Thermo-mechanical Buckling
of
Composite Plates and Shells

by:

Prof. M. R. Eslami

Professor and Fellow of the Academy of sciences,

Mechanical Engineering Department,

Amirkabir University of Technology (Tehran Polytechnic)

Amirkabir University Press, 424 Hafez Avenue, Tehran, Iran

P.O. Box 15875-4413

WEBSITE: http://publication.aut.ac.ir

Amirkabir University of Technology
(Tehran Polytechnic)

Thermo-mechanical Buckling of Composite Plates and Shells, First Edition.

Published by Amirkabir University Press, 424 Hafez Avenue, Tehran, Iran, P.O. Box 15875-4413.

National library of Iran

ISBN: 978-964-463-384-3

Thermo-mechanical Buckling of Composite Plates and Shells, First Edition,

Eslami, M. R.

Series in Solid Mechanics, Composite materials and simulation

WEBSITE: http://publication.aut.ac.ir

Date of Publication: July 2010

Contents

4

List of Figures

List of Tables

Chapter 1

Buckling of Rectangular Plates

1   Isotropic and Composite Plates

1.1   Introduction

Thermoelastic buckling of plates with various geometries and boundary conditions are studied by a number of investigators. Some of the results are given by Gossard et al. [1] and Klosner and Forrey [2]. Bargmann [3] studied the thermal buckling of an initially stress-free elastic simply supported rectangular plate subjected to a general temperature field. The paper presents a close form approximate solution. The temperature field is assumed to be sufficiently general so as to produce within the plate a combined action of the plane compressive, tensile, and shear thermal stresses. Chen and Chen [4] studied thermal buckling of perfect laminated composite rectangular plates subjected to uniform temperature change. The displacement equations of equilibrium are used and the Galerkin method is employed to determine the critical buckling temperature. Thermal buckling behavior of laminated plates subjected to a nonuniform temperature field is investigated by finite element method by Chen and Chen [5]. Tauchert [6] considered thermal buckling behavior of rectangular antisymmetric angle-ply laminates subjected to a uniform temperature rise. Transverse shear deformation is accounted for by employing the thermoelastic version of the Reissner-Mindlin theory. He derived exact solution for the buckling temperature.

Most of the investigations mentioned above are concerned with the simply supported, loosely or rigidly clamped plates, and plates with stiffened edges elastically supported against the rotation. Tauchert [6,7] studied the thermal buckling of the plates with two unsupported edges.

Analysis are limited to the linear case in reference [6]. The assumption of the geometrically perfect plate for the thermal buckling analysis of the plates with analytical approximate solution method is used in the given references. The thermal buckling analysis of geometrically imperfect rectangular plates with different thermal loads are given by Eslami et al. [8]. In this paper, the isotropic and composite plates are considered and an approximate solution method is employed resulting into close form solution.

In this section, a rectangular plate with two cases of boundary conditions is considered. In the first case, one edge is free and the other three edges are simply supported, and in the other case four edges are simply supported. The equilibrium and stability equations are obtained and employed to compute the critical thermoelastic buckling loads of thin rectangular isotropic and orthotropic plates with various boundary conditions under thickness thermal loading, one edge direction temperature difference, and critical uniform temperature rise of a geometrically perfect plate. Closed form solutions based on approximate techniques are presented.

1.2   Basic equations

A thin plate of thickness h is considered. The normal and shear strains at a distance z from the middle plane of the plate are, Eslami [8]

$$\epsilon_x = \epsilon_x^0 + zk_x$$
$$\epsilon_y = \epsilon_y^0 + zk_y$$
$$\epsilon_{xy} = \epsilon_{xy}^0 + zk_{xy} \qquad (1.1\text{-}1)$$

where ϵ_x and ϵ_y are the normal strains, ϵ_{xy} is the shear strain, and k_{ij} are the curvatures. The superscript "0" refers to the strains at the middle surface of the plate. The indices x and y refer to the coordinates system. The general strain-displacement relations may be simplified to give the following terms for the strains at the middle surface and the curvatures in terms of the displacement components, Eslami et al. [8]

$$\epsilon_x^0 = u_{,x} + \frac{1}{2}w_{,x}^2$$
$$\epsilon_y^0 = v_{,y} + \frac{1}{2}w_{,y}^2$$
$$\epsilon_{xy}^0 = u_{,y} + v_{,x} + w_{,x}w_{,y}$$
$$k_x = -w_{,xx}$$
$$k_y = -w_{,yy}$$
$$k_{xy} = -2w_{,xy} \qquad (1.1\text{-}2)$$

where u, v, and w are the middle plane displacements and $(,)$ indicates a partial derivative. The Hooke's law in terms of forces and moments per unit length for orthotropic rectangular plate is, Jones [9]

$$
\begin{pmatrix} N_x \\ N_y \\ N_{xy} \end{pmatrix} = \begin{pmatrix} A_{11} & A_{12} & A_{16} \\ A_{12} & A_{22} & A_{26} \\ A_{16} & A_{26} & A_{66} \end{pmatrix} \begin{pmatrix} \epsilon_x^0 \\ \epsilon_y^0 \\ \epsilon_{xy}^0 \end{pmatrix}
$$

$$
+ \begin{pmatrix} B_{11} & B_{12} & B_{16} \\ B_{12} & B_{22} & B_{26} \\ B_{16} & B_{26} & B_{66} \end{pmatrix} \begin{pmatrix} -w_{,xx} \\ -w_{,yy} \\ -2w_{,xy} \end{pmatrix}
$$

$$
- \begin{pmatrix} N_x^T \\ N_y^T \\ N_{xy}^T \end{pmatrix}
$$

$$
\begin{pmatrix} M_x \\ M_y \\ M_{xy} \end{pmatrix} = \begin{pmatrix} B_{11} & B_{12} & B_{16} \\ B_{12} & B_{22} & B_{26} \\ B_{16} & B_{26} & B_{66} \end{pmatrix} \begin{pmatrix} \epsilon_x^0 \\ \epsilon_y^0 \\ \epsilon_{xy}^0 \end{pmatrix}
$$

$$
+ \begin{pmatrix} D_{11} & D_{12} & D_{16} \\ D_{12} & D_{22} & D_{26} \\ D_{16} & D_{26} & D_{66} \end{pmatrix} \begin{pmatrix} -w_{,xx} \\ -w_{,yy} \\ -2w_{,xy} \end{pmatrix}
$$

$$
- \begin{pmatrix} M_x^T \\ M_y^T \\ M_{xy}^T \end{pmatrix} \tag{1.1-3}
$$

Here, A_{ij}, B_{ij}, and D_{ij} are the stiffness coefficients for the plate and are defined in terms of the transformed reduced stiffness coefficients $(\acute{Q}_{ij})_k$ for the individual layers, $k = 1, 2, 3 \ldots, N$, as by $(Tsai 1985)$

$$
(A_{ij}, B_{ij}, D_{ij}) = \sum_{k=1}^{N} \int_{z_{k-1}}^{z_k} (1, z, z^2)(\acute{Q}_{ij})_k dz \quad (i, j = 1, 2, 6) \tag{1.1-4}
$$

Also appearing in Eq. (1.1-3) are the thermal forces and moments defined as

$$
\begin{pmatrix} N_x^T \\ N_y^T \\ N_{xy}^T \end{pmatrix} = \sum_{k=1}^{N} \int_{z_{k-1}}^{z_k} \begin{pmatrix} \acute{Q}_{11} & \acute{Q}_{12} & \acute{Q}_{16} \\ \acute{Q}_{12} & \acute{Q}_{22} & \acute{Q}_{26} \\ \acute{Q}_{16} & \acute{Q}_{26} & \acute{Q}_{66} \end{pmatrix}_k
$$

$$
\times \begin{pmatrix} \alpha_x \\ \alpha_y \\ \alpha_{xy} \end{pmatrix}_k T dz
$$

$$
\begin{pmatrix} M_x^T \\ M_y^T \\ M_{xy}^T \end{pmatrix} = \sum_{k=1}^{N} \int_{z_{k-1}}^{z_k} \begin{pmatrix} \acute{Q}_{11} & \acute{Q}_{12} & \acute{Q}_{16} \\ \acute{Q}_{12} & \acute{Q}_{22} & \acute{Q}_{26} \\ \acute{Q}_{16} & \acute{Q}_{26} & \acute{Q}_{66} \end{pmatrix}_k
$$

$$
\times \begin{pmatrix} \alpha_x \\ \alpha_y \\ \alpha_{xy} \end{pmatrix}_k zT dz
$$

The total potential energy function of the orthotropic rectangular plate is, Brush and Almroth [10]

$$U = \frac{1}{2} \int \int (\{\epsilon^0\}^T [A]^T \{\epsilon^0\} + 2\{\epsilon^0\}^T [B]^T \{k\}$$

$$-2\{\epsilon^0\}^T \{N^T\} + \{k\}^T [D]^T \{k\} - 2\{k\}^T \{M^T\}$$

$$+ \int_{-h/2}^{h/2} \{\alpha\}^T [\acute{Q}]\{\alpha\} T^2 dz) dx dy \qquad (1.1\text{-}5)$$

Assuming that the orthotropic rectangular plate is under thermal stress alone, the total potential energy is a function of the displacement components and their derivatives and may be written as

$$U = \int \int \int F(u, v, w, u_{,x}, u_{,y}, v_{,x}, v_{,y}, w_{,x}, w_{,y}, w_{,xx}$$

$$, w_{,yy}, w_{,xy}) dx dy dz \qquad (1.1\text{-}6)$$

Minimizing the functional of potential energy leads to the Euler equations, Tauchert [6]

$$\frac{\partial F}{\partial u} - \frac{\partial}{\partial x} \frac{\partial F}{\partial u_{,x}} - \frac{\partial}{\partial y} \frac{\partial F}{\partial u_{,y}} = 0$$

$$\frac{\partial F}{\partial v} - \frac{\partial}{\partial x} \frac{\partial F}{\partial v_{,x}} - \frac{\partial}{\partial y} \frac{\partial F}{\partial v_{,y}} = 0$$

$$\frac{\partial F}{\partial w} - \frac{\partial}{\partial x} \frac{\partial F}{\partial w_{,x}} - \frac{\partial}{\partial y} \frac{\partial F}{\partial w_{,y}} + \frac{\partial^2}{\partial x^2} \frac{\partial F}{\partial w_{,xx}} + \frac{\partial^2}{\partial x \partial y} \frac{\partial F}{\partial w_{,xy}}$$

$$+ \frac{\partial^2}{\partial y^2} \frac{\partial F}{\partial w_{,yy}} = 0 \qquad (1.1\text{-}7)$$

Upon substitution of Eqs. (1.1-2) to (1.1-5) into (1.1-6) and using Eqs. (1.1-8), the equilibrium equations for general thin orthotropic rectangular plate are obtained as, Eslami et al. [8]

$$N_{x,x} + N_{xy,y} = 0$$

$$N_{xy,x} + N_{y,y} = 0$$

$$M_{x,xx} + 2M_{xy,xy} + M_{y,yy} + N_x w_{,xx}$$

$$+ 2N_{xy} w_{,xy} + N_y w_{,yy} = 0 \qquad (1.1\text{-}8)$$

The stability equations of thin orthotropic rectangular plates may be derived by the force summation method, Eslami et al. [8]

$$N_{x1,x} + N_{xy1,y} = 0$$

$$N_{xy1,x} + N_{y1,y} = 0$$

$$M_{x1,xx} + 2M_{xy1,xy} + M_{y1,yy} + N_{x0} w_{1,xx}$$

$$+ 2N_{xy0} w_{1,xy} + N_{y0} w_{1,yy} = 0 \qquad (1.1\text{-}9)$$

where in these equations N_{ij1} and M_{ij1} are related to the linear strain components, and N_{ij0} and as M_{ij0} represent the prebuckling thermal forces and moments

$$
\begin{pmatrix} N_{x0} \\ N_{y0} \\ N_{xy0} \end{pmatrix} = \begin{pmatrix} A_{11} & A_{12} & A_{16} \\ A_{12} & A_{22} & A_{26} \\ A_{16} & A_{26} & A_{66} \end{pmatrix} \begin{pmatrix} \epsilon_{x1} \\ \epsilon_{y1} \\ \epsilon_{xy1} \end{pmatrix}
$$
$$
+ \begin{pmatrix} B_{11} & B_{12} & B_{16} \\ B_{12} & B_{22} & B_{26} \\ B_{16} & B_{26} & B_{66} \end{pmatrix} \begin{pmatrix} k_{x0} \\ k_{y0} \\ k_{xy0} \end{pmatrix}
$$
$$
\begin{pmatrix} M_{x0} \\ M_{y0} \\ M_{xy0} \end{pmatrix} = \begin{pmatrix} B_{11} & B_{12} & B_{16} \\ B_{12} & B_{22} & B_{26} \\ B_{16} & B_{26} & B_{66} \end{pmatrix} \begin{pmatrix} \epsilon_{x1} \\ \epsilon_{y1} \\ \epsilon_{xy1} \end{pmatrix}
$$
$$
+ \begin{pmatrix} D_{11} & D_{12} & D_{16} \\ D_{12} & D_{22} & D_{26} \\ D_{16} & D_{26} & D_{66} \end{pmatrix} \begin{pmatrix} k_{x0} \\ k_{y0} \\ k_{xy0} \end{pmatrix}
$$
$$
\begin{pmatrix} N_{x1} \\ N_{y1} \\ N_{xy1} \end{pmatrix} = \begin{pmatrix} A_{11} & A_{12} & A_{16} \\ A_{12} & A_{22} & A_{26} \\ A_{16} & A_{26} & A_{66} \end{pmatrix} \begin{pmatrix} \epsilon_{x1} \\ \epsilon_{y1} \\ \epsilon_{xy1} \end{pmatrix}
$$
$$
+ \begin{pmatrix} B_{11} & B_{12} & B_{16} \\ B_{12} & B_{22} & B_{26} \\ B_{16} & B_{26} & B_{66} \end{pmatrix} \begin{pmatrix} k_{x1} \\ k_{y1} \\ k_{xy1} \end{pmatrix}
$$
$$
\begin{pmatrix} M_{x1} \\ M_{y1} \\ M_{xy1} \end{pmatrix} = \begin{pmatrix} B_{11} & B_{12} & B_{16} \\ B_{12} & B_{22} & B_{26} \\ B_{16} & B_{26} & B_{66} \end{pmatrix} \begin{pmatrix} \epsilon_{x1} \\ \epsilon_{y1} \\ \epsilon_{xy1} \end{pmatrix}
$$
$$
+ \begin{pmatrix} D_{11} & D_{12} & D_{16} \\ D_{12} & D_{22} & D_{26} \\ D_{16} & D_{26} & D_{66} \end{pmatrix} \begin{pmatrix} k_{x1} \\ k_{y1} \\ k_{xy1} \end{pmatrix}
$$

$$(1.1\text{-}10)$$

and $\epsilon_{x1}, \epsilon_{y1}, ..., k_{x1}, k_{y1}, ...$ are

$$
\begin{aligned}
\epsilon_{x1} = e_{x1} = u_{1,x} && k_{x1} = -w_{1,xx} \\
\epsilon_{y1} = e_{y1} = v_{1,y} && k_{y1} = -w_{1,yy} \\
\epsilon_{xy1} = e_{xy1} = u_{1,y} + v_{1,x} && k_{xy} = -2w_{1,xy}
\end{aligned}
$$
$$(1.1\text{-}11)$$

For simplicity, it is assumed that ply orientations are 0^0 or 90^0, and thus

$$A_{16} = A_{26} = B_{16} = B_{26} = D_{16} = D_{26} = 0 \qquad (1.1\text{-}12)$$

The stability equations in terms of the displacement components are obtained by substitution of Eqs. (1.1-11) into (1.1-10) with the aid of Eqs. (1.1-12) and (1.1-13)

$$A_{11}u_{1,xx} + (A_{12} + A_{66})v_{1,xy} - B_{11}w_{1,xxx} - (B_{12}$$

$$+2B_{66})w_{1,xyy} + A_{66}u_{1,yy} = 0$$
$$(A_{66} + A_{12})u_{1,xy} + A_{66}v_{1,xx} - (2B_{66} + B_{12})w_{1,xxy}$$
$$+A_{22}v_{1,yy} - B_{22}w_{1,yyy} = 0$$
$$B_{11}u_{1,xxx} + (B_{12} + 2B_{66})u_{1,xyy} + B_{22}v_{1,yyy} + (B_{12}$$
$$+2B_{66})v_{1,xxy} - D_{11}w_{1,xxxx} - 2(D_{12} + 2D_{66})w_{1,xxyy}$$
$$-D_{22}w_{1,yyyy} + N_{x0}w_{1,xx} + 2N_{xy0}w_{1,xy} + N_{y0}w_{1,yy} = 0$$

$$(1.1\text{-}13)$$

These equations are related to the thermal forces through the prebuckling terms, such as N_{x0}, through Eqs. (1.1-11).

1.3   Plates with one free and three simply supported edges

Consider a rectangular plate with simply supported edges along $x = 0, x = a$, and $y = 0$ and free edge along $y = b$. The boundary conditions are, Timoshenko and Gere [11]

$$u = w = M_x = N_{xy} = 0 \qquad (\text{at} \quad x = 0 \text{ and } \quad x = a)$$
$$v = w = M_y = N_{xy} = N_y = 0 \qquad (\text{at} \quad y = 0)$$
$$M_y = M_{yx} = Q_y = N_y = 0 \quad (\text{at free edge} \quad y = b)$$

$$(1.1\text{-}14)$$

The lateral deflection is approximated as, Timoshenko and Gere [11]

$$w_1 = F(y)\sin\frac{m\pi}{a}x \qquad (1.1\text{-}15)$$

Consider a rectangular isotropic plate. For this case $B_{ij} = 0$ and thus the last of Eqs. (1.1-14) is function of w_1 only. This equation is used to obtained the critical temperature for various cases of thermal loading.

A. Critical uniform temperature rise

Consider a rectangular isotropic plate of dimensions a and b and thickness h. The initial uniform temperature of the plate is assumed to be T_i. If the plate is simply supported along the edges $x = 0$ and $x = a$ and the displacement in x-direction is prevented, the temperature may be uniformly raised to a final value T_f such that the plate buckles. To find the critical $\Delta T = T_f - T_i$, the prebuckling stresses are, Eslami et al. [8]

$$N_{x0} = -N_{x0}^T = -\frac{E\alpha}{1-\nu}\Delta T$$
$$N_{y0} = 0$$
$$N_{xy0} = 0 \qquad\qquad (1.1\text{-}16)$$

Substitution of Eqs. (1.1-16) and (1.1-17) into the last Equation of (1.1-14) yields a homogeneous differential equation of fourth order. Solving this equation for $F(y)$ yields

$$F(y) = C_1 \cosh \alpha y + C_2 \sinh \alpha y + C_3 \cos \beta y + C_4 \sin \beta y \qquad (1.1\text{-}17)$$

where α and β are defined as

$$\alpha = \sqrt{(\frac{m\pi}{a})^2 + (\frac{m\pi}{a})\sqrt{(\bar{N}_{x0}^T/D)}}$$

$$\beta = \sqrt{-(\frac{m\pi}{a})^2 + (\frac{m\pi}{a})\sqrt{(\bar{N}_{x0}^T/D)}}$$

$$(1.1\text{-}18)$$

Considering the boundary conditions related to y, gives

$$F(0) = 0$$
$$F''(0) = 0$$
$$F''(b) - \nu(\frac{m\pi}{a})^2 = 0$$
$$F'''(b) - (\frac{m\pi}{a})^2(2 - \nu)F'(b) = 0 \qquad (1.1\text{-}19)$$

Introducing the conditions (1.1-20) into Eq. (1.1-18) yields a system of four homogeneous equations for $C_1, C_2, C_3,$ and C_4. The determinant of the coefficients is set equal to zero which gives

$$\alpha(\beta^2 + \nu(\frac{m\pi}{a})^2)^2 \tan \beta b = \beta(\alpha^2 - \nu(\frac{m\pi}{a})^2)^2 \tanh \alpha b \qquad (1.1\text{-}20)$$

Substituting α and β from Eq. (1.1-19) gives

$$[-m_0^2 + m_0\sqrt{\bar{N}_{x0}^T}]^{1/2}[(1 - \nu)m_0^2 + m_0\sqrt{\bar{N}_{x0}^T}]^2$$
$$. \tanh [m_0^2 + m_0\sqrt{\bar{N}_{x0}^T}]^{1/2}$$
$$= [m_0^2 + m_0\sqrt{\bar{N}_{x0}^T}]^{1/2}[(\nu - 1)m_0^2 + m_0\sqrt{\bar{N}_{x0}^T}]^2$$
$$. \tan [m_0^2 + m_0\sqrt{\bar{N}_{x0}^T}]^{1/2}$$

$$(1.1\text{-}21)$$

where m_0 and $\bar{N}_{x0}^T$ are defined as

$$m_0 = \frac{m\pi b}{a}$$
$$\bar{N}_{x0}^T = \frac{b^2 N_{x0}^T}{D} \qquad (1.1\text{-}22)$$

Minimizing ΔT with respect to m from Eq. (1.1-22) gives the critical temperature difference as

$$\Delta T_{cr} = \frac{\pi^2 h^2}{12\alpha(1+\nu)}\left(\frac{1}{a^2} + \frac{0.456}{b^2}\right) \qquad (1.1\text{-}23)$$

This relation is identical with the relation given by Tauchert [6] and Thornton [12].

B. Temperature gradient through the thickness

Assume a linear temperature variation across the plate thickness as

$$T(z) = \Delta T \frac{(z + \frac{h}{2})}{h} \qquad (1.1\text{-}24)$$

where z measures from the middle plane of the plate and ΔT is the difference of the temperature between $z = \frac{-h}{2}$ and $z = \frac{h}{2}$. For plates with the given boundary condition, prebuckling forces are, Eslami et al. [8]

$$N_{x0} = -N_{x0}^T = \frac{E\alpha h}{1-\nu}\left(\frac{\Delta T}{2}\right)$$
$$N_{y0} = 0$$
$$N_{xy0} = 0 \qquad (1.1\text{-}25)$$

With a similar method, the critical temperature difference across the thickness is obtained by setting the determinant of the coefficient to zero and therefore the critical temperature for isotropic plate is obtained by minimizing ΔT with respect to m, as

$$\Delta T_{cr} = \frac{\pi^2 h^2}{6\alpha(1+\nu)}\left(\frac{1}{a^2} + \frac{0.456}{b^2}\right) \qquad (1.1\text{-}26)$$

C. Linear variation along the x-direction

Consider a rectangular isotropic plate of dimensions a and b under temperature difference along the x-direction and with one free edge and three simply supported edges, where the displacement of the edges in the x-directions are prevented. Assume a linear temperature variation along the x-direction

$$T(x) = \Delta T \frac{x}{a} \qquad (1.1\text{-}27)$$

where ΔT is the difference between $x = 0$ and $x = a$. The prebuckling forces are, Eslami et al. [8]

$$N_{x0} = -N_{x0}^T = k_1 x$$
$$N_{y0} = 0$$
$$N_{xy0} = 0 \tag{1.1-28}$$

where in this equation k_1 is defined as

$$k_1 = \frac{E\alpha h \Delta T}{a(1-\nu)} \tag{1.1-29}$$

To calculate the critical temperature ΔT, the displacement equation of stability (1.1-14) is used and Galerkin's method with the help of Eq. (1.1-16) is employed to determine the critical buckling temperature. We call the left-hand side of the Eq. (1.1-14) by R. For the given boundary conditions and considering Eq. (1.1-29), Galerkin's method leads to the following equations:

$$R = D\nabla^4 w_1 + (k_1 x)w_{1,xx} = 0$$
$$\int_0^b \int_0^a R(\sin \frac{m\pi}{a}x)dxdy = 0 \tag{1.1-30}$$

Equation (1.1-31) results into four homogeneous equations for the constant coefficients C_1, C_2, C_3, and C_4. With a similar method, the critical temperature difference in the x-direction is

$$\Delta T_{cr} = \frac{\pi^2 h^2}{6\alpha(1+\nu)}(\frac{1}{a^2} + \frac{0.456}{b^2}) \tag{1.1-31}$$

1.4   Plates with four simply supported edges

Consider a rectangular plate with four edges simply supported. The displacements in x and y-directions are prevented, so the boundary conditions are, Leissa and Narita [13]

$$u = w = M_x = 0 \quad \text{(For} \quad x = 0 \quad \text{and} \quad x = a)$$
$$v = w = M_y = 0 \quad \text{(For} \quad y = 0 \quad \text{and} \quad y = b) \tag{1.1-32}$$

where a and b are the plate dimensions. The displacement components for a rectangular orthotropic plate satisfying the simply supported edges conditions are considered as

$$u_1 = A \cos \frac{m\pi}{a} x \sin \frac{n\pi}{b} y$$
$$v_1 = B \sin \frac{m\pi}{a} x \cos \frac{n\pi}{b} y$$
$$w_1 = C \sin \frac{m\pi}{a} x \sin \frac{n\pi}{b} y \qquad (1.1\text{-}33)$$

Substitution of these assumed solutions into the stability equations (1.1-14), and taking the determinant of the matrix of the coefficients equal to zero leads to the critical temperature.

A. Critical uniform temperature rise

Consider a rectangular orthotropic plate of dimensions a, b, and thickness h with four edges simply supported. The initial uniform temperature of the plate is assumed to be T_i. If the plate is simply supported and the displacements in x and y-directions are prevented, the temperature may be uniformly raised to a final value T_f such that the plate buckles. To find the critical $\Delta T = T_f - T_i$, the prebuckling stresses are, Eslami et al. [8]

$$\begin{pmatrix} N_{x0} \\ N_{y0} \\ N_{xy0} \end{pmatrix} = - \int_{-h/2}^{h/2} [Q]\{\alpha\}\Delta T dz = -[A]\{\alpha\}\Delta T \qquad (1.1\text{-}34)$$

or

$$N_{x0} = -(A_{11}\alpha_x + A_{12}\alpha_y)\Delta T$$
$$N_{y0} = -(A_{12}\alpha_x + A_{22}\alpha_y)\Delta T$$
$$N_{xy0} = -A_{66}\alpha_{xy}\Delta T = 0 \qquad (1.1\text{-}35)$$

Substitution of Eqs. (1.1-34) and (1.1-35) into Eq. (1.1-14) yields a system of three homogeneous equations for A, B, and C, i.e.,

$$[K_{ij}] \begin{pmatrix} A \\ B \\ C \end{pmatrix} = 0 \qquad (1.1\text{-}36)$$

in which K_{ij} is a symmetric matrix with the components

$$K_{11} = -A_{11}(\frac{m\pi}{a})^2 - A_{66}(\frac{n\pi}{b})^2$$
$$K_{12} = -A_{12}(\frac{m\pi}{a})(\frac{n\pi}{b}) - A_{66}(\frac{m\pi}{a})(\frac{n\pi}{b})$$
$$K_{13} = B_{11}(\frac{m\pi}{a})^3 + (B_{12} + 2B_{66})(\frac{m\pi}{a})(\frac{n\pi}{b})^2$$

$$K_{22} = -A_{66}(\frac{m\pi}{a})^2 - A_{22}(\frac{n\pi}{b})^2$$

$$K_{23} = B_{22}(\frac{n\pi}{b})^3 + (B_{12} + 2B_{66})(\frac{m\pi}{a})^2(\frac{n\pi}{b})$$

$$K_{33} = -D_{11}(\frac{m\pi}{a})^4 - 2(D_{12} + D_{66})(\frac{m\pi}{a})^2(\frac{n\pi}{b})^2$$

$$-D_{22}(\frac{n\pi}{b})^4 + (A_{11}\alpha_x + A_{12}\alpha_y)(\frac{m\pi}{a})^2\Delta T$$

$$+(A_{12}\alpha_x + A_{22}\alpha_y)(\frac{n\pi}{b})^2\Delta T \qquad (1.1\text{-}37)$$

The critical value of the ΔT is found by setting $\mid K_{ij} \mid = 0$

$$(\Delta T) = \frac{\delta + D_{11}(\frac{m\pi}{a})^4 + 2(D_{12} + 2D_{66})(\frac{m\pi}{a})^2(\frac{n\pi}{b})^2 + D_{22}(\frac{n\pi}{b})^4}{\{(A_{11}\alpha_x + A_{12}\alpha_y)(\frac{m\pi}{a})^2 + (A_{12}\alpha_x + A_{22}\alpha_y)(\frac{n\pi}{b})^2\}} \qquad (1.1\text{-}38)$$

in which δ is

$$\delta = \frac{K_{11}K_{23}^2 + K_{13}^2 K_{22} - 2K_{12}K_{13}K_{23}}{K_{11}K_{22} - K_{12}^2} \qquad (1.1\text{-}39)$$

Equation (1.1-39) may be used to determine the critical temperature of isotropic and symmetric orthotropic thin plates. For isotropic thin plates,the constants A_{ij}, B_{ij} , and D_{ij} are simplified.Substituting the simplified constants in Eqs. (1.1-39) and minimizing ΔT with respect to m and n, the critical temperature difference is obtained by $m = n = 1$, so

$$\Delta T_{cr} = \frac{\pi^2 h^2}{12(1+\nu)\alpha}(\frac{1}{a^2} + \frac{1}{b^2}) \qquad (1.1\text{-}40)$$

This relation is identical with the relation given by Tauchert [14] and Thornton [12].

For symmetric orthotropic thin plates coefficients $B_{ij} = 0$, so $\delta = 0$ and Eq. (1.1-39) is the same relation given by Leissa and Narita [13]. Minimizing ΔT with respect to m and n yields the following conditions for buckling of symmetric orthotropic thin plate

$$c_3 > 0$$

$$c_3 > c_1{}^2 D_{11}{}^2 \qquad (1.1\text{-}41)$$

The critical uniform temperature which causes thermal buckling is one of the two following cases

$$\Delta T_{cr} = \frac{\pi^2}{b^2}\frac{D_{11}c_4^2 + 2(D_{12} + 2D_{66})c_4 + D_{22}}{c_1 c_4 + c_2}$$

$$\text{For} \qquad n = 1 \, , \, m = \frac{a}{b}\sqrt{c_4} \qquad (1.1\text{-}42)$$

$$\Delta T_{cr} = \frac{\pi^2}{a^2}\frac{D_{11}c_4^2 + 2(D_{12} + 2D_{66})c_4 + D_{22}}{c_4(c_1 c_4 + c_2)}$$

$$\text{For} \qquad m = 1 \, , \, n = \frac{b}{a}\frac{1}{\sqrt{c_4}} \qquad (1.1\text{-}43)$$

where the parameters c_i are

$$c_1 = A_{11}\alpha_x + A_{12}\alpha_y$$
$$c_2 = A_{12}\alpha_x + A_{22}\alpha_y$$
$$c_3 = c_2^2 D_{11}^2 - 2c_1 c_2 D_{11} D_{12} - 4c_1 c_2 D_{11} D_{66} + c_1^2 D_{22} D_{11}$$
$$c_4 = \frac{-c_1 D_{11} + \sqrt{c_3}}{c_1 D_{11}} \tag{1.1-44}$$

Equations (1.1-43) and (1.1-44) gives identical value for ΔT_{cr}.

B. Temperature gradient through the thickness

Assume linear temperature variation across the plate thickness as

$$T(z) = \Delta T \frac{(z + \frac{h}{2})}{h} \tag{1.1-45}$$

where z is measured from the middle plane of the plate. For simply supported edges, the prebuckling forces in the orthotropic plate are, Eslami et al. [8]

$$N_{x0} = -(B_{11}\frac{\alpha_x}{h} + B_{12}\frac{\alpha_y}{h} + A_{11}\frac{\alpha_x}{2} + A_{12}\frac{\alpha_y}{2})\Delta T$$
$$N_{y0} = -(B_{12}\frac{\alpha_x}{h} + B_{22}\frac{\alpha_y}{h} + A_{12}\frac{\alpha_x}{2} + A_{12}\frac{\alpha_y}{2})\Delta T$$
$$N_{xy0} = 0 \tag{1.1-46}$$

With a similar method, the critical temperature difference across the thickness is obtained by setting the determinant of the coefficient matrix to zero

$$\Delta T = \frac{P_1}{P_2}$$
$$P_1 = \delta + D_{11}(\frac{m\pi}{a})^4 + 2(D_{12} + 2D_{66})(\frac{m\pi}{a})^2(\frac{n\pi}{b})^2$$
$$+ D_{22}(\frac{n\pi}{b})^4$$
$$P_2 = (B_{11}\frac{\alpha_x}{h} + B_{12}\frac{\alpha_y}{h} + A_{11}\frac{\alpha_x}{2} + A_{12}\frac{\alpha_y}{2})(\frac{m\pi}{a})^2$$
$$+ (B_{12}\frac{\alpha_x}{h} + B_{22}\frac{\alpha_y}{h} + A_{12}\frac{\alpha_x}{2} + A_{22}\frac{\alpha_y}{2})(\frac{n\pi}{b})^2 \tag{1.1-47}$$

Equation (1.1-48) may be used to determine the critical temperature of isotropic and symmetric orthotropic thin plates. For isotropic thin plates,the constants A_{ij}, B_{ij}, and D_{ij} are simplified.Substituting the simplified constants in Eqs. (1.1-48) and minimizing ΔT with respect to m and n, the critical temperature difference is obtained by $m = n = 1$, so

$$\Delta T_{cr} = \frac{\pi^2 h^2}{6(1+\nu)\alpha}\left(\frac{1}{a^2} + \frac{1}{b^2}\right) \tag{1.1-48}$$

The critical ΔT for symmetric orthotropic plate is one of two following cases

$$\Delta T_{cr} = \frac{2\pi^2}{b^2}\frac{D_{11}c_4^2 + 2(D_{12} + 2D_{66})c_4 + D_{22}}{c_1c_4 + c_2}$$

$$\text{For } n = 1 , \ m = \frac{a}{b}\sqrt{c_4} \tag{1.1-49}$$

$$\Delta T_{cr} = \frac{2\pi^2}{a^2}\frac{D_{11}c_4^2 + 2(D_{12} + 2D_{66})c_4 + D_{22}}{c_4(c_1c_4 + c_2)}$$

$$\text{For } m = 1 , \ n = \frac{b}{a}\frac{1}{\sqrt{c_4}} \tag{1.1-50}$$

Equations (1.1-50) and (1.1-51) give identical value for ΔT_{cr}.

C. Linear variation along the x-direction

Consider a rectangular orthotropic plate of dimensions a and b under temperature difference along the x-direction and with simply supported edges, where the displacement of the edges along the x and y-directions are prevented. Assume a linear temperature variation along the x-direction

$$T(x) = \Delta T\frac{x}{a} \tag{1.1-51}$$

where $\Delta T = T(a) - T(0)$. The prebuckling forces are, Eslami et al. [8]

$$N_{x0} = -(A_{11}\alpha_x + A_{12}\alpha_y)\Delta T\frac{x}{a}$$

$$N_{y0} = -(A_{12}\alpha_x + A_{22}\alpha_y)\Delta T\frac{x}{a}$$

$$N_{xy0} = 0 \tag{1.1-52}$$

To calculate the critical temperature ΔT, the displacement equations of stability (1.1-14) are used and Galerkin's method with the help of Eq. (1.1-34) is employed to determine the critical buckling temperature. We call the left-hand side of the Eqs. (1.1-14) by $R1, R2$, and $R3$ respectively. For the simply supported boundary conditions, considering Eqs. (1.1-34), the Galerkin's method leads to the following equations

$$\int_0^a \int_0^b R1(\cos\frac{m\pi}{a}x\sin\frac{n\pi}{b}y)dxdy = 0$$

$$\int_0^a \int_0^b R2(\sin\frac{m\pi}{a}x\cos\frac{n\pi}{b}y)dxdy = 0$$

$$\int_0^a \int_0^b R3(\sin\frac{m\pi}{a}x\sin\frac{n\pi}{b}y)dxdy = 0 \qquad (1.1\text{-}53)$$

Equations (1.1-54) result into three homogeneous equations for the constant coefficients A, B, and C. With a similar method, the critical temperature difference in the x-direction is

$$\Delta T_{cr} = \frac{P_1}{P_2}$$

$$P_1 = \delta + [D_{11}(\frac{m\pi}{a})^4 + 2(D_{12}+2D_{66})(\frac{m\pi}{a})^2(\frac{n\pi}{b})^2$$

$$+ D_{22}(\frac{n\pi}{b})^4]\frac{a}{2}$$

$$P_2 = \frac{a}{4}\{(A_{11}\alpha_x + A_{12}\alpha_y)(\frac{m\pi}{a})^2$$

$$+ (A_{12}\alpha_x + A_{22}\alpha_y)(\frac{n\pi}{b})^2\} \qquad (1.1\text{-}54)$$

Equations (1.1-55) can be used for isotropic and symmetric orthotropic plate. For isotropic thin plates,the constants A_{ij}, B_{ij}, and D_{ij} are simplified.Substituting the simplified constants in Eqs. (1.1-54) and minimizing ΔT with respect to m and n, the critical temperature difference is obtained by $m = n = 1$, so

$$\Delta T = \frac{h^2\pi^2}{6(1+\nu)\alpha}[\frac{1}{a^2} + \frac{1}{b^2}] \qquad (1.1\text{-}55)$$

The critical ΔT for symmetric orthotropic plate is one of the two following cases

$$\Delta T_{cr} = \frac{2\pi^2}{b^2}\frac{D_{11}c_4^2 + 2(D_{12}+2D_{66})c_4 + D_{22}}{c_1c_4 + c_2}$$

$$\text{For } n=1, \ m = \frac{a}{b}\sqrt{c_4} \qquad (1.1\text{-}56)$$

$$\Delta T_{cr} = \frac{2\pi^2}{a^2}\frac{D_{11}c_4^2 + 2(D_{12}+2D_{66})c_4 + D_{22}}{c_4(c_1c_4 + c_2)}$$

$$\text{For } m=1, \ n = \frac{b}{a}\frac{1}{\sqrt{c_4}} \qquad (1.1\text{-}57)$$

Equations (57) and (58) give identical value for ΔT_{cr}.

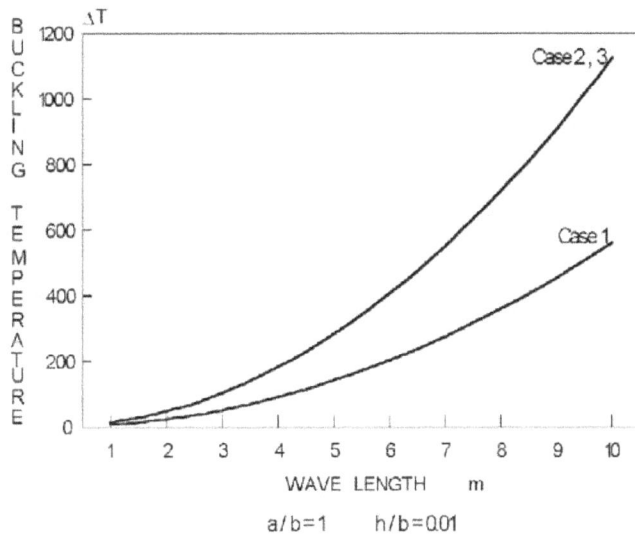

Figure 1.1-1: Temperature for a square plate with different thermal load. Case 1:$T_i \longrightarrow T_i + \Delta$T , Case 2: $T = T(z)$, Case 3: $T = T(x)$.

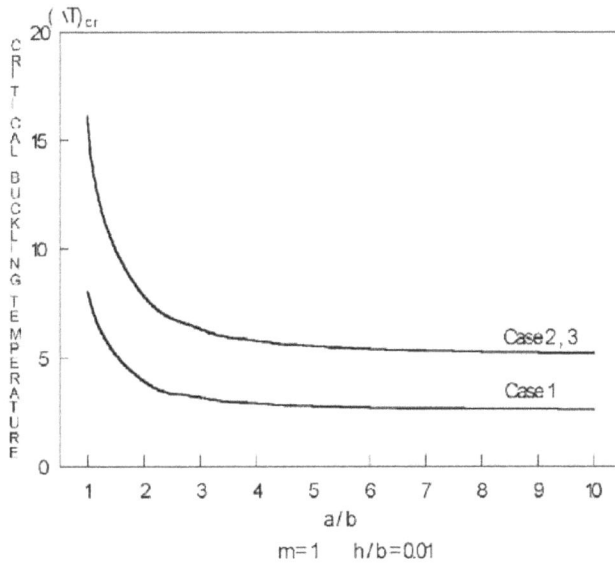

Figure 1.1-2: Critical buckling temperature for an isotropic plate in different thermal load. Case 1: $T_i \longrightarrow T_i + \Delta$T, Case 2: $T = T(z)$, Case 3: $T = T(x)$.

1.5   Results and discussion

Figure (1.1-1) shows the buckling temperature for the uniform temperature rise of a plate versus the wave number along the x-direction for different wave numbers along the y-axis.In this figure case 1 corresponds to the uniform temperature rise, where cases 2 and 3 correspond to the gradient through the thickness and linear temperature variation along the x-direction. The curves show that the lowest critical buckling is reached for $m = 1$. Also the case has been checked for the other types of the thermal loading, and it is found that the lowest buckling temperature occurs at $m = 1$.

Figure (1.1-2) is the plot of critical buckling temperature rise along the x-direction versus aspect ratio for different thermal loads. It is noted that as the aspect ratio is increased, the critical buckling temperature rise is decreased. This curve shows that if aspect ratio reaches to one, the plate is more stable and the buckling happens at higher temperature differences.

Figure (1.1-3) shows the buckling temperature for the uniform temperature rise of an orthotropic plate (material specification is $T300/N5208$), with ply angles [0/90], versus aspect ratio for uniform temperature rise thermal load. It is noted that as the aspect ratio is increased, the critical buckling temperature rise is increased. The case has been checked for other types of thermal loading and it is found that with increasing aspect ratio, thermal buckling load increases.

2   Higher-Order Theory

2.1   Introduction

Chen and Chen [4] studied thermal buckling of laminated composite rectangular plates based on a first-order displacement field and subjected to a uniform temperature change. The displacement equations of equilibrium are used and the Galerkin method is employed to determined the critical buckling temperature. Thermal buckling behavior of laminated plates subjected to a non-uniform temperature field and based on a first-order displacement is investigated by the finite-element method [5].

Tauchert considered thermal buckling behaviors of rectangular antisymmetric angle-ply laminates and based on a first-order displacement field obtained the buckling thermal loads of a plate subjected to uniform temperature rise [6,7,14]. He derived exact solutions for the buckling temperature of thin [14] and thick [6,7] simply supported perfect plates. Chang and Leu [6] studied thermal buckling analysis of antisymmetric angle-ply laminates based on a higher-order displacement field.

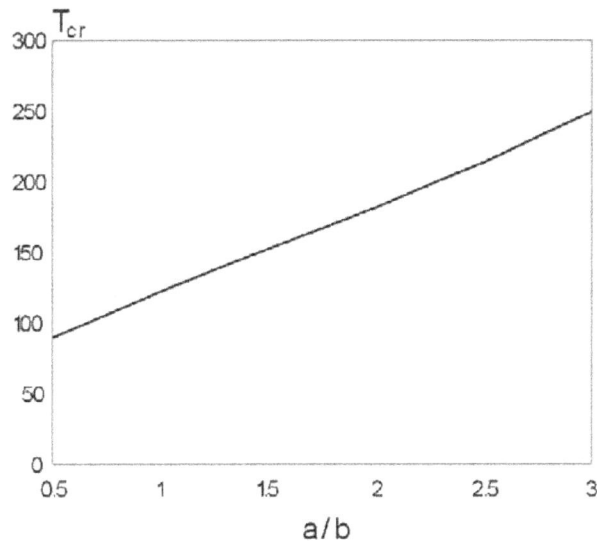

Figure 1.1-3: Buckling temperature for an orthotropic plate.

A higher-order deformation theory, which accounts for transverse shear and transverse normal strains, is derived for the thermal buckling analysis of antisymmetric angle-ply laminates that are simply supported and subjected to a uniform temperature rise. In this paper, they assumed a linear strain-displacement relations to derive the equilibrium equations. To introduce the higher-order theory, they assumed higher-order displacements of up to the third order for u and v and second order for w.

In this section the equilibrium and stability equations are obtained and employed to compute the critical thermoelastic buckling loads of thin rectangular orthotropic plates. The equilibrium equations are obtained using a nonlinear strain-displacement relations and the stability equations are obtained through the force-summation method based on the linearized strain-displacement relations. To introduce the higher-order theory, the displacement components u and v are approximated by the third order polynomials, and the lateral deflection w is approximated by just one term. Thus, while the transverse shear strains are nonzero, the normal lateral strain and stress are assumed to be zero. The eigenvalue solution of the stability equations based on the pre-assumed displacement fields are obtained to present the thermal buckling loads. In addition to the thermal buckling load of the uniform temperature rise given by Chang and Leu [15], the gradient through the thickness thermal buckling load and the one edge direction buckling temperature difference are also derived and presented.

2.2   Basic equations

A thin plate of thickness h is considered. The general strain-displacement relations are [8]

$$\epsilon_{xx} = u_{,x} + \frac{1}{2}w_{,x}^2$$

$$\epsilon_{yy} = v_{,y} + \frac{1}{2}w_{,y}^2$$

$$\epsilon_{xy} = u_{,y} + v_{,x} + w_{,x}w_{,y}$$

$$\epsilon_{xz} = u_{,z} + w_{,x} + w_{,x}w_{,z}$$

$$\epsilon_{yz} = v_{,z} + w_{,y} + w_{,y}w_{,z} \qquad (1.2\text{-}1)$$

where ϵ_{xx} and ϵ_{yy} are the normal strains, and ϵ_{xy}, ϵ_{xz}, and ϵ_{yz}, are the shear strains respectively. The indices x, y, and z refer to the coordinates system. Also u, v, and w are the displacements in the x, y, and z directions, respectively, and $(,)$ indicates a partial derivative.

We now assume the following higher-order displacements [16]

$$u(x, y, z) = u_0(x, y) + z\phi_0(x, y) + z^2\phi_0'(x, y) + z^3\phi_0''(x, y)$$

$$v(x, y, z) = v_0(x, y) + z\psi_0(x, y) + z^2\psi_0'(x, y) + z^3\psi_0''(x, y)$$

$$w(x, y, z) = w_0(x, y) \qquad (1.2\text{-}2)$$

where u_0, v_0, and w_0 denote the displacements of a point (x, y) on the midplane, and ϕ_0 and ψ_0 are the rotations of normal to midplane about the y and x axes, respectively. The functions $\phi_0', \phi_0'', \psi_0'$, and ψ_0'' will be determined using the conditions that transverse shear stresses, σ_{xz} and σ_{yz} vanish on the plate top and bottom surfaces. These are equivalent to $\epsilon_{xz} = \epsilon_{yz} = 0$ at $z = \pm\frac{h}{2}$, which yields [16]

$$\phi_0' = \psi_0' = 0$$

$$\phi_0'' = \frac{-4}{3h^2}(\phi_0 + w_{0,x})$$

$$\psi_0'' = \frac{-4}{3h^2}(\psi_0 + w_{0,y}) \qquad (1.2\text{-}3)$$

Using the conditions (1.2-3), reduce Eqs. (1.2-2) to [16]

$$u = u_0 + z\phi_0 - \frac{4z^3}{3h^2}(\phi_0 + w_{0,x})$$

$$v = v_0 + z\psi_0 - \frac{4z^3}{3h^2}(v1 + w_{0,y})$$

$$w = w_0 \qquad (1.2\text{-}4)$$

The strains associated with the displacements in Eq. (1.2-2) are

$$
\begin{pmatrix} \epsilon_{xx} \\ \epsilon_{yy} \\ \epsilon xy \end{pmatrix} = \begin{pmatrix} \epsilon_{xx}^0 \\ \epsilon_{yy}^0 \\ \epsilon_{xy}^0 \end{pmatrix} + z \begin{pmatrix} k_{xx}^0 \\ k_{yy}^0 \\ k_{xy}^0 \end{pmatrix} + z^3 \begin{pmatrix} k_{xx}^2 \\ k_{yy}^2 \\ k_{xy}^2 \end{pmatrix}
$$

$$
\begin{pmatrix} \epsilon_{xz} \\ \epsilon_{yz} \end{pmatrix} = \begin{pmatrix} \epsilon_{xz}^0 \\ \epsilon_{yz}^0 \end{pmatrix} + z^2 \begin{pmatrix} k_{xz}^1 \\ k_{yz}^1 \end{pmatrix} \qquad (1.2\text{-}5)
$$

where

$$
\begin{pmatrix} k_{xx}^0 \\ k_{yy}^0 \\ k_{xy}^0 \end{pmatrix} = \begin{pmatrix} \phi_{0,x} \\ \psi_{0,y} \\ \phi_{0,y} + \psi_{0,x} \end{pmatrix}
$$

$$
\begin{pmatrix} k_{xx}^2 \\ k_{yy}^2 \\ k_{xy}^2 \end{pmatrix} = \begin{pmatrix} \frac{-4}{3h^2}(\phi_{0,x} + w_{0,xx}) \\ \frac{-4}{3h^2}(\psi_{0,y} + w_{0,yy}) \\ \frac{-4}{3h^2}(\phi_{0,y} + \psi_{0,x} + 2w_{0,xy}) \end{pmatrix}
$$

$$
\begin{pmatrix} k_{xz}^1 \\ k_{yz}^1 \end{pmatrix} = \begin{pmatrix} \frac{-4}{h^2}(\phi_0 + w_{0,x}) \\ \frac{-4}{h^2}(\psi_0 + w_{0,y}) \end{pmatrix}
$$

$$
\begin{pmatrix} \epsilon_{xx}^0 \\ \epsilon_{yy}^0 \\ \epsilon_{xy}^0 \end{pmatrix} = \begin{pmatrix} u_{0,x} + \frac{1}{2}w_{0,x}^2 \\ v_{0,y} + \frac{1}{2}w_{0,y}^2 \\ u_{0,y} + v_{0,x} + w_{0,x}w_{0,y} \end{pmatrix}
$$

$$
\begin{pmatrix} \epsilon_{xz}^0 \\ \epsilon_{yz}^0 \end{pmatrix} = \begin{pmatrix} \phi_0 + w_{0,x} \\ \psi_0 + w_{0,y} \end{pmatrix} \qquad (1.2\text{-}6)
$$

The constitutive equations may be expressed in terms of stresses and strains in the plate coordinates as [16]

$$
\begin{pmatrix} \sigma_{xx} \\ \sigma_{yy} \\ \sigma_{xy} \end{pmatrix} = \begin{pmatrix} \acute{Q}_{11} & \acute{Q}_{12} & \acute{Q}_{16} \\ \acute{Q}_{12} & \acute{Q}_{22} & \acute{Q}_{26} \\ \acute{Q}_{16} & \acute{Q}_{26} & \acute{Q}_{66} \end{pmatrix} \left(\begin{pmatrix} \epsilon_{xx}^0 \\ \epsilon_{yy}^0 \\ \epsilon_{xy}^0 \end{pmatrix} + z \begin{pmatrix} k_{xx}^0 \\ k_{yy}^0 \\ k_{xy}^0 \end{pmatrix} + z^3 \begin{pmatrix} k_{xx}^2 \\ k_{yy}^2 \\ k_{xy}^2 \end{pmatrix} - T \begin{pmatrix} \alpha_{xx} \\ \alpha_{yy} \\ \alpha_{xy} \end{pmatrix} \right)
$$

$$
\begin{pmatrix} \sigma_{xz} \\ \sigma_{yz} \end{pmatrix} = \begin{pmatrix} \acute{Q}_{44} & \acute{Q}_{45} \\ \acute{Q}_{45} & \acute{Q}_{55} \end{pmatrix} \left(\begin{pmatrix} \epsilon_{xz}^0 \\ \epsilon_{yz}^0 \end{pmatrix} + z^2 \begin{pmatrix} k_{xz}^1 \\ k_{yz}^1 \end{pmatrix} - T \begin{pmatrix} \alpha_{xz} \\ \alpha_{yz} \end{pmatrix} \right) \qquad (1.2\text{-}7)
$$

in which T is the absolute temperature, and α_{ij} are linear thermal expansion coefficients. $\acute{Q}_{ij}$ are the plane-stress-reduced elastic constants in the material axes of the layer [8]. The stress resultants $N_{ij}, M_{ij}, P_{ij}, Q_{ij}$, and R_{ij} are

$$
\left(\begin{pmatrix} N_{xx} \\ N_{yy} \\ N_{xy} \end{pmatrix}, \begin{pmatrix} M_{xx} \\ M_{yy} \\ M_{xy} \end{pmatrix}, \begin{pmatrix} P_{xx} \\ P_{yy} \\ P_{xy} \end{pmatrix} \right) == \int_{\frac{-h}{2}}^{\frac{h}{2}} \begin{pmatrix} \sigma_{xx} \\ \sigma_{yy} \\ \sigma_{xy} \end{pmatrix} (1, z, z^3) dz
$$

$$
\left(\begin{pmatrix} Q_{xz} \\ Q_{yz} \end{pmatrix}, \begin{pmatrix} R_{xz} \\ R_{yz} \end{pmatrix} \right) = \int_{\frac{-h}{2}}^{\frac{h}{2}} \begin{pmatrix} \sigma_{xz} \\ \sigma yz \end{pmatrix} (1, z^2) dz \qquad (1.2\text{-}8)
$$

Substitution Eq. (1.2-7) into Eq. (1.2-8), the stress resultants are obtained as

$$
\begin{pmatrix} N_{xx} \\ N_{yy} \\ N_{xy} \end{pmatrix} = \begin{pmatrix} A_{11} & A_{12} & A_{16} \\ A_{12} & A_{22} & A_{26} \\ A_{16} & A_{26} & A_{66} \end{pmatrix} \begin{pmatrix} \epsilon_{xx}^0 \\ \epsilon_{yy}^0 \\ \epsilon_{xy}^0 \end{pmatrix} + \begin{pmatrix} B_{11} & B_{12} & B_{16} \\ B_{12} & B_{22} & B_{26} \\ B_{16} & B_{26} & B_{66} \end{pmatrix} \begin{pmatrix} k_{xx}^0 \\ k_{yy}^0 \\ k_{xy}^0 \end{pmatrix}
$$

$$
+ \begin{pmatrix} E_{11} & E_{12} & E_{16} \\ E_{12} & E_{22} & E_{26} \\ E_{16} & E_{26} & E_{66} \end{pmatrix} \begin{pmatrix} k_{xx}^2 \\ k_{yy}^2 \\ k_{xy}^2 \end{pmatrix} - \begin{pmatrix} N_{xx}^T \\ N_{yy}^T \\ N_{xy}^T \end{pmatrix}
$$

$$
\begin{pmatrix} M_{xx} \\ M_{yy} \\ M_{xy} \end{pmatrix} = \begin{pmatrix} B_{11} & B_{12} & B_{16} \\ B_{12} & B_{22} & B_{26} \\ B_{16} & B_{26} & B_{66} \end{pmatrix} \begin{pmatrix} \epsilon_{xx}^0 \\ \epsilon_{yy}^0 \\ \epsilon_{xy}^0 \end{pmatrix} + \begin{pmatrix} D_{11} & D_{12} & D_{16} \\ D_{12} & D_{22} & D_{26} \\ D_{16} & D_{26} & D_{66} \end{pmatrix} \begin{pmatrix} k_{xx}^0 \\ k_{yy}^0 \\ k_{xy}^0 \end{pmatrix}
$$

$$
+ \begin{pmatrix} F_{11} & F_{12} & F_{16} \\ F_{12} & F_{22} & F_{26} \\ F_{16} & F_{26} & F_{66} \end{pmatrix} \begin{pmatrix} k_{xx}^2 \\ k_{yy}^2 \\ k_{xy}^2 \end{pmatrix} - \begin{pmatrix} M_{xx}^T \\ M_{yy}^T \\ M_{xy}^T \end{pmatrix}
$$

$$
\begin{pmatrix} P_{xx} \\ P_{yy} \\ P_{xy} \end{pmatrix} = \begin{pmatrix} E_{11} & E_{12} & E_{16} \\ E_{12} & E_{22} & E_{26} \\ E_{16} & E_{26} & E_{66} \end{pmatrix} \begin{pmatrix} \epsilon_{xx}^0 \\ \epsilon_{yy}^0 \\ \epsilon_{xy}^0 \end{pmatrix} + \begin{pmatrix} F_{11} & F_{12} & F_{16} \\ F_{12} & F_{22} & F_{26} \\ F_{16} & F_{26} & F_{66} \end{pmatrix} \begin{pmatrix} k_{xx}^0 \\ k_{yy}^0 \\ k_{xy}^0 \end{pmatrix}
$$

$$
+ \begin{pmatrix} H_{11} & H_{12} & H_{16} \\ H_{12} & H_{22} & H_{26} \\ H_{16} & H_{26} & H_{66} \end{pmatrix} \begin{pmatrix} k_{xx}^2 \\ k_{yy}^2 \\ k_{xy}^2 \end{pmatrix} - \begin{pmatrix} P_{xx}^T \\ P_{yy}^T \\ P_{xy}^T \end{pmatrix}
$$

$$
\begin{pmatrix} Q_{xz} \\ Q_{yz} \end{pmatrix} = \begin{pmatrix} A_{44} & A_{45} \\ A_{45} & A_{55} \end{pmatrix} \begin{pmatrix} \epsilon_{xz}^0 \\ \epsilon_{yz}^0 \end{pmatrix} + \begin{pmatrix} D_{44} & D_{45} \\ D_{45} & D_{55} \end{pmatrix} \begin{pmatrix} k_{xz}^1 \\ k_{yz}^1 \end{pmatrix} - \begin{pmatrix} Q_{xz}^T \\ Q_{yz}^T \end{pmatrix}
$$

$$
\begin{pmatrix} R_{xz} \\ R_{yz} \end{pmatrix} = \begin{pmatrix} D_{44} & D_{45} \\ D_{45} & D_{55} \end{pmatrix} \begin{pmatrix} \epsilon_{xz}^0 \\ \epsilon_{yz}^0 \end{pmatrix} + \begin{pmatrix} F_{44} & F_{45} \\ D_{45} & D_{55} \end{pmatrix} \begin{pmatrix} k_{xz}^1 \\ k_{yz}^1 \end{pmatrix} - \begin{pmatrix} R_{xz}^T \\ R_{yz}^T \end{pmatrix}
$$

$$(1.2\text{-}9)$$

where A_{kl}, B_{kl}, etc., are the plate stiffness, defined by

$$
(A_{kl}, B_{kl}, D_{kl}, E_{kl}, F_{kl}, H_{kl}) = \int_{\frac{-h}{2}}^{\frac{h}{2}} \acute{Q}_{kl}(1, z, z^2, z^3, z^4, z^6) dz \quad k, l = 1, 2, 6
$$

$$
(A_{kl}, D_{kl}, F_{kl}) = \int_{\frac{-h}{2}}^{\frac{h}{2}} \acute{Q}_{kl}(1, z^2, z^4) dz \quad k, l = 4, 5 \tag{1.2-10}
$$

The total potential energy function is [10]

$$
U = \frac{1}{2} \int \int \int \{\sigma\}^T (\{\epsilon\} - T\{\alpha\}) dx dy dz \tag{1.2-11}
$$

Upon substitution of Eqs. (1.2-5) and (1.2-7) into Eq. (1.2-11), the total potential energy function of an orthotropic plate is obtained as

$$
U = \frac{1}{2} \int \int \int_{\frac{-h}{2}}^{\frac{h}{2}} (\{\epsilon_{ij}^0\}^T + z\{k_{ij}^0\}^T + z^3\{k_{ij}^2\}^T - T\{\alpha_{ij}\}^T)_{i,j=x,y} [\acute{Q}]^T (\{\epsilon_{ij}^0\}
$$

$$+z\{k_{ij}^0\} + z^3\{k_{ij}^2\} - T\{\alpha_{ij}\})_{i,j=x,y} dxdydz$$

$$+\frac{1}{2}\int\int\int_{\frac{-h}{2}}^{\frac{h}{2}}(\{\epsilon_{ij}^0\}^T + z^2\{k_{ij}^1\}^T - T\{\alpha_{ij}\}^T)_{i=x,y;j=z}[\acute{Q}]^T(\{\epsilon_{ij}^0\}$$

$$+z^2\{k_{ij}^1\} - T\{\alpha_{ij}\})_{i=x,y;j=z} dxdydz \tag{1.2-12}$$

Upon substitution from left-hand side of Eqs. (1.2-6) through (1.2-11) into (1.2-12) and using Euler equations [8], the equilibrium equations for general thin orthotropic rectangular plate are obtained as [10]

$$N_{xx,x} + N_{xy,y} = 0$$

$$N_{xy,x} + N_{yy,y} = 0$$

$$Q_{xz,x} + Q_{yz,y} - \frac{4}{h^2}(R_{xz,x} + R_{yz,y}) + \frac{4}{3h^2}(P_{xx,xx} + 2P_{xy,xy} + P_{yy,yy})$$

$$+(w_{0,x}N_{xx})_{,x} + (w_{0,y}N_{xy})_{,x} + (w_{0,y}N_{yy})_{,y} + (w_{0,x}N_{xy})_{,y} = 0$$

$$M_{xx,x} + M_{xy,y} - Q_{xz} + \frac{4}{h^2}R_{xz} - \frac{4}{3h^2}(P_{xx,x} + P_{xy,y}) = 0$$

$$M_{xy,x} + M_{yy,y} - Q_{yz} + \frac{4}{h^2}R_{yz} - \frac{4}{3h^2}(P_{xy,x} + P_{yy,y}) = 0 \tag{1.2-13}$$

The stability equations of thin orthotropic rectangular plates derived by the force summation method. Let us assume that the state of equilibrium of a general orthotropic rectangular plate under load be defined in terms of the displacements components $\bar{u}_0$, $\bar{v}_0$, $\bar{w}_0$, ϕ_0, and $\bar{\psi}_0$. The displacement components of a neighboring state of stable equilibrium will differ by u_1, v_1, w_1, ϕ_1, ψ_1, with respect to the equilibrium position. Thus the total displacements of a neighboring state are

$$u_0 = \bar{u}_0 + u_1$$

$$v_0 = \bar{v}_0 + v_1$$

$$w_0 = \bar{w}_0 + w_1$$

$$\phi_0 = \bar{\phi}_0 + \phi_1$$

$$\psi_0 = \bar{\psi}_0 + \psi_1 \tag{1.2-14}$$

Similarly, the stress resultant components of a neighboring state may be related to the state of equilibrium as

$$N_{ij} = N_{ij0} + \Delta N_{ij}$$

$$M_{ij} = M_{ij0} + \Delta M_{ij}$$

$$P_{ij} = P_{ij0} + \Delta P_{ij}$$

$$Q_{ij} = Q_{ij0} + \Delta Q_{ij}$$

$$R_{ij} = R_{ij0} + \Delta R_{ij} \tag{1.2-15}$$

Now, the stability equations may be obtained by substitution of Eqs. (1.2-14) and (1.2-15) into (1.2-13) which nonlinear terms with the subscript (1.2-1) will be ignored because they are small compared to the linear terms. The remaining terms form the stability equations of a thin orthotropic rectangular plate under general load as follows [15]

$$N_{xx1,x} + N_{xy1,y} = 0$$

$$N_{xy1,x} + N_{yy1,y} = 0$$

$$Q_{xz1,x} + Q_{yz1,y} - \frac{4}{h^2}(R_{xz1,x} + R_{yz1,y}) + \frac{4}{3h^2}(P_{xx1,xx} + 2P_{xy1,xy} + P_{yy1,yy})$$

$$+(w_{1,x}N_{xx0})_{,x} + (w_{1,y}N_{xy0})_{,x} + (w_{1,y}N_{yy0})_{,y} + (w_{1,x}N_{xy0})_{,y} = 0$$

$$M_{xx1,x} + M_{xy1,y} - Q_{xz1} + \frac{4}{h^2}R_{xz1} - \frac{4}{3h^2}(P_{xx1,x} + P_{xy1,y}) = 0$$

$$M_{xy1,x} + M_{yy1,y} - Q_{yz1} + \frac{4}{h^2}R_{yz1} - \frac{4}{3h^2}(P_{xy1,x} + P_{yy1,y}) = 0 \qquad (1.2\text{-}16)$$

Calling the prebuckling modes by N_{ij0}, M_{ij0}, P_{ij0}, Q_{ij0}, and R_{ij0}, and the buckling modes by N_{ij1}, M_{ij1}, P_{ij1}, Q_{ij1}, and R_{ij1}, it follows that [8]

$$\begin{pmatrix} N_{xx0} \\ N_{yy0} \\ N_{xy0} \end{pmatrix} = \begin{pmatrix} A_{11} & A_{12} & A_{16} \\ A_{12} & A_{22} & A_{26} \\ A_{16} & A_{26} & A_{66} \end{pmatrix} \begin{pmatrix} \epsilon_{xx0}^0 \\ \epsilon_{yy0}^0 \\ \epsilon_{xy0}^0 \end{pmatrix} + \begin{pmatrix} B_{11} & B_{12} & B_{16} \\ B_{12} & B_{22} & B_{26} \\ B_{16} & B_{26} & B_{66} \end{pmatrix} \begin{pmatrix} k_{xx0}^0 \\ k_{yy0}^0 \\ k_{xy0}^0 \end{pmatrix}$$

$$+ \begin{pmatrix} E_{11} & E_{12} & E_{16} \\ E_{12} & E_{22} & E_{26} \\ E_{16} & E_{26} & E_{66} \end{pmatrix} \begin{pmatrix} k_{xx0}^2 \\ k_{yy0}^2 \\ k_{xy0}^2 \end{pmatrix} - \begin{pmatrix} N_{xx0}^T \\ N_{yy0}^T \\ N_{xy0}^T \end{pmatrix}$$

$$\begin{pmatrix} M_{xx0} \\ M_{yy0} \\ M_{xy0} \end{pmatrix} = \begin{pmatrix} B_{11} & B_{12} & B_{16} \\ B_{12} & B_{22} & B_{26} \\ B_{16} & B_{26} & B_{66} \end{pmatrix} \begin{pmatrix} \epsilon_{xx0}^0 \\ \epsilon_{yy0}^0 \\ \epsilon_{xy0}^0 \end{pmatrix} + \begin{pmatrix} D_{11} & D_{12} & D_{16} \\ D_{12} & D_{22} & D_{26} \\ D_{16} & D_{26} & D_{66} \end{pmatrix} \begin{pmatrix} k_{xx0}^0 \\ k_{yy0}^0 \\ k_{xy0}^0 \end{pmatrix}$$

$$+ \begin{pmatrix} F_{11} & F_{12} & F_{16} \\ F_{12} & F_{22} & F_{26} \\ F_{16} & F_{26} & F_{66} \end{pmatrix} \begin{pmatrix} k_{xx0}^2 \\ k_{yy0}^2 \\ k_{xy0}^2 \end{pmatrix} - \begin{pmatrix} M_{x0}^T \\ M_{y0}^T \\ M_{xy0}^T \end{pmatrix}$$

$$\begin{pmatrix} P_{xx0} \\ P_{yy0} \\ P_{xy0} \end{pmatrix} = \begin{pmatrix} E_{11} & E_{12} & E_{16} \\ E_{12} & E_{22} & E_{26} \\ E_{16} & E_{26} & E_{66} \end{pmatrix} \begin{pmatrix} \epsilon_{xx0}^0 \\ \epsilon_{yy0}^0 \\ \epsilon_{xy0}^0 \end{pmatrix} + \begin{pmatrix} F_{11} & F_{12} & F_{16} \\ F_{12} & F_{22} & F_{26} \\ F_{16} & F_{26} & F_{66} \end{pmatrix} \begin{pmatrix} k_{xx0}^0 \\ k_{yy0}^0 \\ k_{xy0}^0 \end{pmatrix}$$

$$+ \begin{pmatrix} H_{11} & H_{12} & H_{16} \\ H_{12} & H_{22} & H_{26} \\ H_{16} & H_{26} & H_{66} \end{pmatrix} \begin{pmatrix} k_{xx0}^2 \\ k_{yy0}^2 \\ k_{xy0}^2 \end{pmatrix} - \begin{pmatrix} P_{x0}^T \\ P_{y0}^T \\ P_{xy0}^T \end{pmatrix}$$

$$\begin{pmatrix} Q_{xz0} \\ Q_{yz0} \end{pmatrix} = \begin{pmatrix} A_{44} & A_{45} \\ A_{45} & A_{55} \end{pmatrix} \begin{pmatrix} \epsilon_{xz0}^0 \\ \epsilon_{yz0}^0 \end{pmatrix} + \begin{pmatrix} D_{44} & D_{45} \\ D_{45} & D_{55} \end{pmatrix} \begin{pmatrix} k_{xz0}^1 \\ k_{yz0}^1 \end{pmatrix} - \begin{pmatrix} Q_{x0}^T \\ Q_{y0}^T \end{pmatrix}$$

$$\begin{pmatrix} R_{xz0} \\ R_{yz0} \end{pmatrix} = \begin{pmatrix} D_{44} & D_{45} \\ D_{45} & D_{55} \end{pmatrix} \begin{pmatrix} \epsilon_{xz0}^0 \\ \epsilon_{yz0}^0 \end{pmatrix} + \begin{pmatrix} F_{44} & F_{45} \\ F_{45} & F_{55} \end{pmatrix} \begin{pmatrix} k_{xz0}^1 \\ k_{yz0}^1 \end{pmatrix} - \begin{pmatrix} Q_{x0}^T \\ Q_{y0}^T \end{pmatrix}$$

$$\begin{pmatrix} N_{xx1} \\ N_{yy1} \\ N_{xy1} \end{pmatrix} = \begin{pmatrix} A_{11} & A_{12} & A_{16} \\ A_{12} & A_{22} & A_{26} \\ A_{16} & A_{26} & A_{66} \end{pmatrix} \begin{pmatrix} \epsilon^0_{xx1} \\ \epsilon^0_{yy1} \\ \epsilon^0_{xy1} \end{pmatrix} + \begin{pmatrix} B_{11} & B_{12} & B_{16} \\ B_{12} & B_{22} & B_{26} \\ B_{16} & B_{26} & B_{66} \end{pmatrix} \begin{pmatrix} k^0_{xx1} \\ k^0_{yy1} \\ k^0_{xy1} \end{pmatrix}$$

$$+ \begin{pmatrix} E_{11} & E_{12} & E_{16} \\ E_{12} & E_{22} & E_{26} \\ E_{16} & E_{26} & E_{66} \end{pmatrix} \begin{pmatrix} k^2_{xx1} \\ k^2_{yy1} \\ k^2_{xy1} \end{pmatrix}$$

$$\begin{pmatrix} M_{xx1} \\ M_{yy1} \\ M_{xy1} \end{pmatrix} = \begin{pmatrix} B_{11} & B_{12} & B_{16} \\ B_{12} & B_{22} & B_{26} \\ B_{16} & B_{26} & B_{66} \end{pmatrix} \begin{pmatrix} \epsilon^0_{xx1} \\ \epsilon^0_{yy1} \\ \epsilon^0_{xy1} \end{pmatrix} + \begin{pmatrix} D_{11} & D_{12} & D_{16} \\ D_{12} & D_{22} & D_{26} \\ D_{16} & D_{26} & D_{66} \end{pmatrix} \begin{pmatrix} k^0_{xx1} \\ k^0_{yy1} \\ k^0_{xy1} \end{pmatrix}$$

$$+ \begin{pmatrix} F_{11} & F_{12} & F_{16} \\ F_{12} & F_{22} & F_{26} \\ F_{16} & F_{26} & F_{66} \end{pmatrix} \begin{pmatrix} k^2_{xx1} \\ k^2_{yy1} \\ k^2_{xy1} \end{pmatrix}$$

$$\begin{pmatrix} P_{xx1} \\ P_{yy1} \\ P_{xy1} \end{pmatrix} = \begin{pmatrix} E_{11} & E_{12} & E_{16} \\ E_{12} & E_{22} & E_{26} \\ E_{16} & E_{26} & E_{66} \end{pmatrix} \begin{pmatrix} \epsilon^0_{xx1} \\ \epsilon^0_{yy1} \\ \epsilon^0_{xy1} \end{pmatrix} + \begin{pmatrix} F_{11} & F_{12} & F_{16} \\ F_{12} & F_{22} & F_{26} \\ F_{16} & F_{26} & F_{66} \end{pmatrix} \begin{pmatrix} k^0_{xx1} \\ k^0_{yy1} \\ k^0_{xy1} \end{pmatrix}$$

$$+ \begin{pmatrix} H_{11} & H_{12} & H_{16} \\ H_{12} & H_{22} & H_{26} \\ H_{16} & H_{26} & H_{66} \end{pmatrix} \begin{pmatrix} k^2_{xx1} \\ k^2_{yy1} \\ k^2_{xy1} \end{pmatrix}$$

$$\begin{pmatrix} Q_{xz1} \\ Q_{yz1} \end{pmatrix} = \begin{pmatrix} A_{44} & A_{45} \\ A_{45} & A_{55} \end{pmatrix} \begin{pmatrix} \epsilon^0_{xz1} \\ \epsilon^0_{yz1} \end{pmatrix} + \begin{pmatrix} D_{44} & D_{45} \\ D_{45} & D_{55} \end{pmatrix} \begin{pmatrix} k^1_{xz1} \\ k^1_{yz1} \end{pmatrix}$$

$$\begin{pmatrix} R_{xz1} \\ R_{yz1} \end{pmatrix} = \begin{pmatrix} D_{44} & D_{45} \\ D_{45} & D_{55} \end{pmatrix} \begin{pmatrix} \epsilon^0_{xz1} \\ \epsilon^0_{yz1} \end{pmatrix} + \begin{pmatrix} F_{44} & F_{45} \\ F_{45} & F_{55} \end{pmatrix} \begin{pmatrix} k^1_{xz1} \\ k^1_{yz1} \end{pmatrix}$$

$$(1.2\text{-}17)$$

The linear strain-displacement relations for the deviation components reduce to

$$\begin{aligned}
\epsilon^0_{xx1} &= u_{1,x} & k^0_{xx1} &= \phi_{1,x} & k^2_{xx1} &= \frac{-4}{3h^2}(\phi_{1,x} + w_{1,xx}) \\
\epsilon^0_{yy1} &= v_{1,x} & k^0_{yy1} &= \psi_{1,x} & k^2_{yy1} &= \frac{-4}{3h^2}(\psi_{1,x} + w_{1,yy}) \\
\epsilon^0_{xy1} &= u_{1,y} + v_{1,x} & k^0_{xx1} &= \phi_{1,y} + \psi_{1,x} & k^2_{xy1} &= \frac{-4}{3h^2}(\phi_{1,y} + \psi_{1,x} + 2w_{1,xy})
\end{aligned}$$

2.3 Stability equations in terms of the displacements

For symmetric (about the mid-plane) cross-ply plates, the following plate stiffness coefficients are identically zero [16]

$$\begin{aligned}
B_{ij} &= E_{ij} = 0 \\
A_{16} &= A_{26} = D_{16} = D_{26} = F_{16} = F_{26} = H_{16} = H_{26} = 0 \\
A_{45} &= D_{45} = F_{45} = 0
\end{aligned} \qquad (1.2\text{-}18)$$

The stability equations in terms of the displacement components are obtained by substitution of Eqs. (1.2-17) into (1.2-16) with the aid of Eqs. (1.2-18) and (1.2-19)

$$A_{11}u_{1,xx} + A_{12}v_{1,xy} + A_{66}(u_{1,yy} + v_{1,xy}) = 0$$

$$A_{66}(u_{1,xy} + v_{1,xx}) + A_{12}u_{1,xy} + A_{22}v_{1,yy} = 0$$

$$A_{44}(\phi_{1,x} + w_{1,xx}) + D_{44}(\frac{-4}{h^2})(\phi_{1,x} + w_{1,xx}) + A_{55}(\psi_{1,y} + w_{1,yy})$$

$$+D_{55}(\frac{-4}{h^2})(\psi_{1,y} + w_{1,yy}) + (\frac{-4}{h^2})\{D_{44}(\phi_{1,x} + w_{1,xx}) + F_{44}(\frac{-4}{h^2})(\phi_{1,x} + w_{0,xx})$$

$$+D_{55}(\psi_{1,y} + w_{1,yy}) + F_{55}(\frac{-4}{h^2})(\psi_{1,y} + w_{1,yy})\}$$

$$(\frac{4}{h^2})\{F_{11}\phi_{1,xxx} + F_{12}\psi_{1,xxy} + H_{11}(\frac{-4}{3h^2})(\phi_{1,xxx} + w_{1,xxxx}) + H_{12}(\frac{-4}{3h^2})$$

$$\times(\psi_{1,xxy} + w_{1,xxyy}) + 2F_{66}(\phi_{1,xyy} + \psi_{1,xxy}) + 2H_{66}(\frac{-4}{3h^2})(\phi_{1,xyy} + \psi_{1,xxy} + 2w_{1,xxyy}$$

$$+F_{12}\phi_{1,xyy} + F_{22}\psi_{1,yyy} + H_{12}(\frac{-4}{3h^2})(\phi_{1,xyy} + w_{1,xxyy} + H_{22}(\frac{-4}{3h^2})(\psi_{1,yyy} + w_{1,yyyy}$$

$$+\frac{h^2}{4}w_{2,yyyy})\} + (N_{xx0}w_{1,x} + N_{xy0}w_{1,y})_{,x} + (N_{yy0}w_{1,y} + N_{xy0}w_{1,x})_{,y} = 0$$

$$D_{11}\phi_{1,xx} + D_{12}\psi_{1,xy} + F_{11}(\frac{-4}{3h^2})(\phi_{1,xx} + w_{1,xxx} + \frac{h^2}{4}w_{2,xxx}) + F_{12}(\frac{-4}{3h^2})(\psi_{1,xy} + w_{1,xyy}$$

$$+\frac{h^2}{4}w_{2,xyy}) + D_{66}(\phi_{1,yy} + \psi_{1,xy}) + F_{66}(\frac{-4}{3h^2})(\phi_{1,yy} + \psi_{1,xy} + 2w_{1,xyy}) - A_{44}(\phi_1 + w_{1,x})$$

$$-D_{44}(\frac{-4}{h^2})(\phi_1 + w_{1,x}) + \frac{4}{h^2}\{D_{44}(\phi_1 + w_{1,x}) + F_{44}(\frac{-4}{h^2})(\phi_1 + w_{1,x})\}$$

$$\frac{-4}{3h^2}\{F_{11}\phi_{1,xx} + F_{12}\psi_{1,xy} + H_{11}(\frac{-4}{3h^2})(\phi_{1,xx} + w_{1,xxx} + \frac{h^2}{4}w_{2,xxx}) + H_{12}(\frac{-4}{3h^2})(\psi_{1,xy}$$

$$+w_{1,xyy}) + F_{66}(\phi_{1,yy} + \psi_{1,xy}) + H_{66}(\frac{-4}{3h^2})(\phi_{1,yy} + \psi_{1,xy} + 2w_{1,xyy} + \frac{h^2}{2}w_{2,xyy})\} = 0$$

$$D_{66}(\phi_{1,xy} + \psi_{1,xx}) + F_{66}(\frac{-4}{3h^2})(\phi_{1,xy} + \psi_{1,xx} + 2w_{1,xxy} + \frac{h^2}{2}w_{2,xxy}) + D_{12}\phi_{1,xy}$$

$$+D_{22}\psi_{1,yy} + F_{12}(\frac{-4}{3h^2})(\phi_{1,xy} + w_{1,xxy} + \frac{h^2}{4}w_{2,xxy}) + F_{22}(\frac{-4}{3h^2})(\psi_{1,yy} + w_{1,yyy})$$

$$-A_{55}(\psi_1 + w_{1,y}) - D_{55}(\frac{-4}{h^2})(\psi_1 + w_{1,y}) + \frac{4}{h^2}\{D_{55}(\psi_1 + w_{1,y}) + F_{55}(\frac{-4}{h^2})(\psi_1 + w_{1,y})\}$$

$$-\frac{4}{3h^2}\{F_{66}(\phi_{1,xy} + \psi_{1,xx}) + H_{66}(\frac{-4}{3h^2})(\phi_{1,xy} + \psi_{1,xx} + 2w_{1,xxy}) + F_{12}\phi_{1,xy} + F_{22}\psi_{1,yy}$$

$$+H_{12}(\frac{-4}{3h^2})(\phi_{1,xy} + w_{1,xxy}) + H_{22}(\frac{-4}{3h^2})(\psi_{1,yy} + w_{1,yyy})\} = 0 \qquad (1.2\text{-}19)$$

These equations are related to the thermal forces through the prebuckling terms, such as N_{xx0}, through Eqs. (1.2-17).

In the next section, three cases of thermal buckling are discussed and the critical temperatures are calculated. The edges are assumed to be simply supported and also the displacement in x and y directions are prevented, so the boundary conditions are [16]

$$w_1(x,0) = w_1(x,b) = w_1(0,y) = w_1(a,y) = 0$$

$$P_{yy1}(x,0) = P_{yy1}(x,b) = P_{xx1}(0,y) = P_{xx1}(a,y) = 0$$
$$M_{yy1}(x,0) = M_{yy1}(x,b) = M_{xx1}(0,y) = M_{xx1}(a,y) = 0$$
$$u_1(x,0) = u_1(x,b) = v_1(0,y) = v_1(a,y) = 0$$
$$\phi_1(x,0) = \phi_1(x,b) = \psi_1(0,y) = \psi_1(a,y) = 0 \qquad (1.2\text{-}20)$$

where a and b are the plate dimensions. The displacement components for a rectangular orthotropic plate satisfying the simply supported edges conditions are considered as [16]

$$w_1 = w_{1mn} \sin \alpha_m x \sin \beta_n y$$
$$u_1 = u_{1mn} \cos \alpha_m x \sin \beta_n y$$
$$v_1 = v_{1mn} \sin \alpha_m x \cos \beta_n y$$
$$\phi_1 = \phi_{1mn} \cos \alpha_m x \sin \beta_n y$$
$$\psi_1 = \psi_{1mn} \sin \alpha_m x \cos \beta_n y \qquad (1.2\text{-}21)$$

where $\alpha_m = \frac{m\pi}{a}$ and $\beta_n = \frac{n\pi}{b}$. Substitution of these assumed solutions into the five stability equations (1.2-20), and taking the determinant of the matrix of the coefficients equal to zero leads to the critical temperature.

A. Critical uniform temperature rise

Consider a rectangular orthotropic plate of dimensions a, b, and thickness h. The initial uniform temperature of the plate is assumed to be T_i. The temperature may be uniformly raised to a final value T_f such that the plate buckles. To find the critical $\Delta T = T_f - T_i$, the prebuckling stresses are [8]

$$N_{ij0} = -N_{ij0}^T = -\int_{\frac{-h}{2}}^{\frac{h}{2}} Q_{kl}\acute{\alpha}_{ij}\Delta T dz = -A_{ij}\acute{\alpha}_{ij}\Delta T \qquad (1.2\text{-}22)$$

or

$$N_{xx0} = -N_{xx0}^T = -(A_{11}\alpha_{xx} + A_{12}\alpha_{yy})\Delta T$$
$$N_{yy0} = -N_{yy0}^T = -(A_{12}\alpha_{xx} + A_{22}\alpha_{yy})\Delta T$$
$$N_{xy0} = -N_{xy0}^T = -A_{66}\alpha_{xy}\Delta T = 0$$

$$(1.2\text{-}23)$$

Substitution of Eqs. (1.2-22) and (1.2-24) into Eq. (1.2-20) yields a system of five homogeneous equations for w_{1mn}, u_{1mn}, v_{1mn}, ϕ_{1mn}, and

ψ_{1mn}, i. e.,

$$[K_{ij}] \begin{pmatrix} u_{1mn} \\ v_{1mn} \\ w_{1mn} \\ \phi_{1mn} \\ \psi_{1mn} \end{pmatrix} = 0 \qquad (1.2\text{-}24)$$

in which K_{ij} is a symmetric matrix with the components [15]

$K_{11} = \alpha_m^2 A_{11} + \beta_n^2 A_{66}$

$K_{12} = \alpha_m \beta_n (A_{12} + A_{66})$

$K_{13} = 0$

$K_{14} = 0$

$K_{15} = 0$

$K_{22} = \beta_n^2 A_{22} + \alpha_m^2 A_{66}$

$K_{23} = 0$

$K_{24} = 0$

$K_{25} = 0$

$K_{33} = \alpha_m^2 A_{44} + \beta_n^2 A_{55} - \dfrac{8}{h^2}(\alpha_m^2 D_{44} + \beta_n^2 D_{55}) + (\dfrac{4}{h^2})^2(\alpha_m^2 F_{44}$

$+\beta_n^2 F_{55}) + (\dfrac{4}{3h^2})^2[\alpha_m^2 H_{11} + 2\alpha_m^2 \beta_n^2(H_{12} + 2H_{66}) + \beta_n^4 H_{22}]$

$-\{(A_{11}\alpha_{xx} + A_{12}\alpha_{yy})\alpha_m^2 + (A_{12}\alpha_{xx} + A_{22}\alpha_{yy})\beta_n^2\}\Delta T$

$K_{34} = \alpha_m A_{44} - \dfrac{8}{h^2}\alpha_m D_{44} - \dfrac{4}{3h^2}[\alpha_m^3 F_{11} + \alpha_m \beta_n^2(F_{12} + 2F_{66})]$

$+(\dfrac{4}{h^2})^2\alpha_m F_{44} + (\dfrac{4}{3h^2})^2[\alpha_m^3 H_{11} + \alpha_m \beta_n^2(H_{12} + 2H_{66})]$

$K_{35} = \beta_n A_{55} - \dfrac{8}{h^2}\beta_n D_{55} - \dfrac{4}{3h^2}[\beta_n^3 F_{22} + \alpha_m^2 \beta_n(F_{12} + 2F_{66})]$

$+(\dfrac{4}{h^2})^2\beta_n F_{55} + (\dfrac{4}{3h^2})^2[\beta_n^3 H_{22} + \alpha_m^2 \beta_n(H_{12} + 2H_{66})]$

$-\{(D_{11}\alpha_{xx} + D_{12}\alpha_{yy})\alpha_m^2 + (D_{12}\alpha_{xx} + D_{22}\alpha_{yy})\beta_n^2\}\Delta T$

$K_{44} = A_{44} + \alpha_m^2 D_{11} + \beta_n^2 D_{66} - \dfrac{8}{h^2}D_{44} + (\dfrac{4}{h^2})^2 F_{44}$

$-\dfrac{8}{3h^2}(\alpha_m^2 F_{11} + \beta_n^2 F_{66}) + (\dfrac{4}{3h^2})^2(\alpha_m^2 H_{11} + \beta_n^2 H_{66})$

$K_{45} = \alpha_m \beta_n(D_{12} + D_{66}) - \dfrac{8}{3h^2}\alpha_m \beta_n(F_{12} + F_{66})$

$+(\dfrac{4}{3h^2})^2\alpha_m \beta_n(H_{12} + H_{66})$

$K_{55} = A_{55} + \alpha_m^2 D_{66} + \beta_n^2 D_{22} - \dfrac{8}{h^2}D_{55} + (\dfrac{4}{h^2})^2 F_{55}$

$-\dfrac{8}{3h^2}(\alpha_m^2 F_{66} + \beta_n^2 F_{22}) + (\dfrac{4}{3h^2})^2(\alpha_m^2 H_{66} + \beta_n^2 H_{22})$

$$(1.2\text{-}25)$$

The critical value of the ΔT is found by setting $\mid K_{ij} \mid = 0$ and is

$$\Delta T = \frac{\delta - \gamma}{(A_{11}\alpha_{xx} + A_{12}\alpha_{yy})\alpha_m^2 + (A_{12}\alpha_{xx} + A_{22}\alpha_{yy})\beta_n^2} \qquad (1.2\text{-}26)$$

in which δ and γ are

$$\delta = \frac{K_{12}(K_{12}K_{33} - K_{13}K_{23}) - K_{13}(K_{12}K_{23} - K_{22}K_{13})}{K_{22}K_{33} - K_{23}^2}$$

$$\gamma = \alpha_m^2 A_{44} + \beta_n^2 A_{55} - \frac{8}{h^2}(\alpha_m^2 D_{44} + \beta_n^2 D_{55}) + (\frac{4}{h^2})^2(\alpha_m^2 F_{44}$$

$$+\beta_n^2 F_{55}) + (\frac{4}{3h^2})^2[\alpha_m^2 H_{11} + 2\alpha_m^2\beta_n^2(H_{12} + 2H_{66}) + \beta_n^4 H_{22}]$$

$$(1.2\text{-}27)$$

Equation (1.2-26) is used to determine the critical temperature of symmetric orthotropic thin plates.

B. Temperature gradient through the thickness

Assume linear temperature variation across the plate thickness as

$$T(z) = \Delta T \frac{(z + \frac{h}{2})}{h} \qquad (1.2\text{-}28)$$

where z measures from the middle plane of the plate. For simply supported edges, the prebuckling forces in the symmetric orthotropic plate are [8]

$$N_{xx0} = -N_{xx0}^T = -(A_{11}\alpha_{xx} + A_{12}\alpha_{yy})\frac{\Delta T}{2}$$

$$N_{yy0} = -N_{yy0}^T = -(A_{12}\alpha_{xx} + A_{22}\alpha_{yy})\frac{\Delta T}{2}$$

$$N_{xy0} = -N_{xy0}^T = 0$$

$$(1.2\text{-}29)$$

With a similar method, the critical temperature difference across the thickness is obtained by setting the determinant of the coefficient to zero, which yields

$$\Delta T = \frac{\delta - \gamma}{(A_{11}\alpha_{xx} + A_{12}\alpha_{yy})\frac{\alpha_m^2}{2} + (A_{12}\alpha_{xx} + A_{22}\alpha_{yy})\frac{\beta_n^2}{2}} \qquad (1.2\text{-}30)$$

Equation (1.2-30) is used to determine the critical temperature of symmetric orthotropic thin plates.

C. Linear variation along the x-direction

Consider a rectangular symmetric orthotropic plate of dimensions a and b under temperature difference across the x-direction and with simply supported edges, where the motion of the edges in the x and y directions are prevented. Assume a linear temperature variation along the x-direction

$$T(x) = \Delta T \frac{x}{a} \qquad (1.2\text{-}31)$$

where $\Delta T = T(a) - T(0)$. The prebuckling forces are [8]

$$N_{xx0} = -N_{xx0}^T = -(A_{11}\alpha_{xx} + A_{12}\alpha_{yy})\Delta T \frac{x}{a}$$

$$N_{yy0} = -N_{yy0}^T = -(A_{12}\alpha_{xx} + A_{22}\alpha_{yy})\Delta T \frac{x}{a}$$

$$N_{xy0} = -N_{xy0}^T = 0$$

$$R_{xy0} = -R_{xy0}^T = 0$$

$$(1.2\text{-}32)$$

To calculate the critical temperature ΔT, the displacement equations of stability (1.2-20) are used and the Galerkin's method with the help of Eq. (1.2-22) is employed to determine the critical buckling temperature. We call the left-hand side of the five Eqs. (1.2-22) by $R1, R2, R3, R4$, and $R5$, respectively. Considering Eqs. (1.2-22), the Galerkin method leads to the following equations

$$\int_0^a \int_0^b R1 \cos \alpha_m x \sin \beta_n y \, dx \, dy = 0$$

$$\int_0^a \int_0^b R2 \sin \alpha_m x \cos \beta_n y \, dx \, dy = 0$$

$$\int_0^a \int_0^b R3 \sin \alpha_m x \sin \beta_n y \, dx \, dy = 0$$

$$\int_0^a \int_0^b R4 \cos \alpha_m x \sin \beta_n y \, dx \, dy = 0$$

$$\int_0^a \int_0^b R5 \sin \alpha_m x \cos \beta_n y \, dx \, dy = 0$$

$$(1.2\text{-}33)$$

Equations (1.2-33) result into five homogeneous equations for the constant coefficients $u_{1mn}, v_{1mn}, w_{1mn}, \phi_{1mn}$, and ψ_{1mn}. With a similar method, the critical temperature difference in the x-direction is

$$\Delta T = \frac{\delta - \gamma}{(A_{11}\alpha_{xx} + A_{12}\alpha_{yy})\frac{\alpha_m^2}{2} + (A_{12}\alpha_{xx} + A_{22}\alpha_{yy})\frac{\beta_n^2}{2}} \qquad (1.2\text{-}34)$$

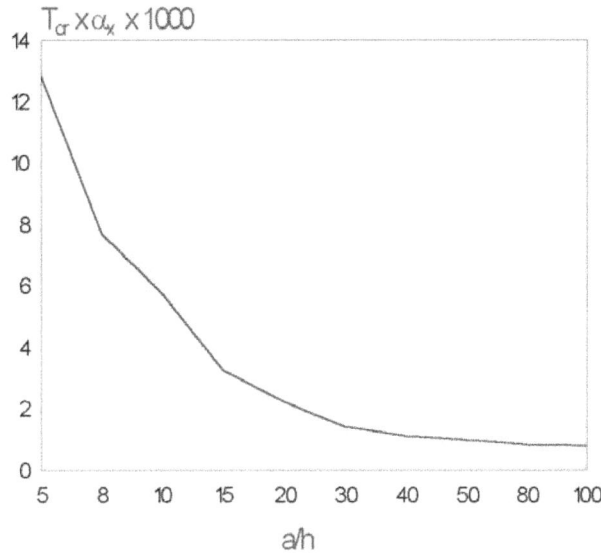

Figure 1.2-1: Thermal buckling load.

Equation (1.2-34) is used to obtain the thermal buckling load of the symmetric orthotropic thin plate.

Buckling temperature for an orthotropic plate.

2.4   Results and discussion

Results for the thermal buckling temperature of the plate for the three cases of loading are given in closed form by Eqs. (1.2-26), (1.2-30), and (1.2-34). The buckling modes m and n appear in the definition of the parameters α_m and β_n. The buckling load is the minimum ΔT for all values of m and n. For the composite materials the lowest buckling load occur at higher buckling modes [17].

Figure (1.2-1) shows the critical buckling temperature, T_{cr}, for the uniform temperature rise of an orthotropic plate (material specification is T300/N5208), with ply angles [0/90/90/0], versus the a/h ratio. The curve shows that with increasing a/h ratio, the critical buckling temperature decreases. This result is qualitatively identical with the result given by Chang and Leu [15], where they have presented the results for an orthotropic plate with [+45/-45] ply angle arrangement. The case has been checked for the other types of the thermal loading and it is found that with increasing a/h ratio, the critical buckling temperature decreases.

Figure (1.2-2) shows the critical buckling temperature for the uni-

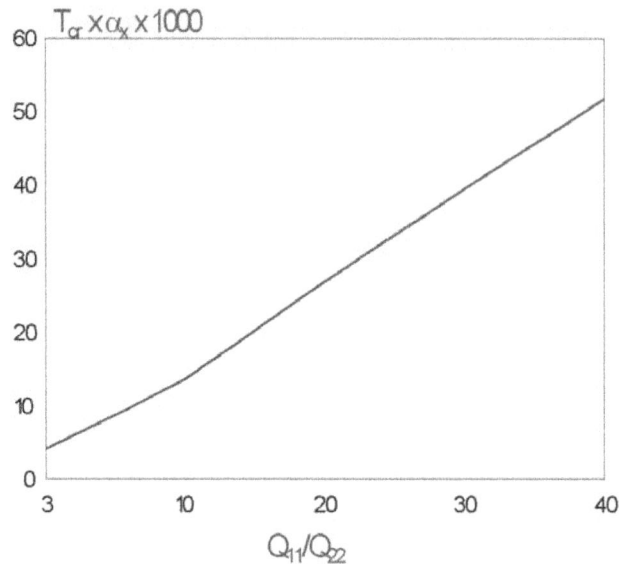

Figure 1.2-2: Thermal buckling load.

form temperature rise of an orthotropic plate (material specification is T300/N5208), with ply angles [0/90/90/0] and $a/h = 10$, versus the $\frac{Q'_{11}}{Q'_{22}}$ ratio. The curve shows that with increasing $\frac{Q'_{11}}{Q'_{22}}$ ratio, the critical buckling temperature increases. This result is qualitatively identical with the result given by Chang and Leu [15]. The case has been checked for the other types of the thermal loading and it is found that with increasing a/h ratio, the critical buckling temperature increases.

3   Imperfect Plates, Higher-Order Theory

3.1   Introduction

Chen and Chen [4] studied thermal buckling of laminated composite rectangular plates based on a first-order displacement field and subjected to a uniform temperature change. The displacement equations of equilibrium are used and the Galerkin method is employed to determined the critical buckling temperature. Thermal buckling behavior of laminated plates subjected to a non-uniform temperature field and based on a first-order displacement is investigated by the finite-element method [5].

Tauchert and Huang and Tauchert considered thermal buckling behaviors of rectangular antisymmetric angle-ply laminates based on a first-order displacement field subjected to a uniform temperature rise

[14,6,7]. They derived exact solutions for the buckling temperature of thin [14] and thick [6,7] simply supported perfect plates. Tauchert et al. compared thermal buckling loads of laminated plates, based on various higher-order theories and classical theory [18]. Chang and Leu [15] studied the thermal buckling of antisymmetric angle-ply laminated rectangular plate based on a higher-order displacement field. A higher-order deformation theory, which accounts for transverse shear and transverse normal strains, is derived for the thermal buckling analysis of antisymmetric angle-ply laminated rectangular plates that are simply supported and subjected to a uniform temperature rise. Using the higher-order displacement field, the three-dimensional Hooke's law, and total potential energy, exact closed-form solutions of the thermal buckling temperature are obtained. Romeo and Frulla [19] studied nonlinear analysis of anisotropic plates with initial imperfections and various boundary conditions subjected to combined bi-axial compression and shear loads. In this paper, nonlinear analysis is developed for symmetric panels under bi-axial compression and shear loads in order to obtain the out-of-plane deflection in the prebuckling range. The nonlinear differential equations are obtained using the principle of stationary potential energy and are expressed in terms of the out-of-plane displacement and the Airy stress function. The equations are solved with Galerkin's method for various boundary conditions. Mossavarali and Eslami [20] computed the critical thermoelastic buckling loads of thin rectangular isotropic and orthotropic plates based on higher-order displacement field. In this paper, the equilibrium and stability equations are obtained and employed to compute the critical thermoelastic buckling loads under thickness thermal loading, one edge direction temperature difference, and the critical uniform initial-final temperature rise for simply supported perfect plates. Eslami et al. [8] studied thermoelastic buckling of isotropic and orthotropic plates with imperfections based on a first-order displacement field. In this paper stability equations are considered and the Koiter model for imperfection are adopted. Thermal buckling of imperfect plates under the three thermal loads which were thickness thermal loading, one edge direction temperature difference, and critical uniform temperature rise, are obtained. Closed form solutions are presented in all cases.

In this section, the equilibrium and stability equations of a plate based on a third-order displacement field for in-plane displacements are obtained using variational formulations. The imperfection model of Koiter is adapted. The critical thermoelastic buckling loads of thin rectangular imperfect isotropic plates under thickness thermal loading, one edge direction temperature difference, and the critical uniform initial-final temperature rise for simply supported imperfect plates are obtained. Closed form solutions are presented in all cases.

3.2   Basic equations

A thin plate of thickness h is considered. The general strain-displacement relations are [15]

$$\epsilon_{xx} = u_{,x} + \frac{1}{2}w_{,x}^2$$

$$\epsilon_{yy} = v_{,y} + \frac{1}{2}w_{,y}^2$$

$$\epsilon_{xy} = u_{,y} + v_{,x} + w_{,x}w_{,y}$$

$$\epsilon_{xz} = u_{,z} + w_{,x} + w_{,x}w_{,z}$$

$$\epsilon_{yz} = v_{,z} + w_{,y} + w_{,y}w_{,z} \qquad (1.3\text{-}1)$$

where ϵ_{xx} and ϵ_{yy} are the normal strains, and ϵ_{xy}, ϵ_{xz}, and ϵ_{yz}, are the shear strains. The indices x, y, and z refer to the coordinates system. Also u, v, and w are the displacements in the x, y, and z directions, respectively, and (,) indicates a partial derivative.

We now assume the following higher-order displacements [16]

$$u(x,y,z) = u_0(x,y) + z\phi_0(x,y) + z^2\acute{\phi}_0(x,y) + z^3\phi_0^{''}(x,y)$$

$$v(x,y,z) = v_0(x,y) + z\psi_0(x,y) + z^2\acute{\psi}_0(x,y) + z^3\psi_0^{''}(x,y)$$

$$w(x,y,z) = w_0(x,y) \qquad (1.3\text{-}2)$$

where u_0, v_0, and w_0 denote the displacements of a point (x,y) on the mid-plane, and ϕ_0 and ψ_0 are the rotations of normal to mid-plane about the y and x axes, respectively. The functions $\acute{\phi}_0, \phi_0^{''}, \acute{\psi}_0$, and $\psi_0^{''}$ will be determined using the conditions that transverse shear stresses, σ_{xz} and σ_{yz} vanish on the plate top and bottom surfaces. These are equivalent to $\epsilon_{xz} = \epsilon_{yz} = 0$ at $z = \pm\frac{h}{2}$, which yields [16]

$$\acute{\phi}_0 = \acute{\psi}_0 = 0$$

$$\phi_0^{''} = \frac{-4}{3h^2}(\phi_0 + w_{0,x})$$

$$\psi_0^{''} = \frac{-4}{3h^2}(\psi_0 + w_{0,y}) \qquad (1.3\text{-}3)$$

Using the conditions (1.3-3), Eqs. (1.3-2) reduce to [16]

$$u = u_0 + z\phi_0 - \frac{4z^3}{3h^2}(\phi_0 + w_{0,x})$$

$$v = v_0 + z\psi_0 - \frac{4z^3}{3h^2}(v1 + w_{0,y})$$

$$w = w_0 \qquad (1.3\text{-}4)$$

The strains associated with the displacements in Eq. (1.3-2) are [16]

$$
\begin{pmatrix} \epsilon_{xx} \\ \epsilon_{yy} \\ \epsilon_{xy} \end{pmatrix} = \begin{pmatrix} \epsilon_{xx}^0 \\ \epsilon_{yy}^0 \\ \epsilon_{xy}^0 \end{pmatrix} + z \begin{pmatrix} k_{xx}^0 \\ k_{yy}^0 \\ k_{xy}^0 \end{pmatrix} + z^3 \begin{pmatrix} k_{xx}^2 \\ k_{yy}^2 \\ k_{xy}^2 \end{pmatrix}
$$

$$
\begin{pmatrix} \epsilon_{xz} \\ \epsilon_{yz} \end{pmatrix} = \begin{pmatrix} \epsilon_{xz}^0 \\ \epsilon_{yz}^0 \end{pmatrix} + z^2 \begin{pmatrix} k_{xz}^1 \\ k_{yz}^2 \end{pmatrix} \tag{1.3-5}
$$

where

$$
\begin{pmatrix} k_{xx}^0 \\ k_{yy}^0 \\ k_{xy}^0 \end{pmatrix} = \begin{pmatrix} \phi_{0,x} \\ \psi_{0,y} \\ \phi_{0,y} + \psi_{0,x} \end{pmatrix}
$$

$$
\begin{pmatrix} k_{xx}^2 \\ k_{yy}^2 \\ k_{xy}^2 \end{pmatrix} = \begin{pmatrix} \frac{-4}{3h^2}(\phi_{0,x} + w_{0,xx}) \\ \frac{-4}{3h^2}(\psi_{0,y} + w_{0,yy}) \\ \frac{-4}{3h^2}(\phi_{0,y} + \psi_{0,x} + 2w_{0,xy}) \end{pmatrix}
$$

$$
\begin{pmatrix} k_{xz}^1 \\ k_{yz}^1 \end{pmatrix} = \begin{pmatrix} \frac{-4}{h^2}(\phi_0 + w_{0,x}) \\ \frac{-4}{h^2}(\psi_0 + w_{0,y}) \end{pmatrix}
$$

$$
\begin{pmatrix} \epsilon_{xx}^0 \\ \epsilon_{yy}^0 \\ \epsilon_{xy}^0 \end{pmatrix} = \begin{pmatrix} u_{0,x} + \frac{1}{2}w_{0,x}^2 \\ v_{0,y} + \frac{1}{2}w_{0,y}^2 \\ u_{0,y} + v_{0,x} + w_{0,x}w_{0,y} \end{pmatrix}
$$

$$
\begin{pmatrix} \epsilon_{xz}^0 \\ \epsilon_{yz}^0 \end{pmatrix} = \begin{pmatrix} \phi_0 + w_{0,x} \\ \psi_0 + w_{0,y} \end{pmatrix} \tag{1.3-6}
$$

The constitutive equations may be expressed in terms of stresses and strains in the plate coordinates as

$$
\begin{pmatrix} \sigma_{xx} \\ \sigma_{yy} \\ \sigma_{xy} \end{pmatrix} = \begin{pmatrix} \acute{Q}_{11} & \acute{Q}_{12} & \acute{Q}_{16} \\ \acute{Q}_{12} & \acute{Q}_{22} & \acute{Q}_{26} \\ \acute{Q}_{16} & \acute{Q}_{26} & \acute{Q}_{66} \end{pmatrix} \left(\begin{pmatrix} \epsilon_{xx}^0 \\ \epsilon_{yy}^0 \\ \epsilon_{xy}^0 \end{pmatrix} + z \begin{pmatrix} k_{xx}^0 \\ k_{yy}^0 \\ k_{xy}^0 \end{pmatrix} + z^3 \begin{pmatrix} k_{xx}^2 \\ k_{yy}^2 \\ k_{xy}^2 \end{pmatrix} - T \begin{pmatrix} \alpha_{xx} \\ \alpha_{yy} \\ \alpha_{xy} \end{pmatrix} \right)
$$

$$
\begin{pmatrix} \sigma_{xz} \\ \sigma_{yz} \end{pmatrix} = \begin{pmatrix} \acute{Q}_{44} & \acute{Q}_{45} \\ \acute{Q}_{45} & \acute{Q}_{55} \end{pmatrix} \left(\begin{pmatrix} \epsilon_{xz}^0 \\ \epsilon_{yz}^0 \end{pmatrix} + z^2 \begin{pmatrix} k_{xz}^1 \\ k_{yz}^1 \end{pmatrix} - T \begin{pmatrix} \alpha_{xz} \\ \alpha_{yz} \end{pmatrix} \right) \tag{1.3-7}
$$

where T is the absolute temperature, and α_{ij} are linear thermal expansion coefficients. Here, $\acute{Q}_{ij}$ are the plane-stress-reduced elastic constants in the material axes of the layer [11]. The stress resultants $N_{ij}, M_{ij}, P_{ij}, Q_{ij}$, and R_{ij} are [16]

$$
\left(\begin{pmatrix} N_{xx} \\ N_{yy} \\ N_{xy} \end{pmatrix}, \begin{pmatrix} M_{xx} \\ M_{yy} \\ M_{xy} \end{pmatrix}, \begin{pmatrix} P_{xx} \\ P_{yy} \\ P_{xy} \end{pmatrix} \right) = \int_{\frac{-h}{2}}^{\frac{h}{2}} \begin{pmatrix} \sigma_{xx} \\ \sigma_{yy} \\ \sigma_{xy} \end{pmatrix} (1, z, z^3) dz
$$

$$
\left(\begin{pmatrix} Q_{xz} \\ Q_{yz} \end{pmatrix}, \begin{pmatrix} R_{xz} \\ R_{yz} \end{pmatrix} \right) = \int_{\frac{-h}{2}}^{\frac{h}{2}} \begin{pmatrix} \sigma_{xz} \\ \sigma_{yz} \end{pmatrix} (1, z^2) dz \tag{1.3-8}
$$

By substitution of Eq. (1.3-7) into Eq. (1.3-8), the stress resultants are obtained as [16]

$$N_{ij} = A_{kl}\epsilon^0_{ij} + B_{kl}k^0_{ij} + E_{kl}k^2_{ij} - N^T_{ij} \quad i,j = x,y \ \ k,l = 1,2,6$$

$$M_{ij} = B_{kl}\epsilon^0_{ij} + D_{kl}k^0_{ij} + F_{kl}k^2_{ij} - M^T_{ij} \quad i,j = x,y \ \ k,l = 1,2,6$$

$$P_{ij} = E_{kl}\epsilon^0_{ij} + F_{kl}k^0_{ij} + H_{kl}k^2_{ij} - P^T_{ij} \quad i,j = x,y \ \ k,l = 1,2,6$$

$$Q_i = A_{kl}\epsilon^0_i + D_{kl}k^1_i - Q^T_i \quad i = xz,yz \ \ k,l = 4,5$$

$$R_i = D_{kl}\epsilon^0_i + F_{kl}k^1_i - R^T_i \quad i = xz,yz \ \ k,l = 4,5 \tag{1.3-9}$$

where A_{ij}, B_{ij}, etc., are the plate stiffnesses, defined by [16]

$$(A_{kl}, B_{kl}, D_{kl}, E_{kl}, F_{kl}, H_{kl}) = \int_{\frac{-h}{2}}^{\frac{h}{2}} \acute{Q}_{kl}(1, z, z^2, z^3, z^4, z^6)dz \quad k,l = 1,2,6$$

$$(A_{kl}, D_{kl}, F_{kl}) = \int_{\frac{-h}{2}}^{\frac{h}{2}} \acute{Q}_{kl}(1, z^2, z^4) \quad k,l = 4,5$$

$$(N^T_{ij}, M^T_{ij}, P^T_{ij}) = \int_{\frac{-h}{2}}^{\frac{h}{2}} \acute{Q}_{kl}(1, z, z^3)(-T\alpha_{ij}) \quad i,j = x,y \ \ k,l = 1,2,6$$

$$(Q^T_i, R^T_i) = \int_{\frac{-h}{2}}^{\frac{h}{2}} \acute{Q}_{kl}(1, z^2)(-T\alpha_i) \quad i = xz,yz \ \ k,l = 4,5 \tag{1.3-10}$$

The total potential energy function of the orthotropic rectangular plate is [10]

$$U = \frac{1}{2} \int \int \int_{\frac{-h}{2}}^{\frac{h}{2}} \{\sigma\}^T (\{\epsilon\} - T\{\alpha\})dxdydz \tag{1.3-11}$$

Upon substitution of Eqs. (1.3-5) and (1.3-7) into Eq. (1.3-11), the total potential energy is obtained as

$$U = \frac{1}{2} \int \int \int_{\frac{-h}{2}}^{\frac{h}{2}} (\{\epsilon^0_{ij}\}^T + z\{k^0_{ij}\}^T + z^3\{k^2_{ij}\}^T - T\{\alpha_{ij}\}^T)_{i,j=x,y}[\acute{Q}]^T (\{\epsilon^0_{ij}\}$$

$$+z\{k^0_{ij}\} + z^3\{k^2_{ij}\} - T\{\alpha_{ij}\})_{i,j=x,y}dxdydz$$

$$+\frac{1}{2} \int \int \int_{\frac{-h}{2}}^{\frac{h}{2}} (\{\epsilon^0_{ij}\}^T + z^2\{k^1_{ij}\}^T - T\{\alpha_{ij}\}^T)_{i=x,y;j=z}[\acute{Q}]^T (\{\epsilon^0_{ij}\}$$

$$+z^2\{k^1_{ij}\} - T\{\alpha_{ij}\})_{i=x,y;j=z}dxdydz \tag{1.3-12}$$

Assuming that the orthotropic rectangular plate is under thermal stress alone, the total potential energy is a function of the displacement components and their derivatives and may be written as

$$U = \int \int \int F(u_0, v_0, w_0, \phi_0, \psi_0, u_{0,x}, u_{0,y}, v_{0,x}, v_{0,y},$$

$$\phi_{0,x}, \phi_{0,y}, \psi_{0,x}, \psi_{0,y}, w_{0,x}, w_{0,y}, w_{0,xx}, w_{0,yy}, w_{0,xy}) dx dy dz$$

$$(1.3\text{-}13)$$

Minimizing the functional of potential energy leads to the Euler equations [10]

$$\frac{\partial F}{\partial u_0} - \frac{\partial}{\partial x}\frac{\partial F}{\partial u_{0,x}} - \frac{\partial}{\partial y}\frac{\partial F}{\partial u_{0,y}} = 0$$

$$\frac{\partial F}{\partial v_0} - \frac{\partial}{\partial x}\frac{\partial F}{\partial v_{0,x}} - \frac{\partial}{\partial y}\frac{\partial F}{\partial v_{0,y}} = 0$$

$$\frac{\partial F}{\partial w_0} - \frac{\partial}{\partial x}\frac{\partial F}{\partial w_{0,x}} - \frac{\partial}{\partial y}\frac{\partial F}{\partial w_{0,y}} + \frac{\partial^2}{\partial x^2}\frac{\partial F}{\partial w_{0,xx}} + \frac{\partial^2}{\partial x \partial y}\frac{\partial F}{\partial w_{0,xy}} + \frac{\partial^2}{\partial y^2}\frac{\partial F}{\partial w_{0,yy}} = 0$$

$$\frac{\partial F}{\partial \phi_0} - \frac{\partial}{\partial x}\frac{\partial F}{\partial \phi_{0,x}} - \frac{\partial}{\partial y}\frac{\partial F}{\partial \phi_{0,y}} = 0$$

$$\frac{\partial F}{\partial \psi_0} - \frac{\partial}{\partial x}\frac{\partial F}{\partial \psi_{0,x}} - \frac{\partial}{\partial y}\frac{\partial F}{\partial \psi_{0,y}} = 0 \qquad (1.3\text{-}14)$$

Upon substitution from Eqs. (1.3-6) through (1.3-11) into (1.3-13) and using Eqs. (1.3-14), the equilibrium equations for a general thin orthotropic rectangular plate are obtained as [16]

$$N_{xx,x} + N_{xy,y} = 0$$

$$N_{xy,x} + N_{yy,y} = 0$$

$$Q_{xz,x} + Q_{yz,y} - \frac{4}{h^2}(R_{xz,x} + R_{yz,y}) + \frac{4}{3h^2}(P_{xx,xx} + 2P_{xy,xy} + P_{yy,yy})$$

$$+(w_{0,x}N_{xx})_{,x} + (w_{0,y}N_{xy})_{,x} + (w_{0,y}N_{yy})_{,y} + (w_{0,x}N_{xy})_{,y} = 0$$

$$M_{xx,x} + M_{xy,y} - Q_{xz} + \frac{4}{h^2}R_{xz} - \frac{4}{3h^2}(P_{xx,x} + P_{xy,y}) = 0$$

$$M_{xy,x} + M_{yy,y} - Q_{yz} + \frac{4}{h^2}R_{yz} - \frac{4}{3h^2}(P_{xy,x} + P_{yy,y}) = 0 \qquad (1.3\text{-}15)$$

The stability equations of thin orthotropic rectangular plates may be derived by the force summation method. Let us assume that the state of equilibrium of a general orthotropic rectangular plate under load be defined in terms of the displacements components u_0, v_0, w_0, ϕ_0, and ψ_0. The displacement components of a neighboring state of stable equilibrium will differ by u_1, v_1, w_1, ϕ_1, ψ_1, with respect to the equilibrium position. Thus the total displacements of a neighboring state are

$$u = u_0 + u_1$$
$$v = v_0 + v_1$$
$$w = w_0 + w_1$$
$$\phi = \phi_0 + \phi_1$$
$$\psi = \psi_0 + \psi_1 \tag{1.3-16}$$

Similarly, the stress resultant components of a neighboring state may be related to the state of equilibrium as

$$N_{ij} = N_{ij0} + \Delta N_{ij}$$
$$M_{ij} = M_{ij0} + \Delta M_{ij}$$
$$P_{ij} = P_{ij0} + \Delta P_{ij}$$
$$Q_{ij} = Q_{ij0} + \Delta Q_{ij}$$
$$R_{ij} = R_{ij0} + \Delta R_{ij} \tag{1.3-17}$$

From the stress-strain relations (1.3-9), using the strain-displacement relations (1.3-6), it follows that

$$N_{xx} = A_{11}(u_{0,x} + \frac{1}{2}w_{0,x}^2) + A_{12}(v_{0,y} + \frac{1}{2}w_{0,y}^2) + A_{16}(u_{0,y} + v_{0,x} + w_{0,x}w_{0,y})$$

$$+ B_{11}\phi_{0,x} + B_{12}\psi_{0,y} + B_{16}(\phi_{0,y} + \psi_{0,x}) + E_{11}(\frac{-4}{3h^2})(\phi_{0,x} + w_{0,xx})$$

$$+ E_{12}(\frac{-4}{3h^2})(\psi_{0,y} + w_{0,yy}) + E_{16}(\frac{-4}{3h^2})(\phi_{0,y} + \psi_{0,x} + 2w_{0,xy}) \tag{1.3-18}$$

Substitution from Eqs. (1.3-17) into (1.3-18) yields

$$N_{xx} = N_{xx0} + \Delta N_{xx0} = A_{11}(u_{0,x} + \phi_{1,x} + \frac{1}{2}w_{0,x}^2 + w_{0,x}w_{1,x} + \frac{1}{2}w_{1,x}^2)$$

$$+ A_{12}(v_{0,y} + v_{1,y} + \frac{1}{2}w_{0,y}^2 + w_{0,y}w_{1,y} + \frac{1}{2}w_{1,y}^2)$$

$$+ A_{16}(u_{0,y} + u_{1,y} + v_{0,x} + v_{1,x} + w_{0,x}w_{0,y} + w_{0,x}w_{1,y} + w_{1,x}w_{0,y} + w_{1,x}w_{1,y})$$

$$+ B_{11}(\phi_{0,x} + \phi_{1,x}) + B_{12}(\psi_{0,y} + \psi_{1,y}) + B_{16}(\phi_{0,y} + \phi_{1,y} + \psi_{0,x} + \psi_{1,x})$$

$$+ E_{11}(\frac{-4}{3h^2})(\phi_{0,x} + \phi_{1,x} + w_{0,xx} + w_{1,xx}) + E_{12}(\frac{-4}{3h^2})(\psi_{0,y} + \psi_{1,y} + w_{0,yy}$$

$$+ w_{1,yy} + E_{16}(\frac{-4}{3h^2})(\phi_{0,y} + \phi_{1,y} + \psi_{0,x} + \psi_{1,x} + 2w_{0,xy} + 2w_{1,xy}) \tag{1.3-19}$$

The right hand side of this equation may be divided into terms related to the state of equilibrium, N_{x_0}, and the terms related to the deviations from the equilibrium state ΔN_{xx0}. In expression for ΔN_{xx} there are, in addition to the linear terms, the terms which are infinitesimal deviations of higher orders which may be neglected. Therefore, if N_{xx0} represents the force pertained to the stable equilibrium and N_{xx1} represent the deviations from the equilibrium state, it follows that

$$N_{xx0} = A_{11}(u_{0,x} + \frac{1}{2}w_{0,x}^2) + A_{12}(v_{0,y} + \frac{1}{2}w_{0,y}^2) + A_{16}(u_{0,y} + v_{0,x} + w_{0,x}w_{0,y})$$

$$+B_{11}\phi_{0,x} + B_{12}\psi_{0,y} + B_{16}(\phi_{0,y} + \psi_{0,x}) + E_{11}(\frac{-4}{3h^2})(\phi_{0,x} + w_{0,xx})$$

$$+E_{12}(\frac{-4}{3h^2})(\psi_{0,y} + w_{0,yy}) + E_{16}(\frac{-4}{3h^2})(\phi_{0,y} + \psi_{0,x} + 2w_{0,xy}) \qquad (1.3\text{-}20)$$

$$N_{xx1} = A_{11}(u_{1,x} + w_{0,x}w_{1,x}) + A_{12}(v_{1,y} + w_{0,y}w_{1,y})$$

$$+A_{16}(u_{1,y} + v_{1,x} + w_{0,x}w_{1,y} + w_{1,x}w_{0,y})$$

$$+B_{11}\phi_{1,x} + B_{12}\psi_{1,y} + B_{16}(\phi_{1,y} + \psi_{1,x})$$

$$+E_{11}(\frac{-4}{3h^2})(\phi_{1,x} + w_{1,xx}) + E_{12}(\frac{-4}{3h^2})(\psi_{1,y} + w_{1,yy})$$

$$+E_{16}(\frac{-4}{3h^2})(\phi_{1,y} + \psi_{1,x} + 2w_{1,xy}) \qquad (1.3\text{-}21)$$

With a similar procedure, other components of forces and moments are found to be related to the displacements of equilibrium state and the deviations from the equilibrium. Denoting the prebuckling modes by N_{ij0}, M_{ij0}, P_{ij0}, Q_{ij0}, and R_{ij0}, and the buckling modes by N_{ij1}, M_{ij1}, P_{ij1}, Q_{ij1}, and R_{ij1}, it follows that

$$N_{ij0} = A_{kl}\epsilon_{ij0}^0 + B_{kl}k_{ij0}^0 + E_{kl}k_{ij0}^2 - N_{ij0}^T \quad i,j = x,y \;\; k,l = 1,2,6$$

$$M_{ij0} = B_{kl}\epsilon_{ij0}^0 + D_{kl}k_{ij0}^0 + F_{kl}k_{ij0}^2 - M_{ij0}^T \quad i,j = x,y \;\; k,l = 1,2,6$$

$$P_{ij0} = E_{kl}\epsilon_{ij0}^0 + F_{kl}k_{ij0}^0 + H_{kl}k_{ij0}^2 - P_{ij0}^T \quad i,j = x,y \;\; k,l = 1,2,6$$

$$Q_{i0} = A_{kl}\epsilon_{i0}^0 + D_{kl}k_{i0}^1 - Q_{i0}^T \quad i = xz,yz \;\; k,l = 4,5$$

$$R_{i0} = D_{kl}\epsilon_{i0}^0 + F_{kl}k_{i0}^1 - R_{i0}^T \quad i = xz,yz \;\; k,l = 4,5$$

$$N_{ij1} = A_{kl}\epsilon_{ij1}^0 + B_{kl}k_{ij1}^0 + E_{kl}k_{ij1}^2 \quad i,j = x,y \;\; k,l = 1,2,6$$

$$M_{ij1} = B_{kl}\epsilon_{ij1}^0 + D_{kl}k_{ij1}^0 + F_{kl}k_{ij1}^2 \quad i,j = x,y \;\; k,l = 1,2,6$$

$$P_{ij1} = E_{kl}\epsilon_{ij1}^0 + F_{kl}k_{ij1}^0 + H_{kl}k_{ij1}^2 \quad i,j = x,y \;\; k,l = 1,2,6$$

$$Q_{i1} = A_{kl}\epsilon_{i1}^0 + D_{kl}k_{i1}^1 \quad i = xz,yz \;\; k,l = 4,5$$

$$R_{i1} = D_{kl}\epsilon_{i1}^0 + F_{kl}k_{i1}^1 - R_{i1}^T \quad i = xz,yz \;\; k,l = 4,5 \qquad (1.3\text{-}22)$$

Equations (1.3-16) are substituted into the nonlinear strain-displacement relations (1.3-6) and, neglecting the higher order derivation terms, the linear strain-displacement relations for the deviation components reduce to

$$\epsilon_{xx1}^0 = u_{1,x} \qquad k_{xx1}^0 = \phi_{1,x} \qquad k_{xx1}^2 = \frac{-4}{3h^2}(\phi_{1,x} + w_{1,xx})$$

$$\epsilon_{yy1}^0 = v_{1,x} \qquad k_{yy1}^0 = \psi_{1,x} \qquad k_{yy1}^2 = \frac{-4}{3h^2}(\psi_{1,x} + w_{1,yy})$$

$$\epsilon_{xy1}^0 = u_{1,y} + v_{1,x} \quad k_{xx1}^0 = \phi_{1,y} + \psi_{1,x} \quad k_{xy1}^2 = \frac{-4}{3h^2}(\phi_{1,y} + \psi_{1,x} + 2w_{1,xy})$$

$$(1.3\text{-}23)$$

Now, the stability equations may be obtained by substitution of Eqs. (1.3-16) and (1.3-17) into (1.3-15) with the help of Eqs. (1.3-22) and (1.3-23). Upon substitution, the terms in the resulting equations with the subscript (0) satisfy the equilibrium condition and therefore drop off the equations. Also, the nonlinear terms with the subscript (1) will be ignored because they are small compared to the linear terms. The remaining terms form the stability equations of a thin orthotropic rectangular plate under general load as follows [16]

$$N_{xx1,x} + N_{xy1,y} = 0$$

$$N_{xy1,x} + N_{yy1,y} = 0$$

$$Q_{xz1,x} + Q_{yz1,y} - \frac{4}{h^2}(R_{xz1,x} + R_{yz1,y}) + \frac{4}{3h^2}(P_{xz1,xx} + 2P_{xy1,xy} + P_{yz1,yy})$$

$$+ (w_{1,x}N_{xx0})_{,x} + (w_{1,y}N_{xy0})_{,x} + (w_{1,y}N_{yy0})_{,y} + (w_{1,x}N_{xy0})_{,y} = 0$$

$$M_{xx1,x} + M_{xy1,y} - Q_{xz1} + \frac{4}{h^2}R_{xz1} - \frac{4}{3h^2}(P_{xx1,x} + P_{xy1,y}) = 0$$

$$M_{xy1,x} + M_{yy1,y} - Q_{yz1} + \frac{4}{h^2}R_{yz1} - \frac{4}{3h^2}(P_{xy1,x} + P_{yy1,y}) = 0 \qquad (1.3\text{-}24)$$

where in these equations N_{ij1}, M_{ij1}, P_{ij1}, Q_{ij1}, and R_{ij1}, are related to the linear strain components as given by Eqs. (1.3-22).

3.3   Thermal buckling of imperfect isotropic plates based on the Koiter model

For isotropic thin plates, coefficients $\acute{Q}_{ij}$, A_{ij}, B_{ij}, etc. are

$$\begin{pmatrix} \acute{Q}_{11} & \acute{Q}_{12} & \acute{Q}_{16} \\ \acute{Q}_{12} & \acute{Q}_{22} & \acute{Q}_{26} \\ \acute{Q}_{16} & \acute{Q}_{26} & \acute{Q}_{66} \end{pmatrix} = \begin{pmatrix} \frac{E}{1-\nu^2} & \frac{E\nu}{1-\nu^2} & 0 \\ \frac{E\nu}{1-\nu^2} & \frac{E}{1-\nu^2} & 0 \\ 0 & 0 & \frac{E}{2(1+\nu)} \end{pmatrix}$$

$$\begin{pmatrix} \acute{Q}_{44} & \acute{Q}_{45} \\ \acute{Q}_{45} & \acute{Q}_{55} \end{pmatrix} = \begin{pmatrix} \frac{E}{2(1+\nu)} & 0 \\ 0 & \frac{E}{2(1+\nu)} \end{pmatrix}$$

$$B_{ij} = E_{ij} = 0$$

$$\begin{pmatrix} A_{11} & A_{12} & A_{16} \\ A_{12} & A_{22} & A_{26} \\ A_{16} & A_{26} & A_{66} \end{pmatrix} = \begin{pmatrix} \frac{Eh}{1-\nu^2} & \frac{Eh\nu}{1-\nu^2} & 0 \\ \frac{Eh\nu}{1-\nu^2} & \frac{Eh}{1-\nu^2} & 0 \\ 0 & 0 & \frac{Eh}{2(1+\nu)} \end{pmatrix}$$

$$\begin{pmatrix} A_{44} & A_{45} \\ A_{45} & A_{55} \end{pmatrix} = \begin{pmatrix} \frac{Eh}{2(1+\nu)} & 0 \\ 0 & \frac{Eh}{2(1+\nu)} \end{pmatrix}$$

$$\begin{pmatrix} D_{11} & D_{12} & D_{16} \\ D_{12} & D_{22} & D_{26} \\ D_{16} & D_{26} & D_{66} \end{pmatrix} = \begin{pmatrix} \frac{Eh^3}{12(1-\nu^2)} & \frac{Eh^3\nu}{12(1-\nu^2)} & 0 \\ \frac{Eh^3\nu}{12(1-\nu^2)} & \frac{Eh^3}{12(1-\nu^2)} & 0 \\ 0 & 0 & \frac{Eh^3}{24(1+\nu)} \end{pmatrix}$$

$$\begin{pmatrix} D_{44} & D_{45} \\ D_{45} & D_{55} \end{pmatrix} = \begin{pmatrix} \frac{Eh^3}{24(1+\nu)} & 0 \\ 0 & \frac{Eh^3}{24(1+\nu)} \end{pmatrix}$$

$$\begin{pmatrix} F_{11} & F_{12} & F_{16} \\ F_{12} & F_{22} & F_{26} \\ F_{16} & F_{26} & F_{66} \end{pmatrix} = \begin{pmatrix} \frac{Eh^5}{90(1-\nu^2)} & \frac{Eh^5\nu}{90(1-\nu^2)} & 0 \\ \frac{Eh^5\nu}{90(1-\nu^2)} & \frac{Eh^5}{90(1-\nu^2)} & 0 \\ 0 & 0 & \frac{Eh^5}{180(1+\nu)} \end{pmatrix}$$

$$\begin{pmatrix} F_{44} & F_{45} \\ F_{45} & F_{55} \end{pmatrix} = \begin{pmatrix} \frac{Eh^5}{180(1+\nu)} & 0 \\ 0 & \frac{Eh^5}{180(1+\nu)} \end{pmatrix}$$

$$\begin{pmatrix} H_{11} & H_{12} & H_{16} \\ H_{12} & H_{22} & H_{26} \\ H_{16} & H_{26} & H_{66} \end{pmatrix} = \begin{pmatrix} \frac{Eh^7}{448(1-\nu^2)} & \frac{Eh^7\nu}{448(1-\nu^2)} & 0 \\ \frac{Eh^7\nu}{448(1-\nu^2)} & \frac{Eh^7}{448(1-\nu^2)} & 0 \\ 0 & 0 & \frac{Eh^7}{896(1+\nu)} \end{pmatrix}$$

$$\begin{pmatrix} H_{44} & H_{45} \\ H_{45} & H_{55} \end{pmatrix} = \begin{pmatrix} \frac{Eh^7}{896(1+\nu)} & 0 \\ 0 & \frac{Eh^7}{896(1+\nu)} \end{pmatrix}$$

$$\begin{pmatrix} \alpha_{xx} \\ \alpha_{yy} \\ \alpha_{xy} \end{pmatrix} = \begin{pmatrix} \alpha \\ \alpha \\ \alpha \end{pmatrix} \tag{1.3-25}$$

With the substitution of Eqs. (1.3-6) and (1.3-25) into Eqs. (1.3-9), the stress resultants for isotropic plates are obtained as

$$\begin{pmatrix} N_{xx} \\ N_{yy} \\ N_{xy} \end{pmatrix} = \begin{pmatrix} A_{11} & A_{12} & A_{16} \\ A_{12} & A_{22} & A_{26} \\ A_{16} & A_{26} & A_{66} \end{pmatrix} \begin{pmatrix} u_{0,x} + \frac{1}{2}w_{0,x}^2 \\ v_{0,y} + \frac{1}{2}w_{0,y}^2 \\ u_{0,y} + v_{0,x} + w_{0,x}w_{0,y} \end{pmatrix}$$
$$- \begin{pmatrix} A_{11} & A_{12} & A_{16} \\ A_{12} & A_{22} & A_{26} \\ A_{16} & A_{26} & A_{66} \end{pmatrix} \begin{pmatrix} \alpha_{xx} \\ \alpha_{yy} \\ \alpha_{xy} \end{pmatrix} T$$

$$\begin{pmatrix} M_{xx} \\ M_{yy} \\ M_{xy} \end{pmatrix} = \begin{pmatrix} D_{11} & D_{12} & D_{16} \\ D_{12} & D_{22} & D_{26} \\ D_{16} & D_{26} & D_{66} \end{pmatrix} \begin{pmatrix} \phi_{0,x} \\ \psi_{0,y} \\ \phi_{0,y} + \psi_{0,x} \end{pmatrix}$$
$$+ \begin{pmatrix} F_{11} & F_{12} & F_{16} \\ F_{12} & F_{22} & F_{26} \\ F_{16} & F_{26} & F_{66} \end{pmatrix} \begin{pmatrix} \frac{-4}{3h^2}(\phi_{0,x} + w_{0,xx}) \\ \frac{-4}{3h^2}(\psi_{0,y} + w_{0,yy}) \\ \frac{-4}{3h^2}(\phi_{0,y} + \psi_{0,x} + 2w_{0,xy}) \end{pmatrix} - \begin{pmatrix} B_{11} & B_{12} & B_{16} \\ B_{12} & B_{22} & B_{26} \\ B_{16} & B_{26} & B_{66} \end{pmatrix} \begin{pmatrix} \alpha_{xx} \\ \alpha_{yy} \\ \alpha_{xy} \end{pmatrix} T$$

$$
\begin{pmatrix} P_{xx} \\ P_{yy} \\ P_{xy} \end{pmatrix} = \begin{pmatrix} F_{11} & F_{12} & F_{16} \\ F_{12} & F_{22} & F_{26} \\ F_{16} & F_{26} & F_{66} \end{pmatrix} \begin{pmatrix} \phi_{0,x} \\ \psi_{0,y} \\ \phi_{0,y} + \psi_{0,x} \end{pmatrix}
$$

$$
+ \begin{pmatrix} H_{11} & H_{12} & H_{16} \\ H_{12} & H_{22} & H_{26} \\ H_{16} & H_{26} & H_{66} \end{pmatrix} \begin{pmatrix} \frac{-4}{3h^2}(\phi_{0,x} + w_{0,xx}) \\ \frac{-4}{3h^2}(\psi_{0,y} + w_{0,yy}) \\ \frac{-4}{3h^2}(\phi_{0,y} + \psi_{0,x} + 2w_{0,xy}) \end{pmatrix} - \begin{pmatrix} E_{11} & E_{12} & E_{16} \\ E_{12} & E_{22} & E_{26} \\ E_{16} & E_{26} & E_{66} \end{pmatrix} \begin{pmatrix} \alpha_{xx} \\ \alpha_{yy} \\ \alpha_{xy} \end{pmatrix} T
$$

$$
\begin{pmatrix} Q_{xz} \\ Q_{yz} \end{pmatrix} = \begin{pmatrix} A_{44} & A_{45} \\ A_{45} & A_{55} \end{pmatrix} \begin{pmatrix} \phi_0 + w_{0,x} \\ \psi_0 + w_{0,y} \end{pmatrix}
$$

$$
+ \begin{pmatrix} D_{44} & D_{45} \\ D_{45} & D_{55} \end{pmatrix} \begin{pmatrix} \frac{-4}{h^2}(\phi_0 + w_{0,x}) \\ \frac{-4}{h^2}(\psi_0 + w_{0,y}) \end{pmatrix} - \begin{pmatrix} A_{44} & A_{45} & A_{45} \\ & A_{55} & \end{pmatrix} \begin{pmatrix} \alpha_{xx} \\ \alpha_{yy} \\ \alpha_{xy} \end{pmatrix} T
$$

$$
\begin{pmatrix} R_{xz} \\ R_{yz} \end{pmatrix} = \begin{pmatrix} D_{44} & D_{45} \\ D_{45} & D_{55} \end{pmatrix} \begin{pmatrix} \phi_0 + w_{0,x} \\ \psi_0 + w_{0,y} \end{pmatrix}
$$

$$
+ [F_{ij}]_{i,j=4,5} \begin{pmatrix} \frac{-4}{h^2}(\phi_0 + w_{0,x}) \\ \frac{-4}{h^2}(\psi_0 + w_{0,y}) \end{pmatrix} - \begin{pmatrix} D_{44} & D_{45} \\ D_{45} & D_{55} \end{pmatrix} \begin{pmatrix} \alpha_{xx} \\ \alpha_{yy} \\ \alpha_{xy} \end{pmatrix} T
$$

$$(1.3\text{-}26)$$

where $\{\alpha\}$ is the matrix of linear thermal expansion coefficient for the isotropic plate. Consider a rectangular plate of homogeneous and isotropic material. The boundary conditions along the edges of the plate are assumed to be simply supported. The edge conditions are thus (a and b are the planform dimensions of the plate) [16]

$$
w(x,0) = w(x,b) = w(0,y) = w(a,y) = 0
$$
$$
P_{yy}(x,0) = P_{yy}(x,b) = P_{xx}(0,y) = P_{xx}(a,y) = 0
$$
$$
M_{yy}(x,0) = M_{yy}(x,b) = M_{xx}(0,y) = M_x(a,y) = 0
$$
$$
\phi(x,0) = \phi(x,b) = \psi(0,y) = \psi(a,y) = 0 \qquad (1.3\text{-}27)
$$

The forces in the x and y-directions which are developed in the plate due to the assumed thermal loading and displacement relations are (since in Eq. (1.3-4) $w = w_0$, therefore Q_{ij0}, R_{ij0} do not appear in the stability equations (1.3-24))

$$
N_{xx0} = -N_{xx0}^T
$$
$$
N_{yy0} = -N_{yy0}^T
$$
$$
N_{xy0} = -N_{xy0}^T = 0
$$

$$(1.3\text{-}28)$$

The geometric imperfection of plate based on Koiter model may be assumed as [10]

$$
w^* = \mu h \sin \frac{m\pi}{a} x \sin \frac{n\pi}{b} y \qquad (1.3\text{-}29)
$$

where the coefficient μ varies between 0 and 1, and the coefficient μh represents amplitude of the imperfection. Since the plate is imperfect, there are initial strains related to the imperfection, which with the consideration of Eqs. (1.3-6) are

$$\epsilon_x^* = \frac{1}{2}w_{,x}^{*\,2} \qquad \epsilon_{xz}^* = w_{,x}^*$$

$$\epsilon_y^* = \frac{1}{2}w_{,y}^{*\,2} \qquad \epsilon_{yz}^* = w_{,y}^*$$

$$\epsilon_{xy}^* = w_{,x}^* w_{,y}^* \tag{1.3-30}$$

The total strains of the middle plane are

$$\epsilon_x^0 = u_{0,x} + \frac{1}{2}(w_0 + w^*)_{,x}^{\,2}$$

$$\epsilon_y^0 = v_{0,y} + \frac{1}{2}(w_0 + w^*)_{,y}^{\,2}$$

$$\epsilon_{xy}^0 = u_{0,y} + v_{0,x} + (w_0 + w^*)_{,x}(w_0 + w^*)_{,y}$$

$$\epsilon_{xz}^0 = \phi_0 + (w_0 + w^*)_{,x}$$

$$\epsilon_x^0 = \psi_0 + (w_0 + w^*)_{,y} \tag{1.3-31}$$

The net strains in the plate are thus

$$\epsilon_{xx} = \epsilon_x^0 - \epsilon_x^* = u_{0,x} + \frac{1}{2}w_{0,x}^{\,2} + w_{0,x}w_{,x}^*$$

$$\epsilon_{yy} = \epsilon_y^0 - \epsilon_y^* = v_{0,y} + \frac{1}{2}w_{0,y}^{\,2} + w_{0,y}w_{,y}^*$$

$$\epsilon_{xy} = \epsilon_{xy}^0 - \epsilon_{xy}^* = u_{0,y} + v_{0,x} + w_{0,x}w_{0,y} + w_{0,x}w_{,y}^* + w_{0,y}w_{,x}^*$$

$$\epsilon_{xz} = \epsilon_{xz}^0 - \epsilon_{xz}^* = \phi_0 + w_{0,x}$$

$$\epsilon_{yz} = \epsilon_{yz}^0 - \epsilon_{yz}^* = \psi_0 + w_{0,y} \tag{1.3-32}$$

Substituting Eqs. (1.3-25) and (1.3-26) into the equilibrium equations (1.3-15), and replacing w_0 by $w_0 + w^*$, the equilibrium equations in terms of the displacement components are obtained as

$$\frac{Eh}{1-\nu^2}\left(\frac{\partial^2 u_0}{\partial x^2} + \frac{\partial(w_0 + w^*)}{\partial x}\frac{\partial^2(w_0 + w^*)}{\partial x^2}\right) + \frac{Eh}{2(1+\nu)}\left(\frac{\partial^2 u_0}{\partial y^2} + \frac{\partial^2 v_0}{\partial x \partial y}\right.$$

$$\left. + \frac{\partial^2(w_0 + w^*)}{\partial x \partial y}\frac{\partial(w_0 + w^*)}{\partial y} + \frac{\partial(w_0 + w^*)}{\partial x}\frac{\partial^2(w_0 + w^*)}{\partial y^2}\right) = 0$$

$$\frac{Eh}{1-\nu^2}\left(\frac{\partial^2 v_0}{\partial y^2} + \frac{\partial(w_0 + w^*)}{\partial y}\frac{\partial^2(w_0 + w^*)}{\partial y^2}\right) + \frac{Eh}{2(1+\nu)}\left(\frac{\partial^2 v_0}{\partial x^2} + \frac{\partial^2 u_0}{\partial x \partial y}\right.$$

$$\left. + \frac{\partial^2(w_0 + w^*)}{\partial x \partial y}\frac{\partial(w_0 + w^*)}{\partial x} + \frac{\partial(w_0 + w^*)}{\partial y}\frac{\partial^2(w_0 + w^*)}{\partial x^2}\right) = 0$$

$$\frac{Eh}{2(1+\nu)}\left(\frac{\partial \phi_0}{\partial x}+\frac{\partial^2(w_0+w^*)}{\partial x^2}+\frac{\partial \psi_0}{\partial y}\right.$$

$$+\frac{\partial^2(w_0+w^*)}{\partial y^2}\left.\right)+\frac{Eh^3}{12.2(1+\nu)}\left(\frac{-4}{h^2}\right)\left(\frac{\partial \phi_0}{\partial x}+\frac{\partial^2(w_0+w^*)}{\partial x^2}\right.$$

$$+\frac{\partial \psi_0}{\partial y}+\frac{\partial^2(w_0+w^*)}{\partial y^2}\left.\right)+\left(\frac{-4}{h^2}\right)\frac{Eh^3}{12.2(1+\nu)}\left(\frac{\partial \phi_0}{\partial x}+\frac{\partial^2(w_0+w^*)}{\partial x^2}+\frac{\partial \psi_0}{\partial y}+\frac{\partial^2(w_0+w^*)}{\partial y^2}\right)$$

$$+\left(\frac{-4}{h^2}\right)^2\frac{Eh^5}{90.2(1+\nu)}\left(\frac{\partial \phi_0}{\partial x}+\frac{\partial^2(w_0+w^*)}{\partial x^2}+\frac{\partial \psi_0}{\partial y}+\frac{\partial^2(w_0+w^*)}{\partial y^2}\right.+\left(\frac{4}{3h^2}\right)\frac{Eh^5}{90(1-\nu^2)}\cdot$$

$$\left(\frac{\partial^3 \phi_0}{\partial x^3}+\nu\frac{\partial \psi_0^3}{\partial x^2 \partial y}+\frac{\partial \psi_0^3}{\partial y^3}+\nu\frac{\partial \phi_0^3}{\partial x \partial y^2}\right)-\left(\frac{4}{3h^2}\right)^2\frac{Eh^7}{448(1-\nu^2)}\left(\frac{\partial^3 \phi_0}{\partial x^3}+\frac{\partial^4(w_0+w^*)}{\partial x^4}\right.$$

$$+\nu\frac{\partial^3 \psi_0}{\partial x^2 \partial y}+\nu\frac{\partial^4(w_0+w^*)}{\partial x^2 \partial y^2}+\frac{\partial^3 \psi_0}{\partial y^3}+\frac{\partial^4(w_0+w^*)}{\partial y^4}+\nu\frac{\partial^3 \phi_0}{\partial x \partial y^2}+\nu\frac{\partial^4(w_0+w^*)}{\partial x^2 \partial y^2}\left.\right)$$

$$+\left(\frac{4}{3h^2}\right)\frac{Eh^5}{90(1+\nu)}\left(\frac{\partial^3 \phi_0}{\partial x \partial y^2}+\frac{\partial^3 \psi_0}{\partial x^2 \partial y}\right)-\left(\frac{4}{3h^2}\right)^2\frac{Eh^7}{448(1-\nu^2)}\left(\frac{\partial^3 \phi_0}{\partial x \partial y^2}+\frac{\partial^3 \psi_0}{\partial x^2 \partial y}\right.$$

$$+2\frac{\partial^4(w_0+w^*)}{\partial x^2 \partial y^2}\left.\right)+N_{xx}\frac{\partial^2(w_0+w^*)}{\partial x^2}+2N_{xy}\frac{\partial^2(w_0+w^*)}{\partial x \partial y}+N_{yy}\frac{\partial^2(w_0+w^*)}{\partial y^2}=0$$

$$\frac{Eh^3}{12(1-\nu^2)}\left(\frac{\partial^2 \phi_0}{\partial x^2}+\nu\frac{\partial^2 \psi_0}{\partial x \partial y}\right)+\frac{Eh^5}{90(1-\nu^2)}\left(\frac{-4}{3h^2}\right)\left(\frac{\partial^2 \phi_0}{\partial x^2}+\frac{\partial^3(w_0+w^*)}{\partial x^3}+\nu\frac{\partial^2 \psi_0}{\partial x \partial y}\right.$$

$$+\nu\frac{\partial^3(w_0+w^*)}{\partial x \partial y^2}\left.\right)+\frac{Eh^3}{12.2(1+\nu)}\left(\frac{\partial^2 \phi_0}{\partial y^2}+\frac{\partial^2 \psi_0}{\partial x \partial y}\right)+\frac{Eh^5}{90.2(1+\nu)}\left(\frac{-4}{3h^2}\right)\left(\frac{\partial^2 \phi_0}{\partial y^2}+\frac{\partial^2 \psi_0}{\partial x \partial y}\right.$$

$$+2\frac{\partial^3(w_0+w^*)}{\partial x \partial y^2}\left.\right)-\frac{Eh}{2(1+\nu)}\left(\phi_0+\frac{\partial(w_0+w^*)}{\partial x}\right)-\frac{Eh^3}{12.2(1+\nu)}\left(\frac{-4}{h^2}\right)\left(\phi_0\right.$$

$$+\frac{\partial(w_0+w^*)}{\partial x}\left.\right)+\frac{Eh^3}{12.2(1+\nu)}\left(\frac{4}{h^2}\right)\left(\phi_0+\frac{\partial(w_0+w^*)}{\partial x}\right.-\frac{Eh^5}{90.2(1+\nu)}\left(\frac{4}{h^2}\right)^2\cdot$$

$$\left(\phi_0+\frac{\partial(w_0+w^*)}{\partial x}\right)-\left(\frac{4}{3h^2}\right)\frac{Eh^5}{90(1-\nu^2)}\left(\frac{\partial^2 \phi_0}{\partial x^2}+\nu\frac{\partial^2 \psi_0}{\partial x \partial y}\right)+\left(\frac{4}{3h^2}\right)^2\frac{Eh^7}{448(1-\nu^2)}\cdot$$

$$\left(\frac{\partial^2 \phi_0}{\partial x^2}+\frac{\partial^3(w_0+w^*)}{\partial x^3}+\nu\frac{\partial^2 \psi_0}{\partial x \partial y}+\nu\frac{\partial^3(w_0+w^*)}{\partial x \partial y^2}\right)-\left(\frac{4}{3h^2}\right)\frac{Eh^5}{90.2(1+\nu)}\left(\frac{\partial^2 \phi_0}{\partial y^2}+\frac{\partial^2 \psi_0}{\partial x \partial y}\right)$$

$$+\left(\frac{4}{3h^2}\right)^2\frac{Eh^7}{448.2(1-\nu^2)}\left(\frac{\partial^2 }{\partial y^2}+\frac{\partial^2 \psi_0}{\partial x \partial y}+2\frac{\partial^3(w_0+w^*)}{\partial x \partial y^2}\right)=0$$

$$\frac{Eh^3}{12.2(1+\nu)}\left(\frac{\partial^2 \phi_0}{\partial x \partial y}+\frac{\partial^2 \psi_0}{\partial x^2}\right)+\frac{Eh^5}{90.2(1+\nu)}\left(\frac{-4}{3h^2}\right)\left(\frac{\partial^2 \phi_0}{\partial x \partial y}+\frac{\partial^2 \psi_0}{\partial x^2}+2\frac{\partial^3(w_0+w^*)}{\partial x^2 \partial y}\right)$$

$$+\frac{Eh^3}{12(1-\nu^2)}\left(\frac{\partial^2 \psi_0}{\partial y^2}+\nu\frac{\partial^2 \phi_0}{\partial x \partial y}\right)+\frac{Eh^5}{90(1-\nu^2)}\left(\frac{-4}{3h^2}\right)\left(\frac{\partial^2 \psi_0}{\partial y^2}+\frac{\partial^3(w_0+w^*)}{\partial y^3}+\nu\frac{\partial^2 \phi_0}{\partial x \partial y}\right.$$

$$\nu\frac{\partial^3(w_0+w^*)}{\partial x^2 \partial y}\left.\right)-\frac{Eh}{2(1+\nu)}\left(\psi_0+\frac{\partial(w_0+w^*)}{\partial y}\right)+\frac{Eh^3}{12.2(1+\nu)}\left(\frac{4}{h^2}\right)\left(\psi_0+\frac{\partial(w_0+w^*)}{\partial y}\right)$$

$$+\frac{Eh^3}{12.2(1+\nu)}\left(\frac{4}{h^2}\right)\left(\psi_0+\frac{\partial(w_0+w^*)}{\partial y}\right)-\frac{Eh^5}{90.2(1+\nu)}\left(\frac{4}{h^2}\right)^2\left(\psi_0+\frac{\partial(w_0+w^*)}{\partial y}\right)$$

$$-\frac{Eh^5}{90.2(1+\nu)}\left(\frac{4}{3h^2}\right)\left(\frac{\partial^2 \phi_0}{\partial x \partial y}+\frac{\partial^2 \psi_0}{\partial x^2}\right)+\frac{Eh^7}{448.2(1-\nu^2)}\left(\frac{4}{3h^2}\right)^2\left(\frac{\partial^2 \phi_0}{\partial x \partial y}+\frac{\partial^2 \psi_0}{\partial x^2}\right.$$

$$+2\frac{\partial^3(w_0+w^*)}{\partial x^2 \partial y}\left.\right)-\frac{Eh^5}{90(1-\nu^2)}\left(\frac{4}{3h^2}\right)\left(\frac{\partial^2 \psi_0}{\partial y^2}+\nu\frac{\partial^2 \phi_0}{\partial x \partial y}\right)+\frac{Eh^7}{448(1-\nu^2)}\left(\frac{4}{3h^2}\right)^2\cdot$$

$$\left(\frac{\partial^2 \psi_0}{\partial y^2}+\frac{\partial^3(w_0+w^*)}{\partial y^3}+\nu\frac{\partial^2 \phi_0}{\partial x \partial y}+\nu\frac{\partial^3(w_0+w^*)}{\partial x^2 \partial y}\right)=0 \tag{1.3-33}$$

These equations describe the equilibrium condition of prebuckling displacements of a plate and are used for imperfection analysis of rectangular plate based on the Koiter model.

To obtain the stability equations for imperfect plate, the force summation method may be used. The displacements ϕ, ψ, and w, and forces N_{xx}, N_{yy}, and N_{xy} are related to their values at stable condition and the neighboring state as given

$$
\begin{aligned}
u_0 &= \acute{u}_0 + u_1 & N_{xx} &= N_{xx0} + N_{xx1} \\
v_0 &= \acute{v}_0 + v_1 & N_{xy} &= N_{xy0} + N_{xy1} \\
w_0 &= \acute{w}_0 + w_1 & N_{yy0} &= N_{yy0} + N_{yy1} \\
\phi_0 &= \acute{\phi_0}0 + \phi_1 & \\
\psi_0 &= \acute{\psi_0}0 + \psi_1 &
\end{aligned}
\tag{1.3-34}
$$

where $\acute{u}_0, \acute{v}_0, \acute{w}_0, \acute{\phi_0}0, \acute{\psi_0}0$, represent the state of equilibrium displacement components and $u_1, v_1, w_1, \phi_1, \psi_1$, represent the displacement components of a neighboring state of stable equilibrium which differ with respect to the equilibrium position. Also $N_{xx0}, N_{xy0}, N_{yy0}$, represent the state of equilibrium force components and $N_{xx1}, N_{xy1}, N_{yy1}$, represent the force components of a neighboring state of stable equilibrium which differ with respect to the equilibrium position.

Substitution of Eqs. (1.3-34) into (1.3-33), ignoring the higher order terms, and replacing N_{xx1}, N_{yy1}, and N_{xy1} by the Airy stress function $\Phi_{,yy}, \Phi_{,xx}$, and $\Phi_{,xy}$, respectively, and replacing $N_{xy0} = 0$ (for the assumed loading conditions), the stability equations are obtained. To obtain a fourth equation relating the dependent functions ϕ_1, ψ_1, w_1, and Φ, the compatibility equation may be used as follows

$$
\epsilon_{x,yy} + \epsilon_{y,xx} = \epsilon_{xy,xy}
\tag{1.3-35}
$$

where ϵ_{xx}, ϵ_{yy}, and ϵ_{xy} are the strains of the middle plane of plate. Substituting for them from Eqs. (1.3-32) and considering $u_0 = \acute{u}_0 + u_1, v_0 = \acute{v}_0 + v_1$, and $w_0 = \acute{w}_0 + w_1$ the following equations for linear strains ignoring the higher order terms are obtained

$$
\begin{aligned}
\epsilon_{xx1} &= u_{1,x} + \frac{1}{2}w_{1,x}\acute{w}_{0,x} + w_{1,x}w^*_{,x} \\
\epsilon_{yy1} &= v_{1,y} + \frac{1}{2}w_{1,y}\acute{w}_{0,y} + w_{1,y}w^*_{,y} \\
\epsilon_{xy1} &= u_{1,y} + v_{1,x} + w_{1,x}\acute{w}_{0,y} + w_{1,y}\acute{w}_{0,x} + w_{1,x}w^*_{,y} + w_{1,y}w^*_{,x}
\end{aligned}
\tag{1.3-36}
$$

Substitution of Eqs. (1.3-36) into Eq. (1.3-35) gives the compatibility equation as

$$\frac{1}{Eh}\nabla^4\Phi - [2w_{1,xy}w'_{0,xy} + 2w_{1,xy}w^*_{,xy} - w_{1,xx}w'_{0,yy} - w_{1,yy}w'_{0,xx} - w_{1,yy}w'_{0,xx}$$
$$-w_{1,xx}w^*_{yy} - w_{1,yy}w^*_{,xx}] = 0 \tag{1.3-37}$$

Equations (1.3-33), (1.3-37) and stability equations which obtained from Eqs. (1.3-33) are the basic equations used to obtained the critical buckling loads of imperfect rectangular plates. Based on these equations, thermal buckling of imperfect rectangular plates of isotropic and homogeneous material are derived.

A. Critical uniform temperature rise

The boundary conditions along the edges of the plate are assumed to be simply supported. The edge conditions are thus [8]

$$u_0(x,0) = u_0(x,b) = 0$$
$$v_0(0,y) = v_0(a,y) = 0$$
$$w_0(x,0) = w_0(x,b) = w_0(0,b) = w_0(a,y) = 0$$
$$\phi_0(x,0) = \phi_0(x,b) = 0$$
$$\psi_0(0,y) = \psi_0(a,y) = 0 \tag{1.3-38}$$

The forces in the x and y directions developed in the plate due to a uniform temperature rise ΔT, are [20]

$$N_{xx0} = N_{yy0} = -N^T_{xx0} = -N^T_{yy0} = \frac{-Eh\alpha}{1-\nu}\Delta T$$
$$N_{xy0} = 0 \tag{1.3-39}$$

The displacement components for a rectangular isotropic plate satisfying the simply supported edges conditions are considered as [8]

$$u_0 = u_{0mn}\cos\alpha_m x \sin\beta_n y$$
$$v_0 = v_{0mn}\sin\alpha_m x \cos\beta_n y$$
$$w_0 = w_{0mn}\sin\alpha_m x \sin\beta_n y$$
$$\phi_0 = \phi_{0mn}\cos\alpha_m x \sin\beta_n y$$
$$\psi_0 = \psi_{0mn}\sin\alpha_m x \cos\beta_n y \tag{1.3-40}$$

Substitution of Eqs. (1.3-29), (1.3-39), and (1.3-40) into Eqs. (1.3-33) and ignoring higher order terms yield a system of equations as

$$K_{11}u_{0mn} + K_{12}v_{0mn} = 0$$

$$K_{12}u_{0mn} + K_{22}v_{0mn} = 0$$

$$K_{33}w_{0mn} + K_{34}\phi_{0mn} + K_{35}\psi_{0mn} = -N_x^T(\alpha_m^2 + \beta_n^2)w_{mn} - N_x^T(\alpha_m2 + \beta_n^2)\mu h$$

$$K_{34}w_{0mn} + K_{44}\phi_{0mn} + K_{55}\psi_{0mn} = 0$$

$$K_{35}w_{0mn} + K_{45}\phi_{0mn} + K_{55}\psi_{0mn} = 0 \tag{1.3-41}$$

in which K_{ij} is a symmetric matrix with the components

$$K_{11} = \frac{Eh}{1-\nu^2}\alpha_m^2 + \frac{Eh}{2(1+\nu)}\beta_n^2$$

$$K_{22} = \left(\frac{Eh}{1-\nu^2} + \frac{Eh}{2(1+\nu)}\right)\alpha_m\beta_n$$

$$K_{33} = \frac{-Eh}{2(1+\nu)}(\alpha_m^2 + \beta_n^2) + \frac{Eh^3}{12(1+\nu)}\left(\frac{4}{h^2}\right)(\alpha_m^2 + \beta_n^2) - \frac{Eh^5}{90.2(1+\nu)}\left(\frac{-4}{h^2}\right)^2(\alpha_m^2 + \beta_n^2)$$

$$-\frac{Eh^7}{448(1-\nu^2)}\left(\frac{4}{3h^2}\right)^2(\alpha_m^4 + 2\nu\alpha_m^2\beta_n^2 + \beta_n^4 + 2\alpha_m^2\beta_n^2)$$

$$K_{34} = \frac{-Eh}{2(1+\nu)}\alpha_m + \frac{Eh^3}{12(1+\nu)}\left(\frac{4}{h^2}\right)\alpha_m - \frac{Eh^5}{90.2(1+\nu)}\left(\frac{4}{h^2}\right)^2\alpha_m + \frac{Eh^5}{90(1-\nu^2)}\left(\frac{4}{3h^2}\right)$$

$$\times(\alpha_m^3 + \nu\alpha_m\beta_n^2) - \frac{Eh^7}{448(1-\nu^2)}\left(\frac{4}{3h^2}\right)^2(\alpha_m^3 + \nu\alpha_m\beta_n^2) + \frac{Eh^5}{90(1+\nu)}\left(\frac{4}{3h^2}\right)\alpha_m\beta_n^2$$

$$-\frac{Eh^7}{448(1-\nu^2)}\left(\frac{4}{3h^2}\right)^2\alpha_m\beta_n^2$$

$$K_{35} = \frac{-Eh}{2(1+\nu)}\beta_n + \frac{Eh^3}{12(1+\nu)}\left(\frac{4}{h^2}\right)\beta_n - \frac{Eh^5}{90.2(1+\nu)}\left(\frac{4}{h^2}\right)^2\beta_n + \frac{Eh^5}{90(1-\nu^2)}\left(\frac{4}{3h^2}\right).$$

$$(\beta_n^3 + \nu\alpha_m^2\beta_n) - \frac{Eh^7}{448(1-\nu^2)}\left(\frac{4}{3h^2}\right)^2(\beta_n^3 + \nu\alpha_m^2\beta_n) + \frac{Eh^5}{90(1+\nu)}\left(\frac{4}{3h^2}\right)\alpha_m^2\beta_n$$

$$-\frac{Eh^7}{448(1-\nu^2)}\left(\frac{4}{3h^2}\right)^2\alpha_m^2\beta_n$$

$$K_{44} = \frac{-Eh^3}{12(1-\nu^2)}\alpha_m^2 + \frac{Eh^5}{90(1-\nu^2)}\left(\frac{4}{3h^2}\right)\alpha_m^2 - \frac{Eh^3}{12.2(1+\nu)}\beta_n^2 + \frac{Eh^5}{90.2(1+\nu)}\left(\frac{4}{3h^2}\right)\beta_n^2$$

$$-\frac{Eh}{2(1+\nu)} + \frac{Eh^3}{12(1+\nu)}\left(\frac{4}{h^2}\right) - \frac{Eh^5}{90.2(1+\nu)}\left(\frac{4}{h^2}\right)^2 + \frac{Eh^5}{90(1-\nu^2)}\left(\frac{4}{3h^2}\right)^2\alpha_m^2$$

$$-\frac{Eh^7}{448(1-\nu^2)}\left(\frac{4}{3h^2}\right)^2\alpha_m^2 + \frac{Eh^5}{90.2(1+\nu)}\left(\frac{4}{3h^2}\right)\beta_n^2$$

$$K_{45} = \frac{-Eh^3}{12(1-\nu^2)}\nu\alpha_m\beta_n + \frac{Eh^5}{90(1-\nu^2)}\left(\frac{4}{3h^2}\right)\nu\alpha_m\beta_n + \frac{Eh^7}{448(1-\nu^2)}\left(\frac{4}{3h^2}\right)^2\nu\alpha_m\beta_n$$

$$+\frac{Eh^5}{90.2(1+\nu)}\left(\frac{4}{3h^2}\right)\alpha_m\beta_n - \frac{Eh^7}{448.2(1-\nu^2)}\left(\frac{4}{3h^2}\right)^2\alpha_m\beta_n$$

$$K_{55} = \frac{-Eh^3}{12.2(1+\nu)}\alpha_m^2 + \frac{Eh^5}{90.2(1+\nu)}\left(\frac{4}{3h^2}\right)\alpha_m^2 - \frac{Eh^3}{12(1-\nu^2)}\beta_n^2 + \frac{Eh^5}{90(1-\nu^2)}\left(\frac{4}{3h^2}\right)\beta_n^2$$

$$\frac{-Eh}{2(1+\nu)} + \frac{Eh^3}{12(1+\nu)}\left(\frac{4}{h^2}\right) - \frac{Eh^5}{90.2(1+\nu)}\left(\frac{4}{h^2}\right)^2 + \frac{Eh^5}{90.2(1+\nu)}\left(\frac{4}{3h^2}\right)\alpha_m^2$$

$$\frac{-Eh^7}{448.2(1-\nu^2)}\left(\frac{4}{3h^2}\right)^2\alpha_m^2 + \frac{Eh^5}{90(1-\nu^2)}\left(\frac{4}{3h^2}\right)\beta_n^2 - \frac{Eh^7}{448(1-\nu^2)}\left(\frac{4}{3h^2}\right)^2\beta_n^2 \qquad (1.3\text{-}42)$$

Solution of Eq. (1.3-41), considering u_{0mn}, v_{0mn}, ϕ_{0mn}, ψ_{0mn}, and w_{0mn} as unknown parameters and w_{mn}^* as known parameter, yields

$$w_{0mn} = \frac{N_{x0}^T(\alpha_m^2 + \beta_n^2)}{\delta - \gamma - N_x^T(\alpha_m^2 + \beta_n^2)}\mu h$$

$$\delta = \frac{K_{12}(K_{12}K_{33} - K_{13}K_{23}) - K_{13}(K_{12}K_{23} - K_{22}K_{13})}{K_{22}K_{33} - K_{23}^2}$$

$$\gamma = K_{33} \qquad (1.3\text{-}43)$$

Consider the following dimensionless parameters

$$\bar{N_{xx0}^T} = \frac{N_{xx0}^T b^2}{D}$$

$$\bar{N_{yy0}^T} = \bar{N_{xx0}^T}$$

$$\Psi = \frac{\Phi}{Eh^3}$$

$$\bar{u}_1 = \frac{u_1}{h}$$

$$\bar{v}_1 = \frac{v_1}{h}$$

$$\bar{w}_0 = \frac{\acute{w}_0}{h}$$

$$\bar{w}_1 = \frac{w_1}{h}$$

$$\bar{w}^* = \frac{w^*}{h}$$

$$\bar{\phi}_1 = \frac{\phi_1}{h}$$

$$\bar{\psi}_1 = \frac{\psi_1}{h} \qquad (1.3\text{-}44)$$

Substituting relations (1.3-44) and (1.3-39) into Eq. (1.3-37) and stability equations, the dimensionless form of the stability and compatibility equations are obtained

$$\frac{Eh}{1-\nu^2}\left(\frac{\partial^2 \bar{u}_1}{\partial x^2} + \frac{\partial \bar{w}_1}{\partial x}\frac{\partial^2 \bar{w}_1}{\partial x^2}\right) = 0$$

$$\frac{Eh}{1-\nu^2}\left(\frac{\partial^2 \bar{v}_1}{\partial y^2} + \frac{\partial \bar{w}_1}{\partial y}\frac{\partial^2 \bar{w}_1}{\partial y^2}\right) = 0$$

$$\frac{Eh}{2(1+\nu)}\left(\frac{\partial\bar{\phi}_1}{\partial x}+\frac{\partial^2\bar{w}_1}{\partial x^2}+\frac{\partial\bar{\psi}_1}{\partial y}+\frac{\partial^2\bar{w}_1}{\partial y^2}\right)+\frac{Eh^3}{12.2(1+\nu)}\left(\frac{-4}{h^2}\right)\left(\frac{\partial\bar{\phi}_1}{\partial x}+\frac{\partial^2\bar{w}_1}{\partial x^2}+\frac{\partial\bar{\psi}_1}{\partial y}+\frac{\partial^2\bar{w}_1}{\partial y^2}\right)$$

$$+\left(\frac{-4}{h^2}\right)\frac{Eh^3}{12.2(1+\nu)}\left(\frac{\partial\bar{\phi}_1}{\partial x}+\frac{\partial^2\bar{w}_1}{\partial x^2}+\frac{\partial\bar{\psi}_1}{\partial y}+\frac{\partial^2\bar{w}_1}{\partial y^2}\right)+\left(\frac{-4}{h^2}\right)^2\frac{Eh^5}{90.2(1+\nu)}\left(\frac{\partial\bar{\phi}_1}{\partial x}+\frac{\partial^2\bar{w}_1}{\partial x^2}\right.$$

$$\left.+\frac{\partial\bar{\psi}_1}{\partial y}+\frac{\partial^2\bar{w}_1}{y^2}\right)+\left(\frac{4}{3h^2}\right)\frac{Eh^5}{90(1-\nu^2)}\left(\frac{\partial^3\bar{\phi}_1}{\partial x^3}+\nu\partial^3\bar{\psi}_1\partial x^2\partial y+\frac{\partial^3\bar{\psi}_1}{\partial y^3}+\frac{\nu\partial^3\bar{\phi}_1}{\partial x\partial y^2}\right)-\left(\frac{4}{3h^2}\right)^2.$$

$$\frac{Eh^7}{448(1-\nu^2)}\left(\frac{\partial^3\bar{\phi}_1}{\partial x^3}+\frac{\partial^4\bar{w}_1}{\partial x^4}+\frac{\nu\partial^3\bar{\psi}_1}{\partial x^2\partial y}+\frac{\nu\partial^4\bar{w}_1}{\partial x^2\partial y^2}+\frac{\partial^3\bar{\psi}_1}{\partial y^3}+\frac{\partial^4\bar{w}_1}{\partial y^4}+\frac{\nu\partial^3\bar{\phi}_1}{\partial x\partial y^2}\right.$$

$$\left.+\frac{\nu\partial^2\bar{w}_1}{\partial x^2\partial y^2}\right)+\left(\frac{4}{3h^2}\right)\frac{Eh^5}{90(1+\nu)}\left(\frac{\partial^3\bar{\phi}_1}{\partial x\partial y^2}+\frac{\partial^3\bar{\psi}_1}{\partial x^2\partial y}\right)-\left(\frac{4}{3h^2}\right)^2\frac{Eh^7}{448(1-\nu^2)}.$$

$$\left(\frac{\partial^3\bar{\phi}_1}{\partial x\partial y^2}+\frac{\partial^3\bar{\psi}_1}{\partial x^2\partial y}+\frac{2\partial^4\bar{w}_1}{\partial x^2\partial y^2}\right)+Eh^3\frac{\partial^2\Psi}{\partial y^2}\left(\frac{\partial^2\bar{w}_0}{\partial x^2}+\frac{\partial^2\bar{w}^*}{\partial x^2}\right)-2Eh^3\frac{\partial^2\Psi}{\partial x\partial y}\left(\frac{\partial^2\bar{w}_0}{\partial x\partial y}+\frac{\partial^2\bar{w}^*}{\partial x\partial y}\right)$$

$$+Eh^3\frac{\partial^2\Psi}{\partial x^2}\left(\frac{\partial^2\bar{w}_0}{\partial y^2}+\frac{\partial^2\bar{w}^*}{\partial y^2}\right)-\frac{Eh^3}{12(1-\nu^2)}N_{xx0}^{\bar{T}}\frac{\partial^2\bar{w}_1}{\partial x^2}-\frac{Eh^3}{(1-\nu^2)}N_{yy0}^{\bar{T}}\frac{\partial^2\bar{w}_1}{\partial y^2}=0$$

$$\frac{Eh^3}{12(1-\nu^2)}\left(\frac{\partial^2\bar{\phi}_1}{\partial x^2}+\frac{\nu\partial^2\bar{\psi}_1}{\partial x\partial y}\right)+\frac{Eh^5}{90(1-\nu^2)}\left(\frac{-4}{3h^2}\right)\left(\frac{\partial^2\bar{\phi}_1}{\partial x^2}+\frac{\partial^3\bar{w}_1}{\partial x^3}+\frac{\nu\partial^2\bar{\psi}_1}{\partial x\partial y}+\frac{\nu\partial^3\bar{w}_1}{\partial x\partial y^2}\right)$$

$$+\frac{Eh^3}{12.2(1+\nu)}\left(\frac{\partial^2\bar{\phi}_1}{\partial y^2}+\frac{\partial^2\bar{\psi}_1}{\partial x\partial y}\right)+\frac{Eh^5}{90.2(1+\nu)}\left(\frac{-4}{3h^2}\right)\left(\frac{\partial^2\bar{\phi}_1}{\partial y^2}+\frac{\partial^2\bar{\psi}_1}{\partial x\partial y}+\frac{2\partial^3\bar{w}_1}{\partial x\partial y^2}\right)$$

$$-\frac{Eh}{2(1+\nu)}\left(\bar{\phi}_1+\frac{\partial\bar{w}_1}{\partial x}\right)-\frac{Eh^3}{12(1+\nu)}\left(\frac{-4}{h^2}\right)\left(\bar{\phi}_1+\frac{\partial\bar{w}_1}{\partial x}\right)$$

$$-\frac{Eh^5}{90.2(1+\nu)}\left(\frac{4}{h^2}\right)^2\left(\bar{\phi}_1+\frac{\partial\bar{w}_1}{\partial x}\right)-\left(\frac{4}{3h^2}\right)\frac{Eh^5}{90(1-\nu^2)}\left(\frac{\partial^2\bar{\phi}_1}{\partial x^2}+\frac{\nu\partial^2\bar{\psi}_1}{\partial x\partial y}\right)$$

$$+\left(\frac{4}{3h^2}\right)^2\frac{Eh^7}{448(1-\nu^2)}\left(\frac{\partial^2\bar{\phi}_1}{\partial x^2}+\frac{\partial^3\bar{w}_1}{\partial x^3}+\frac{\nu\partial^2\bar{\psi}_1}{\partial x\partial y}+\frac{\nu\partial^3\bar{w}_1}{\partial x\partial y^2}\right)-\frac{Eh^5}{90.2(1+\nu)}\left(\frac{4}{3h^2}\right).$$

$$\left(\frac{\partial^2\bar{\phi}_1}{\partial y^2}+\frac{\partial^2\bar{\psi}_1}{\partial x\partial y}\right)+\frac{Eh^7}{448.2(1-\nu^2)}\left(\frac{4}{3h^2}\right)^2\left(\frac{\partial^2\bar{\phi}_1}{\partial y^2}+\frac{\partial^2\bar{\psi}_1}{\partial x\partial y}+\frac{2\partial^3\bar{w}_1}{\partial x\partial y^2}\right)=0$$

$$\frac{Eh^3}{12.2(1+\nu)}\left(\frac{\partial^2\bar{\phi}_1}{\partial x\partial y}+\frac{\partial^2\bar{\psi}_1}{\partial x^2}\right)+\frac{Eh^5}{90.2(1+\nu)}\left(\frac{-4}{3h^2}\right)\left(\frac{\partial^2\bar{\phi}_1}{\partial x\partial y}+\frac{\partial^2\bar{\psi}_1}{\partial x^2}+\frac{2\partial^3\bar{w}_1}{\partial x^2\partial y}\right)$$

$$+\frac{Eh^3}{12(1-\nu^2)}\left(\frac{\partial^2\bar{\psi}_1}{\partial y^2}+\frac{\nu\partial^2\bar{\phi}_1}{\partial x\partial y}\right)+\frac{Eh^5}{90(1-\nu^2)}\left(\frac{-4}{3h^2}\right)\left(\frac{\partial^2\bar{\psi}_1}{\partial y^2}+\frac{\partial^3\bar{w}_1}{\partial y^3}+\frac{\nu\partial^2\bar{\phi}_1}{\partial x\partial y}+\frac{\nu\partial^3\bar{w}_1}{\partial x^2\partial y}\right)$$

$$-\frac{Eh}{2(1+\nu)}\left(\bar{\psi}_1+\frac{\partial\bar{w}_1}{\partial y}\right)+\frac{Eh^3}{12(1+\nu)}\left(\frac{4}{h^2}\right)\left(\bar{\psi}_1+\frac{\partial\bar{w}_1}{\partial y}\right)$$

$$-\frac{Eh^5}{90.2(1+\nu)}\left(\frac{4}{h^2}\right)^2\left(\bar{\psi}_1+\frac{\partial\bar{w}_1}{\partial y}\right)-\frac{Eh^5}{90.2(1+\nu)}\left(\frac{4}{3h^2}\right)\left(\frac{\partial^2\bar{\phi}_1}{\partial x\partial y}+\frac{\partial^2\bar{\psi}_1}{\partial x^2}\right)+\frac{Eh^7}{448.2(1-\nu^2)}.$$

$$\left(\frac{4}{3h^2}\right)^2\left(\frac{\partial^2\bar{\phi}_1}{\partial x\partial y}+\frac{\partial^2\bar{\psi}_1}{\partial x^2}+\frac{2\partial^3\bar{w}_1}{\partial x^2\partial y}\right)-\frac{Eh^5}{90(1-\nu^2)}\left(\frac{4}{3h^2}\right)\left(\frac{\partial^2\bar{\psi}_1}{\partial y^2}+\frac{\nu\partial^2\bar{\phi}_1}{\partial x\partial y}\right)+\frac{Eh^7}{448(1-\nu^2)}.$$

$$\left(\frac{4}{3h^2}\right)^2\left(\frac{\partial^2\bar{\psi}_1}{\partial y^2}+\frac{\partial^3\bar{w}_1}{\partial y^3}+\frac{\nu\partial^2\bar{\phi}_1}{\partial x\partial y}+\frac{\nu\partial^3\bar{w}_1}{\partial x^2\partial y}\right)=0$$

$$\nabla^4\Psi-2\bar{w}_{1,xy}\bar{w}_{0,xy}-2\bar{w}_{1,xy}\bar{w}^*{}_{,xy}+\bar{w}_{1,xx}\bar{w}_{0,yy}+\bar{w}_{1,yy}\bar{w}_{0,xx}+\bar{w}_{1,yy}\bar{w}_{0,xx}$$

$$+\bar{w}_{1,xx}\bar{w}^*{}_{yy}+\bar{w}_{1,yy}\bar{w}^*{}_{,xx}=0 \tag{1.3-45}$$

To solve the system of Eqs. (1.3-45), with the consideration of the boundary conditions (1.3-27), the approximate solutions may be considered as

$$\bar{u}_1 = \bar{u}_{1mn} \cos \frac{m\pi}{a} x \sin \frac{n\pi}{b} y$$

$$\bar{v}_1 = \bar{v}_{1mn} \sin \frac{m\pi}{a} x \cos \frac{n\pi}{b} y$$

$$\bar{w}_0 = \bar{w}_{0mn} \sin \frac{m\pi}{a} x \sin \frac{n\pi}{b} y \qquad \bar{\phi}_1 = \bar{\phi}_{1mn} \cos \frac{m\pi}{a} x \sin \frac{n\pi}{b} y$$

$$\bar{w}^* = \bar{w}^*_{mn} \sin \frac{m\pi}{a} x \sin \frac{n\pi}{b} y \qquad \bar{\psi}_1 = \bar{\psi}_{1mn} \sin \frac{m\pi}{a} x \cos \frac{n\pi}{b} y$$

$$\bar{w}_1 = \bar{w}_{1mn} \sin \frac{m\pi}{a} x \sin \frac{n\pi}{b} y \qquad \Psi = E_{mn} \sin \frac{m\pi}{a} x \sin \frac{n\pi}{b} y$$

$$(1.3\text{-}46)$$

where $\bar{u}_{1mn}$, $\bar{v}_{1mn}$, $\bar{w}_{0mn}$, $\bar{w}_{1mn}$, $\bar{w}^*_{mn}$, $\bar{\phi}_{1mn}$, $\bar{\psi}_{1mn}$, and E_{mn} are constant coefficients that depend on m and n. The system of Eqs. (1.3-46) are made orthogonal with respect to the approximate solutions (1.3-46) according to the Galerkin method

$$\int_0^a \int_0^b [\frac{Eh}{1-\nu^2}(\frac{\partial^2 \bar{u}_1}{\partial x^2} + \frac{\partial \bar{w}_1}{\partial x}\frac{\partial^2 \bar{w}_1}{\partial x^2})]\bar{u}_{1mn} \cos \frac{m\pi}{a} x \sin \frac{n\pi}{b} y \, dx \, dy = 0$$

$$\int_0^a \int_0^b [\frac{Eh}{1-\nu^2}(\frac{\partial^2 \bar{v}_1}{\partial y^2} + \frac{\partial \bar{w}_1}{\partial y}\frac{\partial^2 \bar{w}_1}{\partial y^2})]\bar{v}_{1mn} \sin \frac{m\pi}{a} x \cos \frac{n\pi}{b} y \, dx \, dy = 0$$

$$\int_0^a \int_0^b [\frac{Eh}{2(1+\nu)}(\frac{\partial \bar{\phi}_1}{\partial x} + \frac{\partial^2 \bar{w}_1}{\partial x^2} + \frac{\partial \bar{\psi}_1}{\partial y} + \frac{\partial^2 \bar{w}_1}{\partial y^2}) + \frac{Eh^3}{12.2(1+\nu)}(\frac{-4}{h^2})(\frac{\partial \bar{\phi}_1}{\partial x} + \frac{\partial^2 \bar{w}_1}{\partial x^2} + \frac{\partial \bar{\psi}_1}{\partial y}$$

$$+ \frac{\partial^2 \bar{w}_1}{\partial y^2}) + (\frac{-4}{h^2})\frac{Eh^3}{12.2(1+\nu)}(\frac{\partial \bar{\phi}_1}{\partial x} + \frac{\partial^2 \bar{w}_1}{\partial x^2} + \frac{\partial \bar{\psi}_1}{\partial y} + \frac{\partial^2 \bar{w}_1}{\partial y^2}) + (\frac{-4}{h^2})^2\frac{Eh^5}{90.2(1+\nu)}(\frac{\partial \bar{\phi}_1}{\partial x}$$

$$+ \frac{\partial^2 \bar{w}_1}{\partial x^2} + \frac{\partial \bar{\psi}_1}{\partial y} + \frac{\partial^2 \bar{w}_1}{y^2}) + (\frac{4}{3h^2})\frac{Eh^5}{90(1-\nu^2)}(\frac{\partial^3 \bar{\phi}_1}{\partial x^3} + \nu\partial^3 \bar{\psi}_1 \partial x^2 \partial y + \frac{\partial^3 \bar{\psi}_1}{\partial y^3} + \frac{\nu\partial^3 \bar{\phi}_1}{\partial x \partial y^2})$$

$$-(\frac{4}{3h^2})^2 \times \frac{Eh^7}{448(1-\nu^2)}(\frac{\partial^3 \bar{\phi}_1}{\partial x^3} + \frac{\partial^4 \bar{w}_1}{\partial x^4} + \frac{\nu\partial^3 \bar{\psi}_1}{\partial x^2 \partial y} + \frac{\nu\partial^4 \bar{w}_1}{\partial x^2 \partial y^2} + \frac{\partial^3 \bar{\psi}_1}{\partial y^3} + \frac{\partial^4 \bar{w}_1}{\partial y^4} + \frac{\nu\partial^3 \bar{\phi}_1}{\partial x \partial y^2}$$

$$+ \frac{\nu\partial^2 \bar{w}_1}{\partial x^2 \partial y^2}) + (\frac{4}{3h^2})\frac{Eh^5}{90(1+\nu)}(\frac{\partial^3 \bar{\phi}_1}{\partial x \partial y^2} + \frac{\partial^3 \bar{\psi}_1}{\partial x^2 \partial y}) - (\frac{4}{3h^2})^2\frac{Eh^7}{448(1-\nu^2)}.$$

$$(\frac{\partial^3 \bar{\phi}_1}{\partial x \partial y^2} + \frac{\partial^3 \bar{\psi}_1}{\partial x^2 \partial y} + \frac{2\partial^4 \bar{w}_1}{\partial x^2 \partial y^2}) + Eh^3\frac{\partial^2 \Psi}{\partial y^2}(\frac{\partial^2 \bar{w}_0}{\partial x^2} + \frac{\partial^2 \bar{w}^*}{\partial x^2}) - 2Eh^3\frac{\partial^2 \Psi}{\partial x \partial y}(\frac{\partial^2 \bar{w}_0}{\partial x \partial y} + \frac{\partial^2 \bar{w}^*}{\partial x \partial y})$$

$$+ Eh^3\frac{\partial^2 \Psi}{\partial x^2}(\frac{\partial^2 \bar{w}_0}{\partial y^2} + \frac{\partial^2 \bar{w}^*}{\partial y^2}) - \frac{Eh^3}{12(1-\nu^2)}N^{\bar{T}}_{xx0}\frac{\partial^2 \bar{w}_1}{\partial x^2} - \frac{Eh^3}{(1-\nu^2)}N^{\bar{T}}_{yy0}\frac{\partial^2 \bar{w}_1}{\partial y^2}]$$

$$\times \bar{w}_{1mn} \sin \frac{m\pi}{a} x \sin \frac{n\pi}{b} y \, dx \, dy = 0 \qquad (1.3\text{-}47)$$

$$\int_0^a \int_0^b [\frac{Eh^3}{12(1-\nu^2)}(\frac{\partial^2 \bar{\phi}_1}{\partial x^2} + \frac{\nu\partial^2 \bar{\psi}_1}{\partial x \partial y}) + \frac{Eh^5}{90(1-\nu^2)}(\frac{-4}{3h^2})(\frac{\partial^2 \bar{\phi}_1}{\partial x^2} + \frac{\partial^3 \bar{w}_1}{\partial x^3} + \frac{\nu\partial^2 \bar{\psi}_1}{\partial x \partial y} + \frac{\nu\partial^3 \bar{w}_1}{\partial x \partial y^2})$$

$$+ \frac{Eh^3}{12.2(1+\nu)}(\frac{\partial^2 \bar{\phi}_1}{\partial y^2} + \frac{\partial^2 \bar{\psi}_1}{\partial x \partial y}) + \frac{Eh^5}{90.2(1+\nu)}(\frac{-4}{3h^2})(\frac{\partial^2 \bar{\phi}_1}{\partial y^2} + \frac{\partial^2 \bar{\psi}_1}{\partial x \partial y} + \frac{2\partial^3 \bar{w}_1}{\partial x \partial y^2})$$

$$- \frac{Eh}{2(1+\nu)}(\bar{\phi}_1 + \frac{\partial \bar{w}_1}{\partial x}) - \frac{Eh^3}{12(1+\nu)}(\frac{-4}{h^2})(\bar{\phi}_1 + \frac{\partial \bar{w}_1}{\partial x})$$

$$-\frac{Eh^5}{90.2(1+\nu)}(\frac{4}{h^2})^2(\bar{\phi}_1+\frac{\partial\bar{w}_1}{\partial x})-(\frac{4}{3h^2})\frac{Eh^5}{90(1-\nu^2)}(\frac{\partial^2\bar{\phi}_1}{\partial x^2}+\frac{\nu\partial^2\bar{\psi}_1}{\partial x\partial y})$$

$$+(\frac{4}{3h^2})^2\frac{Eh^7}{448(1-\nu^2)}(\frac{\partial^2\bar{\phi}_1}{\partial x^2}+\frac{\partial^3\bar{w}_1}{\partial x^3}+\frac{\nu\partial^2\bar{\psi}_1}{\partial x\partial y}+\frac{\nu\partial^3\bar{w}_1}{\partial x\partial y^2}-\frac{Eh^5}{90.2(1+\nu)}(\frac{4}{3h^2}).$$

$$(\frac{\partial^2\bar{\phi}_1}{\partial y^2}+\frac{\partial^2\bar{\psi}_1}{\partial x\partial y})+\frac{Eh^7}{448.2(1-\nu^2)}(\frac{4}{3h^2})^2(\frac{\partial^2\bar{\phi}_1}{\partial y^2}+\frac{\partial^2\bar{\psi}_1}{\partial x\partial y}+\frac{2\partial^3\bar{w}_1}{\partial x\partial y^2})]$$

$$\times\bar{\phi}_{mn}\cos\frac{m\pi}{a}x\sin\frac{n\pi}{b}ydxdy=0 \qquad (1.3\text{-}48)$$

$$\int_0^a\int_0^b[\frac{Eh^3}{12.2(1+\nu)}(\frac{\partial^2\bar{\phi}_1}{\partial x\partial y}+\frac{\partial^2\bar{\psi}_1}{\partial x^2})+\frac{Eh^5}{90.2(1+\nu)}(\frac{-4}{3h^2})(\frac{\partial^2\bar{\phi}_1}{\partial x\partial y}+\frac{\partial^2\bar{\psi}_1}{\partial x^2}+\frac{2\partial^3\bar{w}_1}{\partial x^2\partial y})$$

$$+\frac{Eh^3}{12(1-\nu^2)}(\frac{\partial^2\bar{\psi}_1}{\partial y^2}+\frac{\nu\partial^2\bar{\phi}_1}{\partial x\partial y})+\frac{Eh^5}{90(1-\nu^2)}(\frac{-4}{3h^2})(\frac{\partial^2\bar{\psi}_1}{\partial y^2}+\frac{\partial^3\bar{w}_1}{\partial y^3}+\frac{\nu\partial^2\bar{\phi}_1}{\partial x\partial y}+\frac{\nu\partial^3\bar{w}_1}{\partial x^2\partial y})$$

$$-\frac{Eh}{2(1+\nu)}(\bar{\psi}_1+\frac{\partial\bar{w}_1}{\partial y})+\frac{Eh^3}{12(1+\nu)}(\frac{4}{h^2})(\bar{\psi}_1+\frac{\partial\bar{w}_1}{\partial y})$$

$$-\frac{Eh^5}{90.2(1+\nu)}(\frac{4}{h^2})^2(\bar{\psi}_1+\frac{\partial\bar{w}_1}{\partial y})-\frac{Eh^5}{90.2(1+\nu)}(\frac{4}{3h^2})(\frac{\partial^2\bar{\phi}_1}{\partial x\partial y}+\frac{\partial^2\bar{\psi}_1}{\partial x^2})+\frac{Eh^7}{448.2(1-\nu^2)}.$$

$$(\frac{4}{3h^2})^2(\frac{\partial^2\bar{\phi}_1}{\partial x\partial y}+\frac{\partial^2\bar{\psi}_1}{\partial x^2}+\frac{2\partial^3\bar{w}_1}{\partial x^2\partial y})-\frac{Eh^5}{90(1-\nu^2)}(\frac{4}{3h^2})(\frac{\partial^2\bar{\psi}_1}{\partial y^2}+\frac{\nu\partial^2\bar{\phi}_1}{\partial x\partial y})+\frac{Eh^7}{448(1-\nu^2)}.$$

$$(\frac{4}{3h^2})^2(\frac{\partial^2\bar{\psi}_1}{\partial y^2}+\frac{\partial^3\bar{w}_1}{\partial y^3}+\frac{\nu\partial^2\bar{\phi}_1}{\partial x\partial y}+\frac{\nu\partial^3\bar{w}_1}{\partial x^2\partial y})]\bar{\psi}_{mn}\sin\frac{m\pi}{a}x\cos\frac{n\pi}{b}ydxdy=0 \quad (1.3\text{-}49)$$

$$\int_0^a\int_0^b[\nabla^4\Psi-2\bar{w}_{1,xy}\bar{w}_{0,xy}-2\bar{w}_{1,xy}\bar{w}^*{}_{,xy}+\bar{w}_{1,xx}\bar{w}_{0,yy}+\bar{w}_{1,yy}\bar{w}_{0,xx}+\bar{w}_{1,yy}\bar{w}_{0,xx}$$

$$+\bar{w}_{1,xx}\bar{w}^*{}_{yy}+\bar{w}_{1,yy}\bar{w}^*{}_{,xx}]E_{mn}\sin\frac{m\pi}{a}x\sin\frac{n\pi}{b}ydxdy=0 \qquad (1.3\text{-}50)$$

The determinant of the system of Eqs. (1.3-47), (1.3-48), (1.3-49), and (1.3-50) for the coefficients $\bar{u}_{1mn}$, $\bar{v}_{1mn}$, $\bar{w}_{1mn}$, $\bar{\phi}_{1mn}$, $\bar{\psi}_{1mn}$, and E_{mn} is set to zero, which yields

$$\bar{N}_{x0}^T=(\gamma-\delta+k_{mn}^{\frac{1}{3}})\frac{b^2}{D(\alpha_m^2+\beta_n^2)} \qquad (1.3\text{-}51)$$

where $D=\frac{Eh^3}{12(1-\nu^2)}$, γ and δ are the same as Eq. (1.3-44), and k_{mn} is

$$k_{mn}=\frac{1024\alpha_m^4\beta_n^4\mu^2h^4(\gamma-\delta)^2}{9m^2n^2\pi^4(\alpha_m^2+\beta_n^2)^2} \qquad (1.3\text{-}52)$$

Equation (1.3-51) is used to obtain the critical uniform temperature rise. Setting $\mu=0$ in Eq. (1.3-51), reduces it to equation given by Mossavarali and Eslami [9] for an isotropic perfect plate. Solving Eq. (1.3-51) for a fixed pair of m and n yields various quantities for $\bar{N}_{x0}^T$. The smallest one for all pairs of m and n gives the critical thermal load, $\bar{N}_{x0}^T$, for thermal buckling.

B. Temperature gradient through the thickness

Consider the same plate with the assumption of linear temperature variation across the plate thickness as

$$T(z) = \Delta T \frac{(z + \frac{h}{2})}{h} \tag{1.3-53}$$

where z measures from the middle plane of the plate and ΔT is the temperature difference between the top and bottom surfaces of the plate. For simply supported edges the prebuckling forces in the isotropic plate are [20]

$$N_{xx0} = N_{yy0} = -N_{x0}^T = -N_{y0}^T = \frac{-E\alpha h}{1 - \nu} \frac{\Delta T}{2}$$
$$N_{xy0} = 0 \tag{1.3-54}$$

With a similar method, the critical thermal load for the case of temperature difference across the thickness is obtained by setting the determinant of the coefficient of Eqs. (1.3-47), (1.3-48), (1.3-49), and (1.3-50) to zero, yielding

$$\bar{N}_{x0}^T = (\gamma - \delta + k_{mn}^{\frac{1}{3}}) \frac{b^2}{D(\alpha_m^2 + \beta_n^2)} \tag{1.3-55}$$

where $\bar{N}_{x0}^T = \frac{-E\alpha h}{1-\nu} \frac{\Delta T}{2} \frac{b^2}{D}$ and k_{mn} is defined in Eqs. (1.3-52).

C. Linear variation along the x-direction

Consider a rectangular isotropic plate of dimensions a and b under temperature difference along the x-direction and with simply supported edges, where the displacement of the edges in the x and y directions are prevented. Assume a linear temperature variation along the direction x

$$T(x) = \Delta T \frac{x}{a} \tag{1.3-56}$$

where $\Delta T = T(a) - T(0)$. For simply supported edges the prebuckling forces in the isotropic plate are [20]

$$N_{xx0} = N_{yy0} = -N_{x0}^T = -N_{y0}^T = \frac{-E\alpha h}{1 - \nu} \Delta T \frac{x}{a}$$
$$N_{xy0} = 0 \tag{1.3-57}$$

The equilibrium equations (1.3-15) are used and the Galerkin's method with the help of Eqs. (1.3-25), (1.3-26), (1.3-29), (1.3-40), and (1.3-57) is employed. We denote the left-hand sides of the Eqs. (1.3-15) by

*R*1, *R*2, *R*3, *R*4, and *R*3 respectively. For simply supported boundary conditions and considering Eqs. (1.3-40), the Galerkin's method leads to the following equations

$$\int_0^b \int_0^a R_1 \cos \alpha_m x \sin \beta_n y dx dy = 0$$

$$\int_0^b \int_0^a R_2 \sin \alpha_m x \cos \beta_n y dx dy = 0$$

$$\int_0^b \int_0^a R_3 \sin \alpha_m x \sin \beta_n y dx dy = 0$$

$$\int_0^b \int_0^a R_4 \cos \alpha_m x \sin \beta_n y dx dy = 0$$

$$\int_0^b \int_0^a R_5 \sin \alpha_m x \cos \beta_n y dx dy = 0 \qquad (1.3\text{-}58)$$

Equations (1.3-58) result in five equations for the constant coefficients u_{0mn}, v_{0mn}, w_{0mn}, ϕ_{0mn}, and ψ_{0mn}. Setting the determinant of coefficients equal to zero, results in the following equations

$$w_{0mn} = \frac{N_{x0}^T(\alpha_m^2 + \beta_n^2)}{\delta - \gamma - N_x^T(\alpha_m^2 + \beta_n^2)} \mu h$$

$$\delta = \frac{K_{12}(K_{12}K_{33} - K_{13}K_{23}) - K_{13}(K_{12}K_{23} - K_{22}K_{13})}{K_{22}K_{33} - K_{23}^2}$$

$$\gamma = K_{33} \qquad (1.3\text{-}59)$$

where K_{ij} coefficients are obtained from Eqs. (1.3-42) and γ, and δ are as per Eqs. (1.3-43). With a similar procedure, and considering Eqs. (1.3-44), (1.3-45), and (1.3-57) instead of (1.3-39), the critical temperature difference in the x-direction is

$$\bar{N}_{x0}^T = (\gamma - \delta + k_{mn}^{\frac{1}{3}})\frac{b^2}{D(\alpha_m^2 + \beta_n^2)} \qquad (1.3\text{-}60)$$

where $\bar{N}_{x0}^T = \frac{-E\alpha h}{1-\nu}\frac{\Delta T}{2}\frac{b^2}{D}$ and other parameters are defined in Eqs. (1.3-53).

3.4   Results and discussion

Table (1.3-1) shows the effects of the imperfection amplitude, μ, on the buckling load, $\bar{N}_{x0}^T$, for a uniform temperature rise of an isotropic imperfect plate based on the first-order [10] and higher-order theories. The table shows that with increasing μ, the critical buckling load, $\bar{N}_{x0}^T$, increases. This means that with increasing μ, stability of the plate increases. Also the difference between higher-order and first-order theories

N_{x0}^T			
μ .	Higher-Order A	First-Order B	$\frac{A-B}{A}$ x 100
.1	19.6893	23.6611	20.17%
.2	19.6927	25.9765	31.91%
.3	19.6956	27.9186	41.75%
.4	19.6981	29.6521	50.53%
.5	19.7004	31.2453	58.60%
.6	19.7026	32.7350	66.15%
.7	19.7047	34.1437	73.28%
.8	19.7067	35.4867	80.07%
.9	19.7085	36.7750	86.59%
1	19.7104	38.0158	92.87%

Table 1.1: $\frac{a}{h} = 70$, $\frac{a}{b} = 1$, $E = 207Gpa$, $\nu = .3$, $\alpha = 2\text{x}10^{-5}$

N_{x0}^T			
$\frac{a}{h}$.	Higher-Order A	First-Order B	$\frac{A-B}{A}$ x 100
10	7.3270	12.5135	70.87%
20	6.8430	11.1943	63.59%
30	6.7365	10.8075	60.43%
40	6.6945	10.6161	58.58%
50	6.6731	10.4992	57.34%
60	6.6605	10.4192	56.43%
70	6.6523	10.3605	55.74%
80	6.6497	10.3151	55.19%

Table 1.2: $\mu = .1$, $E = 207Gpa$, $\nu = .3$, $\alpha = 2\text{x}10^{-5}$

$\bar{N}^T_{x0}$			
$\frac{a}{h}$.	Higher-Order A	First-Order B	$\frac{A-B}{A}$ x 100
5	8.052	13.41	66.5%
10	7.011	10.75	53.3%
20	6.721	10.084	50.0%
30	6.67	9.96	44.3%
40	6.65	9.92	49.1%
50	6.64	9.90	49.1%
60	6.63	9.88	49.1%
70	6.63	9.88	49.1%
80	6.63	9.87	48.9%

Table 1.3: $\mu = 0$, $E = 207Gpa$, $\nu = .3$, $\alpha = 2x10^{-5}$

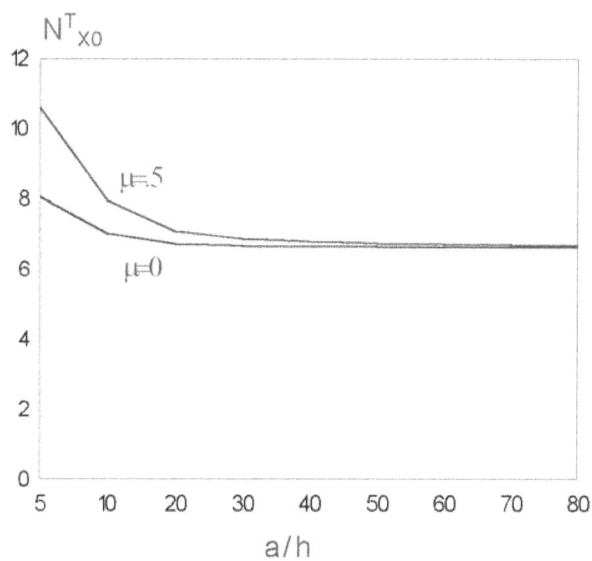

Figure 1.3-1: Thermal buckling load.

varies from 20% to 92%. The case has been checked for the other types
of thermal loading and it is found that with increasing μ, the critical
buckling load, $\bar{N}_{x0}^T$, increases.

Table (1.3-2) shows the effects of the ratio of edge length to thick-
ness, $\left(\frac{a}{h}\right)$, on the buckling load, $\bar{N}_{x0}^T$, for a uniform temperature rise of
an isotropic imperfect plate based on the first-order [8] and higher-order
theories. The table shows that with increasing $\frac{a}{h}$ ratio, the critical buckling load, $\bar{N}_{x0}^T$, decreases. Also the difference between higher-order and
first-order theories varies from 55% to 70%. The case has been checked
for the other types of thermal loading and it is found that with increasing
$\frac{a}{h}$ ratio, the critical buckling load, $\bar{N}_{x0}^T$, decreases. To validate the results
given in tables (1.3-1) and (1.3-2), we may refer to the results represented
by Chang and Leu [15].

Table (1.3-3) shows the effects of the ratio of edge length to thick-
ness, $\left(\frac{a}{h}\right)$, on the buckling load, $\bar{N}_{x0}^T$, for a uniform temperature rise of a
perfect isotropic plate based on the first-order [10] and higher-order the-
ories. The table shows that with increasing $\frac{a}{h}$ ratio, the critical buckling
load, $\bar{N}_{x0}^T$, decreases. This result is qualititavely identical with the result
given by Chang and Leu [15], where they have presented the results for
an orthotropic plate with [+45/-45] ply angle arrangement. Also the difference between higher-order and first-order theories of this paper varies
from 49% to 66%. Results given by Chang and Leu [15] indicate that
this difference varies from 87% to 88%. The case has been checked for
the other types of thermal loading and it is found that with increasing $\frac{a}{b}$
ratio, the critical buckling load, $\bar{N}_{x0}^T$, decreases.

Figure (1.3-1) shows the effects of the ratio of edge length to thick-
ness, $\left(\frac{a}{h}\right)$, on the buckling load, $\bar{N}_{x0}^T$, for a uniform temperature rise of
an imperfect isotropic plate with different imperfection amplitude coeffi-
cient, μ. The figure shows that with increasing $\left(\frac{a}{h}\right)$ ratio, critical buckling
load decreases. Also the figure shows that with increasing μ, the critical
buckling load increases. The case has been checked for the other types of
thermal loading and it is found that with increasing $\frac{a}{h}$ ratio, the critical
buckling load, $\bar{N}_{x0}^T$, decreases.

Figure (1.3-2) shows the effects of the imperfection amplitude coefficient, (μ), on the buckling load, $\bar{N}_{x0}^T$, for a uniform temperature rise of
an imperfect isotropic plate with different $\frac{a}{b}$ ratio. The figure shows that
with increasing $\left(\frac{a}{b}\right)$ ratio, critical buckling load decreases. Also the figure
shows that with increasing μ, the critical buckling load increases. The
case has been checked for the other types of thermal loading and it is
found that with increasing (μ), the critical buckling load, $\bar{N}_{x0}^T$, increases.

4 References

1 - Gossard, M.L., Seide, P., and Roberts, W.M., *Thermal Buckling of Plates*, NACA TN 2771, 1952.

2 - Klosner, J.M., and Forrey, M.J., "Buckling of Simply Supported Plates under Arbitrary Symmetrical Temperature Distributions", Aeronaut. Sci. Vol. 25, pp. 181-184, 1958.

3 - Bargmann, H.W., "Thermal Buckling of Elastic Plates", J. Thermal Stresses, Vol. 8, pp. 71-98, 1985.

4 - Chen, L.W., and Chen, L.Y., "Thermal Buckling of Laminated Composite Plates", J. Thermal Stresses, Vol. 10, pp. 345-356, 1987.

5 - Chen, L.W., and Chen, L.Y., "Thermal Buckling Analysis of Composite Laminated Plates by the Finite Element Method", J. Thermal Stresses, Vol. 12, pp. 41-56, 1989.

6 - Tauchert, T.R., "Thermal Stresses in Plates-Static Problems, in *Thermal Stresses I*, R.B. Hetnarski (ed.), Elsevier, Amsterdam, pp. 23-141, 1986.

7 - Tauchert, T.R., "Thermal Buckling of Thick Antisymmetric Angle-Ply Laminates", J. Thermal Stresses, Vol. 10, pp. 113-124, 1987.

8 - Eslami, M.R., Mossavarali, A., and Peydaye Saheli, Gh., "Thermoelastic Buckling of Isotropic and Orthotropic Plates with Imperfectins", J. Thermal stresses, Vol. 23, No. 9, pp. 853-872, 2000.

9 - Jones, R.M., *Mechanics of Composite Materials* , McGraw-Hill, New York, 1975.

10 - Brush, D.O., and Almroth, B.O., *Buckling of Beams, Plates, and Shells*, Mc Graw-Hill, New York, 1975.

11 - Timoshenko, S.P., and Gere, J.M., *Theory of Plates and Shells*, 2nd ed., McGraw-Hill, New York, 1959.

12 - Thornton, E.A., "Thermal Buckling of Plates and Shells", Appl. Mech. Rev., Vol. 46, No. 10, pp. 485-506, 1993.

13 - Leissa, A.W., and Narita, Y., "Buckling Studies for Simply Sup-

ported Symmetrically Laminated Rectangular Plates", Int. J. Mech. Sci., Vol. 32, No. 11, pp. 909-924, 1990.

14 - Tauchert, T.R., and Huang, N.N., "Thermal Buckling and Post-buckling Behavior of Antisymmetric Angle-Ply Laminates", Proc. Int. Symp. Compos. Mater. Struct., T.T. Loo and C.T. Sun (eds.), Technomic, Lancaster PA, pp. 357-362, 1986.

15 - Chang, J.S., and Leu, S.Y., "Thermal Buckling Analysis of Anti-symmetric Angle-Ply Laminates Based on a Higher-Order Displacements Field", J. Composite Sci. Tech., Vol. 41, pp. 109-128, 1991.

16 - Reddy, J.N., "A Simple Higher-Order Theory for Laminated Composite Plates", J. Applied Mech., Vol. 51, pp. 745-752, 1984.

17 - Eslami, M.R., and Javaheri, M.R., "Thermal and Mechanical Buckling of Composite Cylindrical Shells", J. Thermal Stresses, Vol. 22, No. 6, pp. 527-545, 1999.

18 - Tauchert, T.R., Jonnalagadda, K.D., and Blandford, G.E., "Thermal Buckling of Laminated Plates Using Higher-Order Deformation Theory", International Conference on Composite Materials, Madrid, Spain, pp. 394-401, 1993.

19 - Romeo, G., and Frulla, G., "Nonlinear Analysis of Anisotropic Plates with Initial Imperfections and Various Boundary Conditions Subjected to Combined Biaxial Compresion and Shear Loads", Int. J. Solids Structures, Vol. 14, No. 6, pp. 763-783, 1994.

20 - Mossavarali, A., and Eslami, M.R., "Thermoelastic Buckling of Orthotropic Plates Based on a Higher-Order Displacement Field", Scientia Iranian, Vol. 8, No. 2, pp. 149-157, 2001.

Chapter 2

Buckling of Circular Plates

1   Isotropic Homogeneous Circular Plates

1.1   Introduction

The axisymmetric buckling of annular plates is investigated by Meissner [1] when the outer boundary is uniformly compressed and the concentric hole is free. Yamaki [2] showed that for other boundary conditions than those assumed by Nadai and Meissner the asymmetric-type buckling may be well associated with the lowest critical compression. Grigolyuk [3] considered thermal buckling of circular plates based on the approximate Bubnov method. Mansfield [4] has discussed the buckling and curling of a heated thin circular plate, when the temperature varies through the thickness of the plate and the edges are restrained.

Klossner and Forray [5] have studied the buckling of circular plates under symmetrical temperature distribution using the Raliegh-Ritz method. Sarkar [6] has solved the buckling problem of a heated thin circular plate of isotropic material under uniform compression and thermal fields $T = T(r, z)$ in the plane of the plate and obtained the critical buckling temperature for plates under different edge conditions and different temperature distributions. Das [7] has studied the behaviour of buckled annular plates under uniform compression and two dimensional temperature distribution $T = T(r, z)$ and obtained the critical buckling compression and critical temperature in two cases,viz, when the plate thickness varied linearly, and inversely as the distance from the center of the plate. De [8] studied the behaviour of buckled annular plates under uniform compression and thermal fields $T = T(r, z)$. He assumed that the thickness of the plate varies as the nth power of the distance from the center, i.e. $h = h_0^n$ and obtained the critical buckling value for the temperature. Hong and Wang [9] considered the effect of prebuckling deformation on the elastic buckling of circular plates. Stavsky [10] considered thermoelastic stability of laminated orthotropic circular plate. He obtained closed type solution for certain thermal buckling problem in the form of Bessel function of first kind and fractional order. Chang and Shiao [11] considered thermal buckling of composite circular plate with a hole, using finite element method and obtained the value of critical temperature.

Ye [12] considered the axisymmetric buckling problem of homogeneous and laminated circular plates, based on the three dimensional elasticity for both prebuckling and perturbed analysis. On the basis of three dimensional consideration, the paper presents a formulation suitable for studying the axisymmetric buckling behavior of thick circular plates with transversely isotropic material properties. Most of the investigations mentioned above are concerned with the thermal buckling of isotropic or composite circular plate for the case of uniform temperature rise. Krizhevsky and Stavsky [13], with the use of Hamilton's variational principle, derived the equations of motion for transversely isotropic laminated annular plates. Linearized vibration and buckling equations are obtained for the annular plates uniformly compressed in the radial direction. A closed-form solution is given for the mode shapes in terms of Bessel power and trigonometric functions.

In this section we derive the general thermoelastic equations and the stability equations from the energy method with the use of calculus of variations and compatible with the theory of buckling. The closed form solution of a solid circular plate of isotropic materials subjected to uniform temperature rise, gradient through the thickness, and the linear temperature variation along the radius of the plate are obtained. The buckling loads considered in this paper are of practical importance, and the proposed buckling formula are useful in the design stage of circular plates. The results are validated with the known data in the literature.

1.2   Basic equations

Consider a circular flat plate of thickness h and radius a under axisymmetric thermoelastic loading. The two-dimensional stress-strain relations are given by

$$\epsilon_{rr} = \frac{1}{E}(\sigma_{rr} - \nu\sigma_{\theta\theta}) + \alpha T$$

$$\epsilon_{\theta\theta} = \frac{1}{E}(\sigma_{\theta\theta} - \nu\sigma_{rr}) + \alpha T$$

$$\epsilon_{r\theta} = \frac{1}{G}\sigma_{r\theta} \qquad (2.1\text{-}1)$$

The plate is assumed to be comparatively thin and planes normal to the median surface are assumed to remain plane after deformation, thus shear deformations normal to the plate are disregarded. Strain components at distance z from the middle plane are then given by

$$\epsilon_{rr} = \bar{\epsilon}_{rr} - zk_{rr}$$
$$\epsilon_{\theta\theta} = \bar{\epsilon}_{\theta\theta} - zk_{\theta\theta}$$
$$\epsilon_{r\theta} = \bar{\epsilon}_{r\theta} - 2zk_{r\theta} \qquad (2.1\text{-}2)$$

Here, $\bar{\epsilon}_{ij}$ are the strain components in the median surface, and k_{ij} are the curvatures. According to the Sanders assumption, the middle plane strains and curvatures for symmetric loading are related to the radial displacement u and the lateral deflection w as

$$\bar{\epsilon}_{rr} = u_{,r} + \frac{1}{2}w_{,r}^2$$
$$\bar{\epsilon}_{\theta\theta} = \frac{u}{r}$$
$$\bar{\epsilon}_{r\theta} = 0$$
$$k_{rr} = w_{,rr}$$
$$k_{\theta\theta} = \frac{1}{r}w_{,r}$$
$$k_{r\theta} = 0 \qquad (2.1\text{-}3)$$

where u and w are the non-zero displacement components and functions of in-plane coordinates (r, θ). Substituting Eqs. (2.1-3) into Eqs. (2.1-2) the following expression for the strain components are obtained

$$\epsilon_{rr} = u_{,r} + \frac{1}{2}w_{,r}^2 - zw_{,rr}$$
$$\epsilon_{\theta\theta} = \frac{u}{r} - \frac{z}{r}w_{,r}$$
$$\epsilon_{r\theta} = 0 \qquad (2.1\text{-}4)$$

The total potential energy for the circular plate may be written as

$$V = \frac{E}{2(1-\nu^2)} \int_0^r \int_0^{2\pi} \int_{-h/2}^{h/2} [\epsilon_{rr}^2 + z^2 k_{rr}^2$$
$$-2z\epsilon_{rr}k_{rr} + \epsilon_{\theta\theta}^2 + z^2 k_{\theta\theta}^2 - 2\epsilon_{\theta\theta}zk_{\theta\theta}$$
$$+2\nu(\epsilon_{rr}\epsilon_{\theta\theta} - \epsilon_{rr}zk_{\theta\theta} - \epsilon_{\theta\theta}zk_{rr} + z^2 k_{rr}k_{\theta\theta})$$
$$-2(1+\nu)\alpha T(\epsilon_{rr} + \epsilon_{\theta\theta} - zk_{rr} - zk_{\theta\theta})$$
$$+\frac{1-\nu}{2}(\epsilon_{r\theta}^2 + 4z^2 k_{r\theta}{}^2 - 4\epsilon_{r\theta}zk_{r\theta}) -$$
$$2(1+\nu)\alpha^2 T^2]rdrd\theta dz \qquad (2.1\text{-}5)$$

Integration with respect to z from $-h/2$ to $+h/2$ gives

$$V = \frac{E}{2(1-\nu^2)} \int_0^r \int_0^{2\pi} [(\epsilon_{rr}^2 + \epsilon_{\theta\theta}^2 + 2\nu\epsilon_{rr}\epsilon_{\theta\theta}$$
$$+\frac{1-\nu}{2}\epsilon_{r\theta}^2)h + \frac{h^3}{12}(k_{rr}^2 + k_{\theta\theta}^2$$
$$+2\nu k_{rr}k_{\theta\theta} + \frac{1-\nu}{2}4k_{r\theta}^2)]rdrd\theta$$
$$-\frac{E}{(1-\nu)}\int_{-h/2}^{h/2}[\alpha T(\epsilon_{rr} + \epsilon_{\theta\theta} - zk_{rr} -$$
$$zk_{\theta\theta}) + \alpha^2 T^2]rdrd\theta \tag{2.1-6}$$

Assuming that the circular plate is under thermal stress alone, the total potential energy is a function of the displacement components and their derivatives and may be written as

$$V = \int\int F(u, w, u_{,r}, u_{,\theta}, v_r, v_\theta, w_r, w_\theta)drd\theta \tag{2.1-7}$$

where

$$F = \frac{E}{2(1-\nu^2)} \int_0^r \int_0^{2\pi} [(\epsilon_{rr}^2 + \epsilon_{\theta\theta}^2 + 2\nu\epsilon_{rr}\epsilon_{\theta\theta}$$
$$+\frac{1-\nu}{2}\epsilon_{r\theta}^2)h + (k_{rr}^2 + k_{\theta\theta}^2 + 2\nu k_{rr}k_{\theta\theta}$$
$$+2(1-\nu)k_{r\theta}^2)(\frac{h^3}{12})]rdrd\theta$$
$$-\frac{E}{(1-\nu)}\int_{-h/2}^{h/2}[\alpha T(\epsilon_{rr} + \epsilon_{\theta\theta} -$$
$$zk_{rr} - zk_{\theta\theta}) + \alpha^2 T^2]rdrd\theta \tag{2.1-8}$$

Extremizing the functional of potential energy leads to the Euler equations

$$\frac{\partial F}{\partial u} - \frac{\partial}{\partial r}\frac{\partial F}{\partial u_{,r}} - \frac{\partial}{\partial\theta}\frac{\partial F}{\partial u_{,\theta}} = 0$$
$$\frac{\partial F}{\partial v} - \frac{\partial}{\partial r}\frac{\partial F}{\partial v_{,r}} - \frac{\partial}{\partial\theta}\frac{\partial F}{\partial v_{,\theta}} = 0$$
$$\frac{\partial F}{\partial w} - \frac{\partial}{\partial r}\frac{\partial F}{\partial w_{,r}} - \frac{\partial}{\partial\theta}\frac{\partial F}{\partial w_{,\theta}} + \frac{\partial^2}{\partial r^2}\frac{\partial F}{\partial w_{,rr}}$$
$$+\frac{\partial^2}{\partial r\partial\theta}\frac{\partial F}{\partial w_{,r\theta}} + \frac{\partial^2}{\partial\theta^2}\frac{\partial F}{\partial w_{,\theta\theta}} = 0 \tag{2.1-9}$$

Substitution from Eqs. (2.1-2) through (2.1-6) into Eq. (2.1-7) and using Eqs. (2.1-9) the equilibrium equations for general thin circular plate are obtained as

$$N_{r,r} + \frac{1}{r}N_{r\theta,\theta} + \frac{N_r - N_\theta}{r} = 0$$

$$N_{r\theta,r} + \frac{1}{r}N_{\theta,\theta} + \frac{2}{r}N_{r\theta} = 0$$

$$(rM_r)_{,rr} + 2(M_{r\theta,r\theta} + \frac{1}{r}M_{r\theta,\theta}) +$$

$$(\frac{1}{r}M_{\theta,\theta\theta} - M_{\theta,r}) - [(rN_{r0}\beta_r + rN_{r\theta 0}\beta_\theta)_{,r}$$

$$+(N_{r\theta 0}\beta_r + N_{\theta 0}\beta_\theta)_{,\theta}] = 0 \qquad (2.1\text{-}10)$$

where N_r, N_θ, $N_{r\theta}$, M_r, M_θ, $M_{r\theta}$, β_r, β_θ, M^T, D and C are

$$N_r = C(\bar\epsilon_{rr} + \nu\bar\epsilon_{\theta\theta}) - \frac{E\alpha}{1-\nu}\int_{-h/2}^{h/2} T dz$$

$$N_{r0} = C(\frac{du_0}{dr} + \frac{\nu u_o}{r}) - \frac{E\alpha}{1-\nu}\int_{-h/2}^{h/2} T dz$$

$$N_\theta = C(\bar\epsilon_{\theta\theta} + \nu\bar\epsilon_{rr}) - \frac{E\alpha}{1-\nu}\int_{-h/2}^{h/2} T dz$$

$$N_{\theta_0} = C(\frac{u_0}{r} + \nu\frac{du_0}{dr}) - \frac{E\alpha}{1-\nu}\int_{-h/2}^{h/2} T dz$$

$$N_{r\theta} = N_{r\theta_0} = \frac{Eh}{2(1+\nu)}\bar\epsilon_{r\theta}$$

$$M_r = -D[\beta_{r,r} + \frac{\nu}{r}(\beta_{\theta,\theta} + \beta_r)] + M^T$$

$$M_\theta = -D[\frac{1}{r}(\beta_{\theta,\theta} + \beta_r) + \nu\beta_{r,r}] + M^T$$

$$M_{r\theta} = -D\frac{1-\nu}{2}[r(\frac{\beta_\theta}{r})_{,r} + \frac{\beta_{r,\theta}}{r}]$$

$$\beta_r = w_{,r}$$

$$\beta_\theta = \frac{1}{r}w_{,\theta}$$

$$M^T = -\frac{E\alpha}{1-\nu}\int_{-h/2}^{h/2} T z dz$$

$$D = \frac{Eh^3}{12(1-\nu^2)}$$

$$C = \frac{Eh}{1-\nu^2} \qquad (2.1\text{-}11)$$

Substitution from Eqs. (2.1-11) into the third of Eqs. (2.1-10) we obtained

$$N_{r,r} + \frac{1}{r}N_{r\theta,\theta} + \frac{N_r - N_\theta}{r} = 0$$

$$N_{r\theta,r} + \frac{1}{r}N_{\theta,\theta} + \frac{2}{r}N_{r\theta} = 0$$

$$D\nabla^4 w + \nabla^2 M^T = N_r \beta_{r,r} + \frac{1}{r}N_\theta(\beta_{\theta,\theta} + \beta_r)$$

$$+2N_{r\theta}\beta_{\theta,r} \tag{2.1-12}$$

Equations (2.1-12) are the nonlinear equilibrium equations for thin circular plate under thermoelastic loading. The linear form of the equilibrium equations (2.1-10) reduce to

$$N_{r,r} + \frac{1}{r}N_{r\theta,\theta} + \frac{N_r - N_\theta}{r} = 0$$

$$N_{r\theta,r} + \frac{1}{r}N_{\theta,\theta} + \frac{2}{r}N_{r\theta} = 0$$

$$D\nabla^4 w + \nabla^2 M^T = 0 \tag{2.1-13}$$

The stability equations of thin circular plates may be derived either by the force summation method or the energy method. If V is the total potential energy of the plate, the expansion of V about the equilibrium state into the Taylor series yields

$$\Delta V = \delta V + \frac{1}{2}\delta^2 V + \frac{1}{6}\delta^3 V + ... \tag{2.1-14}$$

The stability of the plate in the neighborhood of equilibrium condition may be determine by the sign of second variation as follows:
1) Stable equilibrium occurs when a small displacement of the system causes the system return to its original position. In this case the original potential energy of the system is a minimum, that is $\delta^2 V > 0$
2) Neutral equilibrium occurs when a small displacement of the system causes the system to remain in its displaced state. In this case the potential energy of the system remains constant, that is $\delta^2 V = 0$
3) Unstable equilibrium occurs when a small displacement of the system causes the system to move farther away from its original position. In this case the original potential energy of the system is a maximum, that is $\delta^2 V < 0$
The condition $\delta^2 V = 0$ is used to derive the stability equations for buckling problem.

Let us assume that u_i^* denotes the displacement component of the equilibrium state and δu_i^* the virtual displacement corresponding to a neighboring state. Denoting by $\bar{\delta}$ the variation with respect to u_i^*, the following rule, know as the Trefftz rule, is stated for the determination of the lowest critical load.

The external load acting on the original configuration is considered to be the critical buckling load if the following variational equation is satisfied

$$\bar{\delta}(\delta^2 V) = 0 \qquad (2.1\text{-}15)$$

This rule provides the governing equations that determine the lowest critical loads. Consider the state of stable equilibrium of a general circular plate under thermal load to be designated by u_0, v_0, w_0. The displacement components of the neighboring state are

$$u = u_0 + u_1$$
$$v = v_0 + v_1$$
$$w = w_0 + w_1 \qquad (2.1\text{-}16)$$

where u_1, v_1, w_1 are arbitrary small increments at displacements. Substituting Eqs. (2.1-16) into Eq. (2.1-8) and collection of second-order terms, we get the second variation of the potential energy as

$$\frac{1}{2}\delta^2 V = \int \int \frac{C}{2}(\epsilon_{rr_1}^2 + \epsilon_{\theta\theta_1}^2 + 2\nu\epsilon_{rr_1}\epsilon_{\theta\theta_1} +$$
$$\frac{(1-\nu)}{2}\epsilon_{r\theta_1}^2)rdrd\theta + \frac{1}{2}\int \int (N_{r_0}\beta_{r1}^2 + N_{\theta_0}\beta_{\theta1}^2$$
$$+2N_{r\theta_0}\beta_{r_1}\beta_{\theta_1})rdrd\theta + \int \int \frac{D}{2}[k_{rr_1}^2 + k_{\theta\theta_1}^2 +$$
$$2\nu k_{rr_1}k_{\theta\theta_1} + 2(1-\nu)k_{r\theta_1}^2]rdrd\theta \qquad (2.1\text{-}17)$$

where the linear strain-displacement relations are

$$\epsilon_{rr_1} = \frac{\partial u_1}{\partial r} \quad \epsilon_{\theta\theta_1} = \frac{1}{r}\frac{\partial v_1}{\partial \theta} + \frac{u_1}{r}$$
$$\epsilon_{r\theta_1} = \frac{1}{r}\frac{\partial u_1}{\partial \theta} + \frac{\partial v_1}{\partial r} - \frac{v_1}{r} \quad k_{rr_1} = \frac{\partial^2 w_1}{\partial r^2}$$
$$k_{\theta\theta_1} = \frac{1}{r}\frac{\partial w_1}{\partial r} \quad k_{r\theta_1} = \frac{1}{r}\frac{\partial^2 w_1}{\partial r\partial \theta} -$$
$$\frac{1}{r^2}\frac{\partial w_1}{\partial \theta} \qquad (2.1\text{-}18)$$

Substituting Eqs. (2.1-18) into Eq. (2.1-17) and using the Euler equations we get the stability equations

$$N_{r_1,r} + \frac{1}{r}N_{r\theta_1,\theta} + \frac{N_{r_1} - N_{\theta_1}}{r} = 0$$

$$N_{r\theta_1,r} + \frac{1}{r}N_{\theta_1,\theta} + \frac{2}{r}N_{r\theta_1} = 0$$

$$(rM_{r_1})_{,rr} + 2(M_{r\theta_1,r\theta} + \frac{1}{r}M_{r\theta_1,\theta}) +$$

$$(\frac{1}{r}M_{\theta_1,\theta\theta} - M_{\theta_1,r}) - [(rN_{r_0}\beta_r + rN_{r\theta_0}\beta_\theta)_{,r}$$

$$+(N_{r\theta_0}\beta_r + N_{\theta_0}\beta_\theta)_{,\theta}] = 0 \qquad (2.1\text{-}19)$$

where N_{r_1}, N_{θ_1}, $N_{r\theta_1}$, N_{r_0}, N_{θ_0}, $N_{r\theta_0}$, $N_{r\theta_0}$, M_{r_1}, M_{θ_1}, $M_{r\theta_1}$ are

$$N_{r_1} = C(\bar\epsilon_{rr_1} + \nu\bar\epsilon_{\theta\theta_1})$$
$$N_{\theta_1} = C(\bar\epsilon_{\theta\theta_1} + \nu\bar\epsilon_{rr_1})$$
$$N_{r\theta_1} = \frac{Eh}{2(1+\nu)}\bar\epsilon_{r\theta_1}$$
$$N_{r_0} = C(\bar\epsilon_{rr} + \nu\bar\epsilon_{\theta\theta}) - N_{r0}^T$$
$$N_{\theta_0} = C(\bar\epsilon_{\theta\theta} + \nu\bar\epsilon_{rr}) - N_{\theta_0}^T$$
$$N_{r\theta_0} = \frac{C(1-\nu)}{2}\bar\epsilon_{r\theta}$$
$$N_{r0}^T = N_{\theta_0}^T = \frac{E}{1-\nu}\int_{-h/2}^{h/2}\alpha T dz$$
$$M_{r_1} = -D[\beta_{r,r} + \frac{\nu}{r}(\beta_{\theta,\theta} + \beta_r)]$$
$$M_{\theta_1} = -D[\frac{1}{r}(\beta_{\theta,\theta} + \beta_r) + \nu\beta_{r,r}]$$
$$M_{r\theta_1} = -D\frac{1-\nu}{2}[r(\frac{\beta_\theta}{r})_{,r} + \frac{\beta_{r,\theta}}{r}]$$

$$(2.1\text{-}20)$$

Substitution from Eqs. (2.1-20) into the third of Eqs. (2.1-19) we obtained the stability equations.

$$N_{r_1,r} + \frac{1}{r}N_{r\theta_1,\theta} + \frac{N_{r_1} - N_{\theta_1}}{r} = 0$$
$$N_{r\theta_1,r} + \frac{1}{r}N_{\theta_1,\theta} + \frac{2}{r}N_{r\theta_1} = 0$$
$$D\nabla^4 w_1 = N_{r_0}\beta_{r_1,r} + \frac{1}{r}N_{\theta_0}(\beta_{\theta_1,\theta} + \beta_{r1}) +$$
$$2N_{r\theta_0}\beta_{\theta_1,r} \qquad (2.1\text{-}21)$$

1.3   Thermal buckling

Let us consider a circular plate in the absence of mechanical loading subjected to a thermal field $T = T(r,z)$. We also limit the discussion to the case of cylindrical symmetry. With this condition, the third of stability equations (2.1-21) becomes:

$$D\nabla^4 w_1 = N_{r0}\frac{\partial^2 w_1}{\partial r^2} + N_{\theta 0}\frac{\partial w_1}{r\partial r} \qquad (2.1\text{-}22)$$

In Eq. (2.1-22) N_{r0} and $N_{\theta 0}$ are the prebuckling thermal forces that must be calculated. From the first of Eqs. (2.1-10) we have

$$N_{\theta 0} = rN_{r0,r} + N_{r0} \qquad (2.1\text{-}23)$$

Substituting Eq. (2.1-23) into Eq. (2.1-22) and one step integrating with respect to r, we get

$$D(\frac{d^3 w_1}{dr^3} + \frac{1}{r}\frac{d^2 w_1}{dr^2} - \frac{1}{r^2}\frac{dw_1}{dr}) = N_{r0}w_{1,r} \qquad (2.1\text{-}24)$$

From Eqs. (2.1-20) we have

$$N_{r0} = C(\frac{du_0}{dr} + \frac{\nu u_o}{r}) - N_{r0}^T$$

$$N_{\theta 0} = C(\frac{u_0}{r} + \nu\frac{du_0}{dr}) - N_{\theta 0}^T \qquad (2.1\text{-}25)$$

Substituting Eqs. (2.1-25) into the first of the equilibrium equations (2.1-10) we get

$$C(r^2\frac{d^2 u_0}{dr^2} + \frac{rdu_0}{dr} - u_0) = r^2\frac{dN_{r0}^T}{dr} \qquad (2.1\text{-}26)$$

If the temperature is a function of the thickness of the plate, or the case of uniform temperature rise, we have

$$N_{r0,r}^T = [\frac{E\alpha}{1-\nu}\int_{-h/2}^{h/2} Tdz]_{,r} = 0 \qquad (2.1\text{-}27)$$

Substituting into Eq. (2.1-26), we get

$$r^2 u_0'' + ru_0' - u_0 = 0 \qquad (2.1\text{-}28)$$

Solution of Eq. (2.1-28) gives

$$u_0 = c_1 + c_2(\frac{1}{r}) \qquad (2.1\text{-}29)$$

Considering a solid circular plate with the following conditions for u, we have :

$$u = 0 \qquad\qquad r = 0$$
$$u = 0 \qquad\qquad r = a \qquad (2.1\text{-}30)$$

The constants of integration $c_1 = c_2 = 0$ and from Eq. (2.1-29)

$$u_0 = 0 \qquad\qquad (2.1\text{-}31)$$

Therefore, from Eq. (2.1-25) we have

$$N_{r0} = -N_{r0}^T \qquad\qquad (2.1\text{-}32)$$

Substituting Eq. (2.1-32) into Eq. (2.1-24) we get

$$r^2\beta'' + r\beta' + \beta(\lambda^2 r^2 - 1) = 0 \qquad\qquad (2.1\text{-}33)$$

where

$$\beta = \frac{dw_1}{dr} \qquad\qquad ; \lambda^2 = \frac{N_{r0}^T}{D} \qquad\qquad (2.1\text{-}34)$$

The solution of Eq. (2.1-33) is

$$\beta = c_3 J_1(\lambda r) + c_4 Y_1(\lambda r) \qquad\qquad (2.1\text{-}35)$$

where J_1 and Y_1 are Bessel functions of first order of the first and second kinds, respectively, and c_3 and c_4 are integration constants. But $\beta = 0$ at $r = 0$, and $Y_1(0) \to \infty$. Therefore $c_4 = 0$ and

$$\beta = c_3 J_1(\lambda r) \qquad\qquad (2.1\text{-}36)$$

For the clamped edge, $\beta = 0$ at $r = a$, where a is the plate radius. Therefore

$$J_1(x) = 0 \qquad\qquad x = \lambda a \qquad\qquad (2.1\text{-}37)$$

and the lowest critical thermal load is expressed in terms of $x_1 = \lambda_1 a$, the first zero of the Bessel function of first kind and order one, as

$$\lambda a = 3.831 \qquad\qquad (2.1\text{-}38)$$

Using the value of λa and substituting into Eqs. (2.1-34) and (2.1-32) gives

$$N_{r0}^T = 14.68\frac{D}{a^2} \qquad\qquad (2.1\text{-}39)$$

From Eq. (2.1-21) for the case of uniform temperature rise, the thermal buckling load of the clamped circular plate reduces to

$$\Delta T_{cr} = \frac{14.68}{12(1+\nu)}(\frac{h}{a})^2\frac{1}{\alpha_T} \qquad\qquad (2.1\text{-}40)$$

For the case of simply supported plate, we similarly obtain

$$\Delta T_{cr} = \frac{4.2}{12(1+\nu)}(\frac{h}{a})^2\frac{1}{\alpha_T} \qquad\qquad (2.1\text{-}41)$$

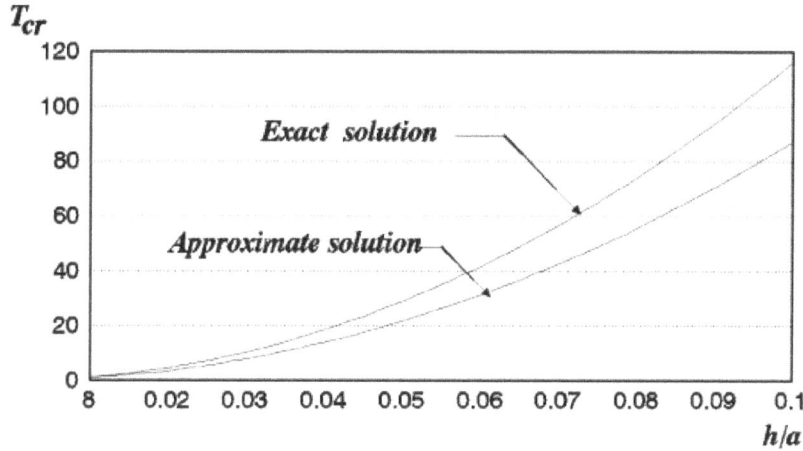

Figure 2.1-1: buckling temperature vs. ratio of h/a for simply supported plate.

For the case of linear temperature variation through the thickness of the plate, the temperature distribution is

$$T(z) = \frac{\Delta T}{h}(z + \frac{h}{2}) \tag{2.1-42}$$

The resulting buckling loads for clamped edge circular plate from Eqs. (2.1-21) and (2.1-32) is

$$\Delta T_{cr} = \frac{29.28}{12(1+\nu)}(\frac{h}{a})^2\frac{1}{\alpha_T} \tag{2.1-43}$$

For simply supported circular plate, the thermal buckling load from Eq. (2.1-35) is

$$\Delta T_{cr} = \frac{8.4}{12(1+\nu)}(\frac{h}{a})^2\frac{1}{\alpha_T} \tag{2.1-44}$$

For the case of linear temperature variation along the direction of radius such that

$$T = \frac{\Delta T}{a}r \tag{2.1-45}$$

The critical thermal buckling load for clamped circular plate from Eqs. (2.1-21), (2.1-24), and (2.1-26) is

$$\Delta T_{cr} = \frac{19.87}{12(1+\nu)}(\frac{h}{a})^2\frac{1}{\alpha_T} \tag{2.1-46}$$

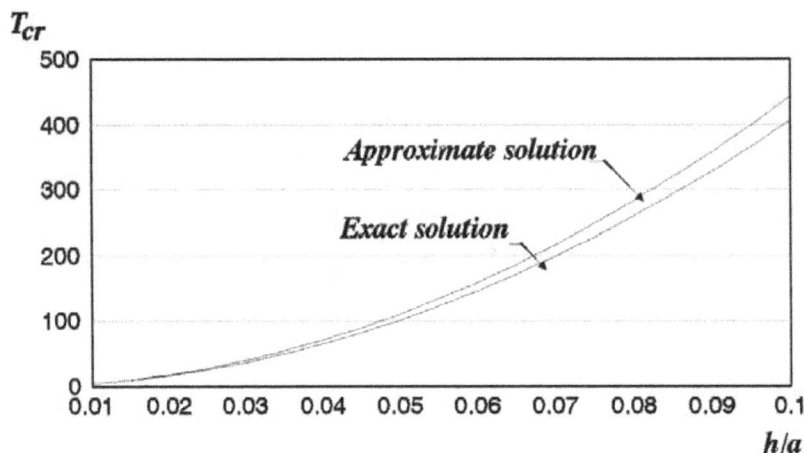

Figure 2.1-2: buckling temperature vs. ratio of h/a.

1.4   Results and discussion

In this section the axisymmetric stability and thermal buckling equations of circular plates subjected to uniform temperature rise, linear gradient through the thickness, and linear temperature variation along the radius are considered. The problem is solved by Nowacki [14] for clamped and Kovalenko [15] for simply supported circular plates. Nowacki [14] and Kovalenko [15] consider the thermal buckling temperature associated with uniform temperature rise. The results of the present paper are compared with the results of Nowacki and Kovalenko and are given in Tables (2.1-1) and (2.1-2).

Table (2.1-1) Comparison of results for uniform temperature rise.

Uniform temperature rise	Simply supported
Kovalenko Solution	$\Delta T_{cr} = \frac{2}{3}\frac{1}{(3+\nu)}\left(\frac{h}{a}\right)^2\frac{1}{\alpha_T}$
Nowacki Solution	
Present section	$\Delta T_{cr} = \frac{4.2}{12(1+\nu)}\left(\frac{h}{a}\right)^2\frac{1}{\alpha_T}$

Uniform temperature rise	Clamped edge
Kovalenko Solution	
Nowacki Solution	$\Delta T_{cr} = \frac{4}{3(1+\nu)}\left(\frac{h}{a}\right)^2\frac{1}{\alpha_T}$
Present section	$\Delta T_{cr} = \frac{14.68}{12(1+\nu)}\left(\frac{h}{a}\right)^2\frac{1}{\alpha_T}$

For $\nu = 0.3$ we have

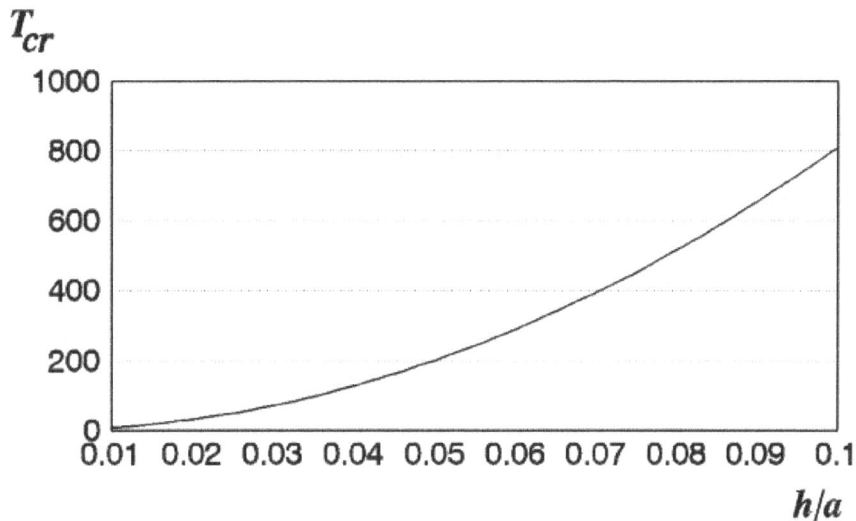

Figure 2.1-3: buckling of circular plate vs. ratio of h/a.

Table (2.1-2) Comparison of results for uniform temperature rise (for $\nu = 0.3$).

Uniform temperature rise	Simply supported
Kovalenko Solution	$\Delta T_{cr} = 0.2(\frac{h}{a})^2 \frac{1}{\alpha_T}$
Nowacki Solution	
Present section	$\Delta T_{cr} = 0.26(\frac{h}{a})^2 \frac{1}{\alpha_T}$

Uniform temperature rise	Clamped edge
Kovalenko Solution	
Nowacki Solution	$\Delta T_{cr} = 1.025(\frac{h}{a})^2 \frac{1}{\alpha_T}$
Present section	$\Delta T_{cr} = 0.94(\frac{h}{a})^2 \frac{1}{\alpha_T}$

The buckling loads for the linear temperature gradient through the thickness and linear temperature along the radius are given in Table (2.1-3) and for $\nu = 0.3$ are given in Table (2.1-4). These types of thermal buckling information are frequently needed in the design stage of circular plates. The temperature variations in circular plate in the z and $r-$ directions may be other than linear distribution, as assumed in this paper. However, with a linear temperature approximation in z and $r-$ directions, the thermal buckling of circular plates can be readily calculated

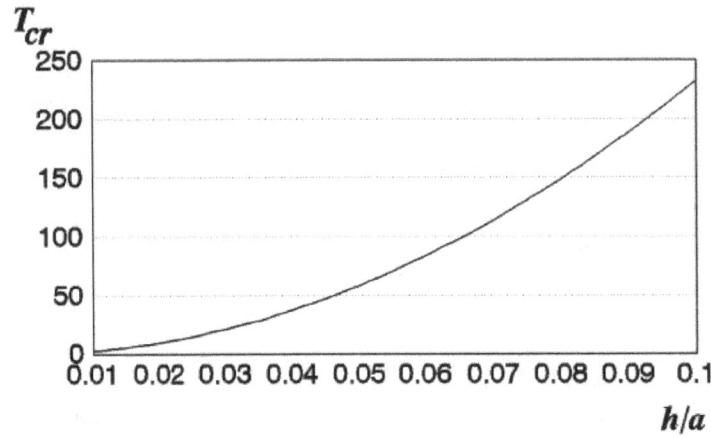

Figure 2.1-4: buckling temperature vs. ratio of h/a.

using Table (2.1-3)

Table (2.1-3) Buckling loads for the linear temperature gradient through the thickness.

Type of load	Clamped edge
$T(z) = \frac{\Delta T}{h}(z + \frac{h}{2})$	$\Delta T_{cr} = \frac{29.28}{12(1+\nu)}\left(\frac{h}{a}\right)^2 \frac{1}{\alpha_T}$
$T = \frac{\Delta T}{a} r$	$\Delta T_{cr} = \frac{19.87}{12(1+\nu)}\left(\frac{h}{a}\right)^2 \frac{1}{\alpha_T}$

Type of load	Simply supported
$T(z) = \frac{\Delta T}{h}(z + \frac{h}{2})$	$\frac{8.4}{12(1+\nu)}\left(\frac{h}{a}\right)^2 \frac{1}{\alpha_T}$
$T = \frac{\Delta T}{a} r$	

For $\nu = 0.3$ we have

Table (2.1-4) Buckling loads for the linear temperature along the radius.

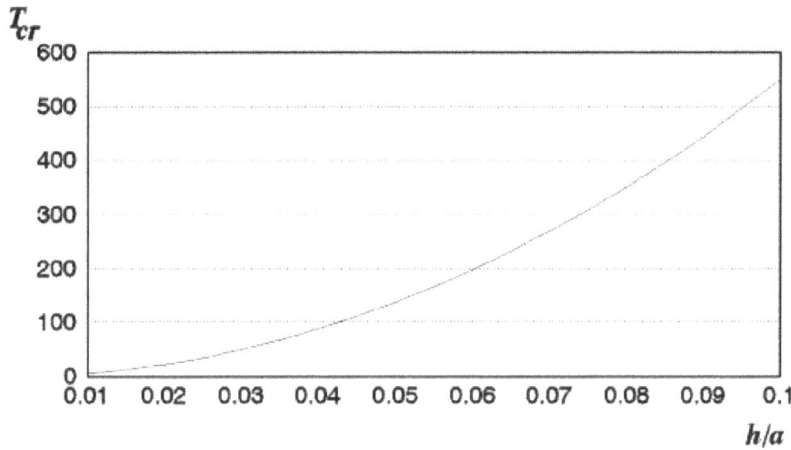

Figure 2.1-5: buckling temperature vs. ratio of h/a.

Type of load	Clamped edge
$T(z) = \frac{\Delta T}{h}(z + \frac{h}{2})$	$\Delta T_{cr} = 1.877(\frac{h}{a})^2 \frac{1}{\alpha_T}$
$T = \frac{\Delta T}{a} r$	$\Delta T_{cr} = 1.273(\frac{h}{a})^2 \frac{1}{\alpha_T}$

Type of load	Simply supported
$T(z) = \frac{\Delta T}{h}(z + \frac{h}{2})$	$0.538(\frac{h}{a})^2 \frac{1}{\alpha_T}$
$T = \frac{\Delta T}{a} r$	

Figures (2.1-1) and (2.1-2) show the buckling temperature T_{cr} versus the ratio of thickness to radius h/a for the case of uniform temperature rise with the simply supported and clamped supported edges. Figures (2.1-3) and (2.1-4) show the buckling temperature T_{cr} vs. ratio thickness to radius h/a for the case of linear gradient through the thickness with the simply supported and clamped supported edges. Figure (2.1-5) shows the buckling temperature T_{cr} vs. ratio thickness to radius h/a for the case of linear temperature variation along the radius with the clamp supported edges.

2   Stability of Orthotropic Circular Plates

2.1   Introduction

In this section we derive the general thermoelastic equilibrium equations and the stability equations for a composite circular plate from the energy method with the use of the calculus of variations and compatible with the theory of buckling. The formulation is derived and is valid for the thermal buckling of non-axisymmetric as well as axisymmetric temperature fields, but the applications are presented for the latter only. The closed-form approximate solutions for circular plates subjected to thermal loads are obtained. The buckling temperature of a solid circular plate under uniform temperature rise, gradient through the thickness, and the linear temperature variation along the radius are obtained. The plate is assumed to be geometrically perfect. The results are validated with the known data in the literature.

2.2   Basic equations

Consider a circular composite circular flat plate under thermoelastic loading. The thermoelastic stress-strain relations may be written for a polar orthotropic as

$$\sigma_{rr} = E_{rr}\epsilon_{rr} + E_{r\theta}\epsilon_{\theta\theta} + A_r T$$
$$\sigma_{\theta\theta} = E_{r\theta}\epsilon_{rr} + E_{\theta\theta}\epsilon_{\theta\theta} + A_\theta T$$
$$\sigma_{r\theta} = G\epsilon_{r\theta} \tag{2.2-1}$$

where E_{ij} are the elastic modulii and the thermal coefficients A_i are expressed in term of the coefficients of thermal expansion [8]

$$A_r = -(E_{rr}\alpha_r + E_{r\theta}\alpha_\theta)$$
$$A_\theta = -(E_{r\theta}\alpha_r + E_{\theta\theta}\alpha_\theta) \tag{2.2-2}$$

It is assumed that both elastic and thermal properties are independent of time and temperature. The plate is assumed to be comparatively thin and planes normal to the median surface are assumed to remain plane after deformation, thus shearing deformation normal to the plate is disregarded as assumed by the Love-Kirchoff hypothesis. Strain components are then given by

$$\epsilon_{rr} = \bar\epsilon_{rr} + zk_{rr}$$
$$\epsilon_{\theta\theta} = \bar\epsilon_{\theta\theta} + zk_{\theta\theta}$$
$$\epsilon_{r\theta} = \bar\epsilon_{r\theta} + 2zk_{r\theta} \tag{2.2-3}$$

Here, $\bar\epsilon_{rr}, \bar\epsilon_{\theta\theta}, \bar\epsilon_{r\theta}$ are the strain components in the median surface, $k_{rr}, k_{\theta\theta}$ and $k_{r\theta}$ are the curvatures which can be expressed in term of the displacement components taking into consideration the first order terms and higher order terms in finite deformation according to the Sanders assumption as

$$\bar{\epsilon}_{rr} = \frac{\partial u}{\partial r} + \frac{1}{2}(\frac{\partial w}{\partial r})^2$$

$$\bar{\epsilon}_{\theta\theta} = \frac{1}{r}\frac{\partial v}{\partial \theta} + \frac{u}{r} + \frac{1}{2}(\frac{1}{r}\frac{\partial w}{\partial \theta})^2$$

$$\bar{\epsilon}_{r\theta} = \frac{1}{r}\frac{\partial u}{\partial \theta} + \frac{\partial v}{\partial r} - \frac{v}{r} + (\frac{1}{r}\frac{\partial w}{\partial r})\frac{\partial w}{\partial \theta} \qquad (2.2\text{-}4)$$

and

$$k_{rr} = -\frac{\partial^2 w}{\partial r^2}$$

$$k_{\theta\theta} = \frac{-1}{r}\frac{\partial w}{\partial r} - \frac{1}{r^2}\frac{\partial^2 w}{\partial \theta^2}$$

$$k_{r\theta} = \frac{-1}{r}(\frac{\partial^2 w}{\partial r\partial \theta}) + \frac{1}{r^2}\frac{\partial w}{\partial \theta} \qquad (2.2\text{-}5)$$

Substituting relations (2.2-4) and (2.2-5) into Eqs. (2.2-3), the following expressions for the strain components are obtained

$$\epsilon_{rr} = \frac{\partial u}{\partial r} + \frac{1}{2}(\frac{\partial w}{\partial r})^2 - z\frac{\partial^2 w}{\partial r^2}$$

$$\epsilon_{\theta\theta} = \frac{1}{r}\frac{\partial v}{\partial \theta} + \frac{u}{r} + \frac{1}{2}(\frac{1}{r}\frac{\partial w}{\partial \theta})^2 - z(\frac{1}{r}\frac{\partial w}{\partial r} + \frac{1}{r^2}\frac{\partial^2 w}{\partial \theta^2})$$

$$\epsilon_{r\theta} = \frac{1}{r}\frac{\partial u}{\partial \theta} + \frac{\partial v}{\partial r} - \frac{v}{r} + (\frac{1}{r}\frac{\partial w}{\partial r})\frac{\partial w}{\partial \theta} + 2z(\frac{-1}{r}\frac{\partial^2 w}{\partial r\partial \theta} + \frac{1}{r^2}\frac{\partial w}{\partial \theta})$$

$$(2.2\text{-}6)$$

The total potential energy function of the orthotropic circular plate is

$$V = \frac{1}{2}\int_0^r \int_0^{2\pi} \int_{-h/2}^{h/2} (\epsilon_{ij} - \alpha T\delta_{ij})\sigma_{ij} r\,dr\,d\theta\,dz \qquad (2.2\text{-}7)$$

or

$$V = \frac{1}{2}\int_0^r \int_0^{2\pi} \int_{-h/2}^{h/2} (\epsilon_{rr}\sigma_{rr} + \epsilon_{\theta\theta}\sigma_{\theta\theta} + \epsilon_{r\theta}\sigma_{r\theta} - \alpha_r T\sigma_{rr} - \alpha_\theta T\sigma_{\theta\theta}) r\,dr\,d\theta\,dz$$

$$(2.2\text{-}8)$$

Upon substitution of Eqs. (2.2-1), (2.2-2) and (2.2-6) into Eq. (2.2-8), and integration with respect to z from $-h/2$ to $h/2$, the total potential energy is obtained as

$$V = \int\int F\,dr\,d\theta \qquad (2.2\text{-}9)$$

where

$$F = \frac{1}{2}E_{rr}[rh(\frac{\partial u}{\partial r})^2 + \frac{rh}{4}(\frac{\partial w}{\partial r})^4 + \frac{rh^3}{12}(\frac{\partial^2 w}{\partial r^2})^2 + rh\frac{\partial u}{\partial r}(\frac{\partial w}{\partial r})^2] + \frac{1}{2}E_{\theta\theta}[$$

$$\frac{h}{r}(\frac{\partial v}{\partial \theta})^2\frac{u^2 h}{r} + \frac{h}{4r^3}(\frac{\partial w}{\partial \theta})^4 + \frac{h^3}{12r}(\frac{\partial w}{\partial r})^2 + \frac{h^3}{12r^3}(\frac{\partial^2 w}{\partial \theta^2})^2 + \frac{h^3}{6r^2}\frac{\partial w}{\partial r}\frac{\partial^2 w}{\partial \theta^2} +$$

$$\frac{2hu}{r}\frac{\partial v}{\partial \theta} + \frac{h}{r^2}\frac{\partial v}{\partial \theta}(\frac{\partial w}{\partial \theta})^2 + \frac{uh}{r^2}(\frac{\partial w}{\partial \theta})^2] + E_{r\theta}[h\frac{\partial v}{\partial \theta}\frac{\partial u}{\partial r} + uh\frac{\partial u}{\partial r} + \frac{h}{2r}(\frac{\partial w}{\partial \theta})^2$$

$$\frac{\partial u}{\partial r} + \frac{h}{2}\frac{\partial v}{\partial \theta}(\frac{\partial w}{\partial r})^2 + \frac{hu}{2}(\frac{\partial w}{\partial r})^2 + \frac{h}{4r}(\frac{\partial w}{\partial r})^2(\frac{\partial w}{\partial \theta})^2 + \frac{h^3}{12}\frac{\partial w}{\partial r}(\frac{\partial^2 w}{\partial r^2}) + \frac{h^3}{12}$$

$$(\frac{\partial^2 w}{\partial r^2})(\frac{\partial^2 w}{\partial \theta^2}) - [r\frac{\partial u}{\partial r} + \frac{r}{2}(\frac{\partial w}{\partial r})^2](E_{rr}\alpha_r \int T dz + E_{r\theta}\alpha_\theta \int T dz) + (E_{rr}$$

$$\alpha_r r\frac{\partial^2 w}{\partial r^2} + E_{r\theta}\alpha_\theta r\frac{\partial^2 w}{\partial r^2}) \int Tz dz - [\frac{\partial v}{\partial \theta} + u + \frac{1}{2r}(\frac{\partial w}{\partial \theta})^2](E_{\theta\theta}\alpha_\theta \int T dz$$

$$+E_{r\theta}\alpha_r \int T dz) + [(E_{\theta\theta}\alpha_\theta + E_{r\theta}\alpha_r)(\frac{\partial w}{\partial r} + \frac{1}{r}\frac{\partial^2 w}{\partial \theta^2})] \int Tz dz + (\frac{1}{2}\alpha_r^2 E_{rr}$$

$$r + \frac{1}{2}\alpha_\theta^2 E_{\theta\theta}r) \int T^2 dz + \alpha_r\alpha_\theta r \int T^2 dz + \frac{1}{2}G[\frac{h}{r}(\frac{\partial u}{\partial \theta})^2 + hr(\frac{\partial v}{\partial r})^2 + \frac{hv^2}{r}$$

$$+\frac{h}{r}(\frac{\partial w}{\partial \theta})^2(\frac{\partial w}{\partial r})^2 + \frac{rh^3}{3}[\frac{1}{r^2}(\frac{\partial^2 w}{\partial r\partial \theta})^2 + \frac{1}{r^4}(\frac{\partial w}{\partial \theta})^2 - \frac{2}{r^3}\frac{\partial w}{\partial \theta}\frac{\partial^2 w}{\partial r\partial \theta}] +$$

$$2h\frac{\partial u}{\partial \theta}\frac{\partial v}{\partial r} - \frac{2vh}{r}\frac{\partial u}{\partial \theta} + \frac{2h}{r}\frac{\partial u}{\partial \theta}\frac{\partial w}{\partial r}\frac{\partial w}{\partial \theta} - 2vh\frac{\partial v}{\partial r} + 2h\frac{\partial v}{\partial r}\frac{\partial w}{\partial r}\frac{\partial w}{\partial \theta} -$$

$$\frac{2vh}{r}\frac{\partial w}{\partial r}\frac{\partial w}{\partial \theta}$$

$$(2.2\text{-}10)$$

Extremizing the functional of potential energy leads to the Euler equations

$$\frac{\partial F}{\partial u} - \frac{\partial}{\partial r}\frac{\partial F}{\partial u_{,r}} - \frac{\partial}{\partial \theta}\frac{\partial F}{\partial u_{,\theta}} = 0$$

$$\frac{\partial F}{\partial v} - \frac{\partial}{\partial r}\frac{\partial F}{\partial v_{,r}} - \frac{\partial}{\partial \theta}\frac{\partial F}{\partial v_{,\theta}} = 0$$

$$\frac{\partial F}{\partial w} - \frac{\partial}{\partial r}\frac{\partial F}{\partial w_{,r}} - \frac{\partial}{\partial \theta}\frac{\partial F}{\partial w_{,\theta}} + \frac{\partial^2}{\partial r^2}\frac{\partial F}{\partial w_{,rr}} + \frac{\partial^2}{\partial r\partial \theta}\frac{\partial F}{\partial w_{,r\theta}} +$$

$$\frac{\partial^2}{\partial \theta^2}\frac{\partial F}{\partial w_{,\theta\theta}} = 0 \qquad\qquad (2.2\text{-}11)$$

Substitution from Eq. (2.2-10) into Eqs. (2.2-11), the equilibrium equations for the general thin orthotropic circular plate are obtained as

$$N_{r,r} + \frac{1}{r}N_{r\theta,\theta} + \frac{N_r - N_\theta}{r} = 0$$

$$N_{r\theta,r} + \frac{1}{r}N_{\theta,\theta} + \frac{2}{r}N_{r\theta} = 0$$

$$M_{r,rr} + \frac{2}{r}M_{r,r} + \frac{2}{r}M_{r\theta,r\theta} + \frac{2}{r^2}M_{r\theta,\theta} + \frac{1}{r^2}M_{\theta,\theta\theta} - \frac{1}{r}M_{\theta,r} + N_r\frac{\partial^2 w}{\partial r^2}$$

$$+N_\theta(\frac{1}{r}\frac{\partial w}{\partial r} + \frac{1}{r^2}\frac{\partial^2 w}{\partial \theta^2}) + 2N_{r\theta}\frac{\partial^2}{\partial r\partial\theta}(\frac{w}{r}) = 0 \qquad (2.2\text{-}12)$$

where N_r, N_θ, $N_{r\theta}$, M_r, M_θ, $M_{r\theta}$ are

$$N_r = E_{rr}[h\frac{\partial u}{\partial r} + \frac{h}{2}(\frac{\partial w}{\partial r})^2] + E_{r\theta}[\frac{h}{r}\frac{\partial v}{\partial \theta} + \frac{hu}{r} + \frac{h}{2r^2}(\frac{\partial w}{\partial \theta})^2]$$
$$+A_r T_0$$

$$N_\theta = E_{r\theta}[\frac{h\partial u}{\partial r} + \frac{h}{2}(\frac{\partial w}{\partial r})^2] + E_{\theta\theta}[\frac{h}{r}\frac{\partial v}{\partial \theta} + \frac{hu}{r} + \frac{h}{2r^2}(\frac{\partial w}{\partial \theta})^2]$$
$$+A_\theta T_0$$

$$N_{r\theta} = G(\frac{h}{r}\frac{\partial u}{\partial \theta} + h\frac{\partial v}{\partial r} - \frac{hv}{r} + \frac{h}{r}\frac{\partial w}{\partial r}\frac{\partial w}{\partial \theta})$$

$$M_r = \frac{-h^3}{12}E_{rr}(\frac{\partial^2 w}{\partial r^2}) - \frac{h^3}{12}E_{r\theta}(\frac{1}{r}\frac{\partial w}{\partial r} + \frac{1}{r^2}\frac{\partial^2 w}{\partial \theta^2}) + A_r T_1$$

$$M_\theta = \frac{-h^3}{12}E_{r\theta}(\frac{\partial^2 w}{\partial r^2}) - \frac{h^3}{12}E_{\theta\theta}(\frac{1}{r}\frac{\partial w}{\partial r} + \frac{1}{r^2}\frac{\partial^2 w}{\partial \theta^2}) + A_\theta T_1$$

$$M_{r\theta} = G\frac{h^3}{6}(\frac{-1}{r}\frac{\partial^2 w}{\partial r\partial\theta} + \frac{1}{r^2}\frac{\partial w}{\partial \theta})$$

$$T_0 = \int_{-h/2}^{h/2} T dz$$

$$T_1 = \int_{-h/2}^{h/2} T z dz \qquad (2.2\text{-}13)$$

Substituting relations (2.2-13) into the third of Eqs. (2.2-12), we obtain

$$N_{r,r} + \frac{1}{r}N_{r\theta,\theta} + \frac{N_r - N_\theta}{r} = 0$$

$$N_{r\theta,r} + \frac{1}{r}N_{\theta,\theta} + \frac{2}{r}N_{r\theta} = 0$$

$$\frac{h^3}{12}E_{rr}(\frac{\partial^4 w}{\partial r^4} + \frac{2}{r}\frac{\partial^3 w}{\partial r^3}) + \frac{h^3}{12}E_{\theta\theta}(\frac{1}{r^3}\frac{\partial w}{\partial r} - \frac{1}{r^2}\frac{\partial^2 w}{\partial r^2} + \frac{2}{r^4}\frac{\partial^2 w}{\partial \theta^2} + \frac{1}{r^4}\frac{\partial^4 w}{\partial \theta^4})$$

$$+\frac{h^3}{6}(E_{r\theta} + 2G)(\frac{1}{r^4}\frac{\partial^2 w}{\partial \theta^2} - \frac{1}{r^3}\frac{\partial^3 w}{\partial r\partial\theta^2} + \frac{1}{r^2}\frac{\partial^4 w}{\partial r^2\partial\theta^2}) + (A_\theta - A_r)\frac{1}{r}\frac{\partial T_1}{\partial r}$$

$$-A_r\frac{\partial^2 T_1}{\partial r^2} - \frac{1}{r^2}A_\theta\frac{\partial^2 T_1}{\partial \theta^2} = N_r\frac{\partial^2 w}{\partial r^2} + N_\theta(\frac{1}{r}\frac{\partial w}{\partial r} + \frac{1}{r^2}\frac{\partial^2 w}{\partial \theta^2}) + 2N_{r\theta}$$

$$\frac{\partial^2}{\partial r\partial\theta}(\frac{w}{r}) \qquad (2.2\text{-}14)$$

The stability equations of thin orthotropic circular plates may be derived either by the force summation method or energy method (variational method). Using the energy method, if V is the total potential energy of the plate, the expansion of V about the equilibrium state into the Taylor series yields

$$\Delta V = \delta V + \frac{1}{2}\delta^2 V + \frac{1}{6}\delta^3 V + \ldots \qquad (2.2\text{-}15)$$

The type of equilibrium depends upon how the system responds when released following a small displacement. The stability of the system in the neighborhood of equilibrium is thus determined by the sign of second variation of its potential energy as follows [16];
1) Stable equilibrium occurs when a small displacement of the system causes the system to return to its original position. In this case, the original potential energy of the system is a minimum, that is $\delta^2 V > 0$.
2) Neutral equilibrium occurs when a small displacement of the system causes the system to remain in its displaced state. In this case, the potential energy of the system remain constant, that is $\delta^2 V = 0$.
3) Unstable equilibrium occurs when a small displacement of the system causes the system to move farther away from its original position. In this case, the original potential energy of the system is a maximum, that is $\delta^2 V < 0$. The condition $\delta^2 V = 0$ is used to derive the stability equations for the buckling problem.

Let us assume that u_i^* denotes the displacement component of the equilibrium state and δu_i^* the virtual displacement corresponding to a neighboring state. Denoting by $(\bar{\delta})$ the variation with respect to u_i^*, the following rule, known as the Trefftz rule, is stated for the determination of the lowest critical load. The external load acting on the original configuration is considered to be the critical buckling load if the following variational equation is satisfied

$$\bar{\delta}(\delta^2 V) = 0 \qquad (2.2\text{-}16)$$

This rule provides the governing equations that determine the lowest critical loads. Consider the state of stable equilibrium of a general circular plate under thermal load to be designated by u_0, v_0, w_0. The displacement components of the neighboring state are

$$u = u_0 + u_1$$
$$v = v_0 + v_1$$
$$w = w_0 + w_1 \qquad (2.2\text{-}17)$$

Here, u_1, v_1, w_1 are arbitrary small increments of displacements. Substituting Eqs. (2.2-17) into Eqs. (2.2-10) and collection of second-order terms, we get the second variation of the potential energy as

$$\frac{1}{2}\delta^2 V = \int\int \left(\frac{1}{2}E_{rr}\left[rh\left(\frac{\partial u_1}{\partial r}\right)^2 + \frac{rh^3}{12}\left(\frac{\partial^2 w_1}{\partial r^2}\right)^2 + rh\frac{\partial u_0}{\partial r}\left(\frac{\partial w_1}{\partial r}\right)^2\right] + \frac{1}{2}E_{\theta\theta}\left[\frac{h}{r}\right.\right.$$

$$(\frac{\partial v_1}{\partial \theta})^2 + \frac{u_1^2 h}{r} + \frac{h^3}{12r}(\frac{\partial w_1}{\partial r})^2 + \frac{h^3}{12r^3}(\frac{\partial^2 w_1}{\partial \theta^2})^2 + \frac{h^3}{6r^2}\frac{\partial w_1}{\partial r}\frac{\partial^2 w_1}{\partial \theta^2} + \frac{2hu_1}{r}\frac{\partial v_1}{\partial \theta}$$

$$+\frac{h}{r^2}\frac{\partial v_0}{\partial \theta}(\frac{\partial w_1}{\partial \theta})^2 + \frac{u_0 h}{r^2}(\frac{\partial w_1}{\partial \theta})^2] + E_{r\theta}[h\frac{\partial v_1}{\partial \theta}\frac{\partial u_1}{\partial r} + u_1 h\frac{\partial u_1}{\partial r} + \frac{h}{2r}(\frac{\partial w_1}{\partial \theta})^2$$

$$\frac{\partial u_0}{\partial r} + \frac{h}{2}\frac{\partial v_0}{\partial \theta}(\frac{\partial w_1}{\partial r})^2 + \frac{hu_0}{2}(\frac{\partial w_1}{\partial r})^2 + \frac{h^3}{12}\frac{\partial w_1}{\partial r}(\frac{\partial^2 w_1}{\partial r^2}) + \frac{h^3}{12r}(\frac{\partial^2 w_1}{\partial r^2})(\frac{\partial^2 w_1}{\partial \theta^2})]$$

$$-\frac{r}{2}(\frac{\partial w_1}{\partial r})^2(E_{rr}\alpha_r \int T dz + E_{r\theta}\alpha_\theta \int T dz) - \frac{1}{2r}(\frac{\partial w_1}{\partial \theta})^2(E_{\theta\theta}\alpha_\theta \int T dz + E_{r\theta}$$

$$\alpha_r \int T dz) + \frac{1}{2}G[\frac{h}{r}(\frac{\partial u_1}{\partial \theta})^2 + hr(\frac{\partial v_1}{\partial r})^2 + \frac{hv_1^2}{r} + \frac{h^3}{3r}(\frac{\partial^2 w_1}{\partial r \partial \theta})^2 + \frac{h^3}{3r^3}(\frac{\partial w_1}{\partial \theta})^2$$

$$+2h\frac{\partial u_1}{\partial \theta}\frac{\partial v_1}{\partial r} - \frac{2v_1 h}{r}\frac{\partial u_1}{\partial \theta} + \frac{2h}{r}\frac{\partial u_0}{\partial \theta}\frac{\partial w_1}{\partial r}\frac{\partial w_1}{\partial \theta} - 2v_1 h\frac{\partial v_1}{\partial r} + 2h\frac{\partial v_0}{\partial r}\frac{\partial w_1}{\partial r}\frac{\partial w_1}{\partial \theta}$$

$$-\frac{2v_0 h}{r}\frac{\partial w_1}{\partial r}\frac{\partial w_1}{\partial \theta} - \frac{2h^3}{3r^2}\frac{\partial w_1}{\partial \theta}\frac{\partial^2 w_1}{\partial r \partial \theta}])dr d\theta \qquad (2.2\text{-}18)$$

Substituting Eq. (2.2-18) into Eqs. (2.2-11) and applying the Euler equations, we arrive at the stability equations as

$$N_{r_1,r} + \frac{1}{r}N_{r\theta_1,\theta} + \frac{N_{r_1} - N_{\theta_1}}{r} = 0$$

$$N_{r\theta_1,r} + \frac{1}{r}N_{\theta_1,\theta} + \frac{2}{r}N_{r\theta_1} = 0$$

$$M_{r_1,rr} + \frac{2}{r}M_{r_1,r} + \frac{2}{r}M_{r\theta_1,r\theta} + \frac{2}{r^2}M_{r\theta_1,\theta} + \frac{1}{r^2}M_{\theta_1,\theta\theta} - \frac{1}{r}M_{\theta_1,r} +$$

$$N_{r_0}\frac{\partial^2 w_1}{\partial r^2} + N_{\theta_0}(\frac{1}{r}\frac{\partial w_1}{\partial r} + \frac{1}{r^2}\frac{\partial^2 w_1}{\partial \theta^2}) + 2N_{r\theta_0}\frac{\partial^2}{\partial r\partial \theta}(\frac{w_1}{r}) = 0 \quad (2.2\text{-}19)$$

where

$$N_{r_1} = E_{rr}h\frac{\partial u_1}{\partial r} + E_{r\theta}[\frac{h}{r}\frac{\partial v_1}{\partial \theta} + \frac{hu_1}{r}]$$

$$N_{\theta_1} = E_{r\theta}\frac{h\partial u_1}{\partial r} + E_{\theta\theta}[\frac{h}{r}\frac{\partial v_1}{\partial \theta} + \frac{hu_1}{r}]$$

$$N_{r\theta_1} = G(\frac{h}{r}\frac{\partial u_1}{\partial \theta} + h\frac{\partial v_1}{\partial r} - \frac{hv_1}{r})$$

$$M_{r_1} = \frac{-h^3}{12}E_{rr}(\frac{\partial^2 w_1}{\partial r^2}) - \frac{h^3}{12}E_{r\theta}(\frac{1}{r}\frac{\partial w_1}{\partial r} + \frac{1}{r^2}\frac{\partial^2 w_1}{\partial \theta^2})$$

$$M_{\theta_1} = \frac{-h^3}{12}E_{r\theta}(\frac{\partial^2 w_1}{\partial r^2}) - \frac{h^3}{12}E_{\theta\theta}(\frac{1}{r}\frac{\partial w_1}{\partial r} + \frac{1}{r^2}\frac{\partial^2 w_1}{\partial \theta^2})$$

$$M_{r\theta_1} = G\frac{h^3}{6}(\frac{-1}{r}\frac{\partial^2 w_1}{\partial r\partial \theta} + \frac{1}{r^2}\frac{\partial w_1}{\partial \theta}) \qquad (2.2\text{-}20)$$

Substituting Eqs. (2.2-20) into the third of Eqs. (2.2-19) we obtain

$$N_{r_1,r} + \frac{1}{r}N_{r\theta_1,\theta} + \frac{N_{r_1} - N_{\theta_1}}{r} = 0$$

$$N_{r\theta_1,r} + \frac{1}{r}N_{\theta_1,\theta} + \frac{2}{r}N_{r\theta_1} = 0$$

$$\frac{h^3}{12}E_{rr}\left(\frac{\partial^4 w_1}{\partial r^4} + \frac{2}{r}\frac{\partial^3 w_1}{\partial r^3}\right) + \frac{h^3}{12}E_{\theta\theta}\left(\frac{1}{r^3}\frac{\partial w_1}{\partial r} - \frac{1}{r^2}\frac{\partial^2 w_1}{\partial r^2} + \frac{2}{r^4}\frac{\partial^2 w_1}{\partial \theta^2} + \frac{1}{r^4}\frac{\partial^4 w_1}{\partial \theta^4}\right)$$

$$+\frac{h^3}{6}(E_{r\theta} + 2G)\left(\frac{1}{r^4}\frac{\partial^2 w_1}{\partial \theta^2} - \frac{1}{r^3}\frac{\partial^3 w_1}{\partial r\partial \theta^2} + \frac{1}{r^2}\frac{\partial^4 w_1}{\partial r^2\partial \theta^2}\right) = N_{r_0}\frac{\partial^2 w_1}{\partial r^2} + N_{\theta_0}$$

$$\left(\frac{1}{r}\frac{\partial w_1}{\partial r} + \frac{1}{r^2}\frac{\partial^2 w_1}{\partial \theta^2}\right) + 2N_{r\theta_0}\frac{\partial^2}{\partial r\partial \theta}\left(\frac{w_1}{r}\right) \tag{2.2-21}$$

2.3   Thermal buckling analysis

To obtain the thermal buckling forces, the prebuckling forces must be obtained from the equilibrium equations and then substituted into the stability equations for the buckling analysis. It is customary to derive the prebuckling forces from the simplified equilibrium equations [16]. The prebuckling forces in terms of the displacement components from Eqs. (2.2-13), neglecting the prebuckling rotations, are

$$N_{r_0} = E_{rr}h\frac{\partial u_0}{\partial r} + E_{r\theta}\left[\frac{h}{r}\frac{\partial v_0}{\partial \theta} + \frac{hu_0}{r}\right] - (E_{rr}\alpha_r + E_{r\theta}\alpha_\theta)T_0$$

$$N_{\theta_0} = E_{r\theta}\frac{h\partial u_0}{\partial r} + E_{\theta\theta}\left[\frac{h}{r}\frac{\partial v_0}{\partial \theta} + \frac{hu_0}{r}\right] - (E_{r\theta}\alpha_r + E_{\theta\theta}\alpha_\theta)T_0$$

$$N_{r\theta_0} = G\left(\frac{h}{r}\frac{\partial u_0}{\partial \theta} + h\frac{\partial v_0}{\partial r} - \frac{hv_0}{r}\right) \tag{2.2-22}$$

Now, we may consider an orthotropic circular plate under axisymmetric temperature distribution. The third of the stability equations (2.2-21), when prebuckling rotations are disregarded, is

$$D_{rr}\left(\frac{d^4 w_1}{dr^4} + \frac{2}{r}\frac{d^3 w_1}{dr^3}\right) + D_{\theta\theta}\left(\frac{1}{r^3}\frac{dw_1}{dr} - \frac{1}{r^2}\frac{d^2 w_1}{dr^2}\right) = N_{r0}\frac{d^2 w_1}{dr^2} + N_{\theta 0}\frac{1}{r}\frac{dw_1}{dr} \tag{2.2-23}$$

where

$$D_{rr} = \frac{h^3}{12}E_{rr} \quad , \quad D_{\theta\theta} = \frac{h^3}{12}E_{\theta\theta} \tag{2.2-24}$$

In equations (2.2-23) N_{r0} and $N_{\theta 0}$ are the prebuckling thermal forces that must be calculated. From the first of the equilibrium equations (2.2-12), we have

$$N_{\theta 0} = rN_{r0,r} + N_{r0} \tag{2.2-25}$$

Substituting Eq. (2.2-25) into Eq. (2.2-23) and with one step integration with respect to r we get

$$r^2\beta'' + r\beta' - d^2\beta = r^2 N_{r0}\beta \tag{2.2-26}$$

where

$$\beta = \frac{dw_1}{dr} \quad , \quad d^2 = \frac{D_{\theta\theta}}{D_{rr}} \tag{2.2-27}$$

Substituting Eqs. (2.2-13) into the first of equilibrium equations (2.2-12) we get

$$r^2\frac{d^2u}{dr^2} + r\frac{du}{dr} - s^2u = kr - \frac{r^2 N_{r_0,r}^T}{A_{rr}} \qquad (2.2\text{-}28)$$

where

$$s^2 = \frac{A_{\theta\theta}}{A_{rr}} \quad , \quad k = \frac{N_{\theta_0}^T - N_{r_0}^T}{A_{rr}} \qquad (2.2\text{-}29)$$

and

$$N_{i_0}^T = \int_{-h/2}^{h/2} A_i T dz \quad , \quad A_{ii} = \int_{-h/2}^{h/2} E_{ii} dz \qquad (2.2\text{-}30)$$

It is noted that for isotropic plates $s = 1$ and $k = 0$. Generally, s may be any positive number and k may take any positive or negative values.

Equation (2.2-28) has general and particular solution. The general solution, when $T = T(z)$, is [8]

$$u_h = c_1 r^s + c_2 r^{-s} \qquad \text{for} \qquad s > 0 \qquad (2.2\text{-}31)$$

where c_1 and c_2 are the constants of integrations. The particular solution is

$$u_p = \frac{k}{1-s^2} r \qquad \text{for} \qquad s \neq 1 \qquad (2.2\text{-}32)$$

or

$$u_p = \frac{k}{2} r \ln r \qquad \text{for} \qquad s = 1 \qquad (2.2\text{-}33)$$

Thus, the complete solution for $s \neq 1$ and $s = 1$ is either of the followings

$$u = c_1 r^s + c_2 r^{-s} + \frac{k}{1-s^2} r \qquad \text{for} \qquad s \neq 1 \quad \text{and} \quad s > 0 \qquad (2.2\text{-}34)$$

and

$$u = c_1 r^s + c_2 r^{-s} + \frac{k}{2} r \ln r \qquad \text{for} \qquad s = 1 \qquad (2.2\text{-}35)$$

The constants of integration should be evaluated knowing the given boundary conditions. For a solid circular plate with restricted boundary's motion in r-direction, the boundary conditions are [8]

$$u = \text{finite} \qquad r = 0$$
$$u = 0 \qquad r = a \qquad (2.2\text{-}36)$$

Substituting the boundary conditions (2.2-36) into Eq. (2.2-34), the constants of integration are obtained and the radial displacement is

$$u = \frac{k}{1-s^2} r[1 - (\frac{r}{a})^{s-1}] \qquad \text{for} \qquad s \neq 1 \qquad (2.2\text{-}37)$$

Substituting into Eq. (2.2-22), the prebuckling radial force is

$$N_{r_0} = \frac{k}{1-s^2}[(A_{rr} + A_{r\theta}) - (sA_{rr} + A_{r\theta})(\frac{r}{a})^{s-1}] + N_{r_0}^T \qquad (2.2\text{-}38)$$

where

$$A_{rr} = E_{rr}h \quad , \quad A_{r\theta} = E_{r\theta}h \qquad (2.2\text{-}39)$$

Substituting Eq. (2.2-38) into Eq. (2.2-26) we get [8]

$$r^2\beta'' + r\beta' + \beta[-\frac{r^2}{D_{rr}}[\frac{k}{1-s^2}(A_{rr} + A_{r\theta}) + N_{r_0}^T] - d^2 +$$
$$\frac{k(sA_{rr} + A_{r\theta})}{D_{rr}(1-s^2)}\frac{r^{s+1}}{a^{s-1}}] = 0 \qquad (2.2\text{-}40)$$

Equation (2.2-40) can be transformed into a Bessel type equation only under further restrictions on the physical parameters s and k. It is noted that for isotropic plates $s = 1$ and $k = 0$, then, equation (2.2-40) transformed into a Bessel type equation in the form

$$r^2\beta'' + r\beta' - \beta(\frac{N_{r_0}^T}{D}r^2 + 1) = 0 \qquad (2.2\text{-}41)$$

which is obtained and solved by Zakerzadeh [17].

In order to obtain some insight into the buckling behavior of composite plates, various combinations of orthotropic layers must be considered. Let us assume [8]

$$s^2 = \frac{A_{\theta\theta}}{A_{rr}} = 9 \qquad (2.2\text{-}42)$$

That is

$$\frac{E_{\theta\theta}}{E_{rr}} = d^2 = 9 \qquad (2.2\text{-}43)$$

In this case, the general solution of Eq. (2.2-40) with the use of series expansion is

$$\beta = c_1 r^3[1 - \frac{1}{16}pr^2 + (\frac{1}{640}p^2 - \frac{1}{40}q)r^4 + 0(r^6)] + c_2[\frac{\ln r(0(r^6))}{r^3}$$
$$+ \frac{-86400 - 1080pr^2 + (1350p^2 - 10800q)r^4 + 0(r^6)}{r^3}] \qquad (2.2\text{-}44)$$

where

$$p = \frac{1}{D_{rr}}[\frac{k}{8}(A_{rr} + A_{r\theta}) - N_{r_0}^T]$$
$$q = \frac{-k(3A_{rr} + A_{r\theta})}{8a^2 D_{rr}} \qquad (2.2\text{-}45)$$

For a solid circular plate $\beta = 0$ at $r = 0$. Therefore $c_2 = 0$ and

$$\beta = c_1 r^3 [1 - \frac{1}{16} p r^2 + (\frac{1}{640} p^2 - \frac{1}{40} q) r^4 + 0(r^6)] \qquad (2.2\text{-}46)$$

This equation gives the critical thermal buckling loads for various boundary conditions of the plates. For the clamped edge $\beta = 0$ at $r = a$, where a is the plate radius. Then we have

$$[1 - \frac{1}{16} p a^2 + (\frac{1}{640} p^2 - \frac{1}{40} q) a^4] = 0 \qquad (2.2\text{-}47)$$

This equation is used to obtain the critical temperature for various cases of thermal loadings. For different values of $s^2 = \frac{A_{\theta\theta}}{A_{rr}}$, Eq. (2.2-40) may be solved to obtain the critical buckling temperature for different boundary conditions. The thermal buckling for different s values are given in the Results and Discussion section.

A. Critical uniform temperature rise

Consider an orthotropic circular plate of radius a and thickness h. The initial uniform temperature of the plate is assumed to be T_i. If the plate is clamped on the edge $r = a$ and the displacement in $r-$direction is prevented, the temperature may be uniformly raised to a final value T_f such that the plate buckles. To find the critical $\Delta T = T_f - T_i$ from Eq. (2.2-30) we have

$$N_{r_0}^T = A_r h \Delta T \quad , \qquad N_{\theta_0}^T = A_\theta h \Delta T \qquad (2.2\text{-}48)$$

and from Eq. $(2.2 - 29)$ and Eq. $(2.2 - 45)$ we have

$$k = \frac{(A_\theta - A_r) \Delta T}{E_{rr}}$$

$$p = [\frac{3}{2} \frac{(A_\theta - A_r)}{h^2} \frac{(E_{rr} + E_{r\theta})}{E_{rr}^2} - \frac{12}{h^2 E_{rr}} A_r] \Delta T$$

$$q = \frac{-3}{2} (A_\theta - A_r) \frac{(3E_{rr} + E_{r\theta})}{E_{rr}^2} \frac{1}{h^2 a^2} \Delta T \qquad (2.2\text{-}49)$$

Substituting these relations into Eq. (2.2-47) we get the quadratic equation, that smallest root of this equation yields the critical temperature difference. The critical temperature for this case is the smallest of the following equations

$$\Delta T_1 = \frac{320}{A_1^2 a^4} [A_4 + \frac{1}{80} (25A_3^2 + 20A_3 A_2 a^2 + 4A_2^2 a^4 - 40A_1 a^2)^{\frac{1}{2}}]$$

$$\Delta T_2 = \frac{320}{A_1^2 a^4} [A_4 - \frac{1}{80} (25A_3^2 + 20A_3 A_2 a^2 + 4A_2^2 a^4 - 40A_1 a^2)^{\frac{1}{2}}]$$

$$(2.2\text{-}50)$$

where

$$A_1 = [\frac{3}{2}(A_\theta - A_r)(E_{rr} + E_{r\theta})E_{rr}^2 - \frac{12A_r E_{rr}}{h^2}]$$

$$A_2 = [\frac{-3}{2h^2 a^2 E_{rr}^2}(A_\theta - A_r)(3E_{rr} + E_{r\theta})]$$

$$A_3 = [\frac{3}{2h^4}(A_\theta - A_r)(E_{rr} + E_{r\theta})E_{rr}^2 - \frac{12A_r E_{rr}}{h^2}]$$

$$A_4 = \frac{1}{16}A_3 a^2 + \frac{1}{40}A_2 a^4 \qquad (2.2\text{-}51)$$

B. Critical gradient through the thickness temperature

Assume a linear temperature variation across the plate thickness as

$$T(z) = \Delta T \frac{(z + \frac{h}{2})}{h} + T_b \qquad (2.2\text{-}52)$$

where z is measured from the middle plane of the plate and $\Delta T = T_t - T_b$ is the difference of the temperature between $z = h/2$ and $z = -h/2$. With a similar method, and with assumption that $T_b = 0$, the prebuckling thermal forces are

$$N_{r0}^T = A_r h \frac{\Delta T}{2} \quad , \quad N_{\theta 0}^T = A_\theta h \frac{\Delta T}{2} \qquad (2.2\text{-}53)$$

and from Eq. (2.2-29) and Eq. (2.2-45) we have

$$k = \frac{(A_\theta - A_r)\Delta T}{2E_{rr}}$$

$$p = [\frac{3}{4}\frac{(A_\theta - A_r)}{h^2}\frac{(E_{rr} + E_{r\theta})}{E_{rr}^2} - \frac{6}{h^2 E_{rr}}A_r]\Delta T$$

$$q = \frac{-3}{4}(A_\theta - A_r)\frac{(3E_{rr} + E_{r\theta})}{E_{rr}^2}\frac{1}{h^2 a^2}\Delta T \qquad (2.2\text{-}54)$$

With the same procedure, the critical temperature across the thickness is obtained by solving Eq. (2.2-40). This gives the critical buckling temperature as the smallest of the following ΔT

$$\Delta T_1 = \frac{320}{A_1^2 a^4}[A_4 + \frac{1}{80}(25A_3^2 + 20A_3 A_2 a^2 + 4A_2^2 a^4 - 40A_1 a^2)^{\frac{1}{2}}]$$

$$\Delta T_2 = \frac{320}{A_1^2 a^4}[A_4 - \frac{1}{80}(25A_3^2 + 20A_3 A_2 a^2 + 4A_2^2 a^4 - 40A_1 a^2)^{\frac{1}{2}}]$$

$$(2.2\text{-}55)$$

where

$$A_1 = [\frac{3}{4}(A_\theta - A_r)(E_{rr} + E_{r\theta})E_{rr}^2 - \frac{6A_r E_{rr}}{h^2}]$$

$$A_2 = [\frac{-3}{4h^2 a^2 E_{rr}^2}(A_\theta - A_r)(3E_{rr} + E_{r\theta})]$$

$$A_3 = [\frac{3}{4h^4}(A_\theta - A_r)(E_{rr} + E_{r\theta})E_{rr}^2 - \frac{6A_r E_{rr}}{h^2}]$$

$$A_4 = \frac{1}{16}A_3 a^2 + \frac{1}{40}A_2 a^4 \tag{2.2-56}$$

C. Critical temperature with linear variation in the $r-$direction

Consider an orthotropic circular plate of radius a and thickness h under temperature difference along the $r-$direction. Assume a linear temperature variation in the $r-$direction as

$$T_r = \Delta T \frac{r}{a} + T_0 \tag{2.2-57}$$

where $\Delta T = T_a - T_0$ is the difference between the temperatures at $r = a$ and $r = 0$. With assumption that $T_0 = 0$, the membrane solution of the equilibrium equations (2.2-12), using Eq. (2.2-28), is

$$N_{r_0}^T = A_r \frac{r}{a} h \Delta T \quad , \quad N_{\theta_0}^T = A_\theta \frac{r}{a} h \Delta T \tag{2.2-58}$$

and

$$r^2 \frac{d^2 u}{dr^2} + r \frac{du}{dr} - s^2 u = p_1 r^2 \tag{2.2-59}$$

where

$$p_1 = \frac{(A_\theta - 2A_r)h\Delta T}{aA_{rr}} \tag{2.2-60}$$

The general homogeneous solution of Eq. (2.2-58) is

$$u_h = c_1 r^s + c_2 r^{-s} \tag{2.2-61}$$

and the particular solution is

$$u_p = \frac{p_1}{4 - s^2} r^2 \tag{2.2-62}$$

Thus, the general solution for the displacement u from Eq. (2.2-59) becomes

$$u = c_1 r^s + c_2 r^{-s} + \frac{p_1}{4 - s^2} r^2 \tag{2.2-63}$$

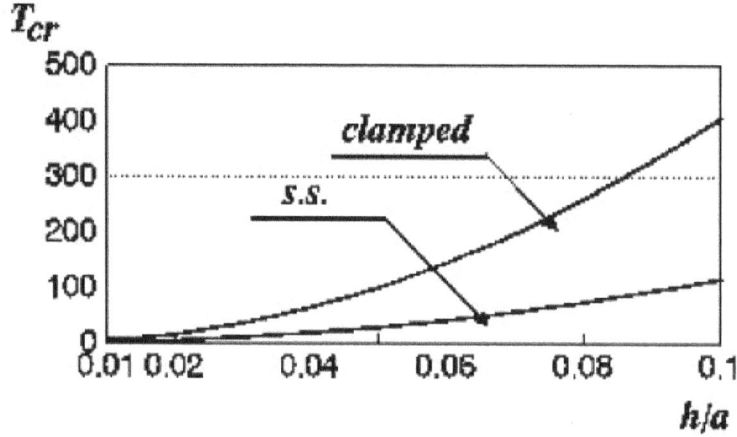

Figure 2.2-1: buckling temperature for clamped and simply supported plates vs. ratio of h/a, for the case uniform temperature rise, $(s = 1, k = 0)$.

Considering a solid circular plate with the following condition for u

$$u = \text{finite} \qquad r = 0$$
$$u = 0 \qquad r = a \qquad (2.2\text{-}64)$$

it is concluded that c_2 vanishes and u takes the form

$$u = \frac{p_1}{4 - s^2}(-a^{2-s}r^s + r^2) \qquad (2.2\text{-}65)$$

Thus, from Eq. (2.2-38) the prebuckling force is

$$N_{r0} = A_{rr}[\frac{p_1}{4 - s^2}(2r - a^{2-s}sr^{s-1})] + A_{r\theta}[\frac{p_1}{4 - s^2}(r - a^{2-s}r^{s-1})] + N_r^T \qquad (2.2\text{-}66)$$

Substituting Eq. (2.2-65) into Eq. (2.2-40) we get

$$r^2\beta'' + r\beta' - d^2\beta = \frac{r^2\beta}{D_r r}[A_{rr}[\frac{p_1}{4 - s^2}(2r - a^{2-s}sr^{s-1})] +$$

$$A_{r\theta}[\frac{p_1}{4 - s^2}(r - a^{2-s}r^{s-1})] + N_r^T] \qquad (2.2\text{-}67)$$

For case $s = 3$ [8], this equation becomes

$$r^2\beta'' + r\beta' + \beta[-d^2 + (\frac{2p_1}{5}\frac{A_{rr}}{D_{rr}} + \frac{p_1}{5}\frac{A_{r\theta}}{D_{rr}} - \frac{A_r}{D_{rr}}\frac{h}{a}\Delta T)r^3$$

$$-(\frac{3p_1}{5a}\frac{A_{rr}}{D_{rr}} + \frac{p_1}{5a}\frac{A_{r\theta}}{D_{rr}})r^4] = 0 \qquad (2.2\text{-}68)$$

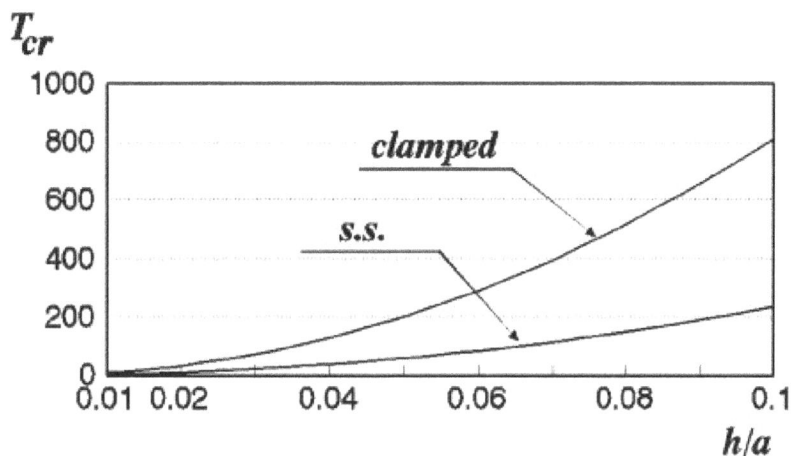

Figure 2.2-2: buckling temperature for clamped and simply supported plates vs. ratio of h/a, for the case linear gradient through the thickness, $(s = 1, k = 0)$.

With the use of series expansions we have

$$
\begin{aligned}
\beta = c_1 r^3 (1 &- \frac{1}{27} B r^3 + \frac{1}{40} A r^4 + (O) r^6) + \\
C_2 [&\frac{(\ln(r)(O(r^6))}{r^3} + \frac{-86400 - 9600 B r^3 + 1080 a r^4 + O r^6}{r^3}]
\end{aligned}
\tag{2.2-69}
$$

For a solid circular plate $\beta = 0$ at $r = 0$. Therefore $c_2 = 0$ and for clamped edge, $\beta = 0$ at $r = a$, then

$$
1 - \frac{1}{27} B a^3 + \frac{1}{40} A a^4 = 0
\tag{2.2-70}
$$

where

$$
\begin{aligned}
A &= \frac{3p_1}{5a} \frac{A_{rr}}{D_{rr}} + \frac{p_1}{5a} \frac{A_{r\theta}}{D_{rr}} \\
B &= \frac{2p_1}{5} \frac{A_{rr}}{D_{rr}} + \frac{p_1}{5} \frac{A_{r\theta}}{D_{rr}} - \frac{A_r}{D_{rr}} \frac{h}{a} \Delta T
\end{aligned}
\tag{2.2-71}
$$

The smallest root of this equation yields the critical temperature for buckling, which is

$$
\Delta T_{cr} = \frac{1}{ha^2(\frac{1}{27} B_1 - \frac{1}{40} A_1})
\tag{2.2-72}
$$

where

$$A_1 = \frac{3}{5D_{rr}}(A_\theta - 2A_r) + \frac{A_{r\theta}}{5D_{rr}A_{rr}}(A_\theta - 2A_r)$$

$$B_1 = \frac{2}{5D_{rr}}(A_\theta - 2A_r) + \frac{A_{r\theta}}{5D_{rr}Arr}(A_\theta - 2A_r) - \frac{A_r}{D_{rr}} \quad (2.2\text{-}73)$$

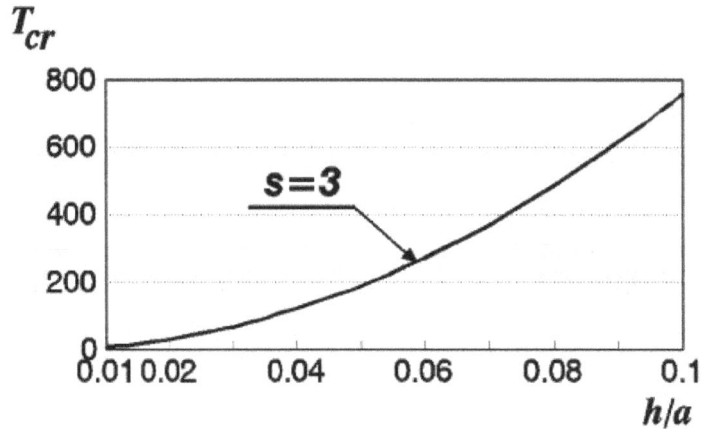

Figure 2.2-3: buckling temperature for clamped supported plate vs. ratio of h/a, for the case uniform temperature rise.

2.4   Results and discussion

Axisymmetric stability and thermal buckling equations of a circular plate subjected to uniform temperature rise, linear gradient through the thickness, and linear temperature variation along the radius are considered. These types of thermal buckling are frequently needed in the design stage of circular plates. To validate the buckling expressions obtained in this section with the known results, the buckling formula obtained for three cases of thermal buckling of composite plates are reduced to those for isotropic plates. For the case of an isotropic circular plate $s = 1$, $k = 0$, and $d = 1$. Substituting these values into Eq. (2.2-40), Eq. (2.2-41) is obtained and the critical buckling temperature reduce to the expression given in Table (2.2-1) and reported in [8, 17].

It is interesting to note that for the homogeneous isotropic circular plates, the buckling temperature is independent of Young's modulus, while for the orthotropic plates the buckling temperature is directly related to the moduli of elasticity $E_{rr}, E_{\theta\theta}$, and $E_{r\theta}$, in addition to its dependency on the coefficients of thermal expansion α_r and α_θ. Figure (2.2-1) shows the buckling temperature of an isotropic solid circular plate

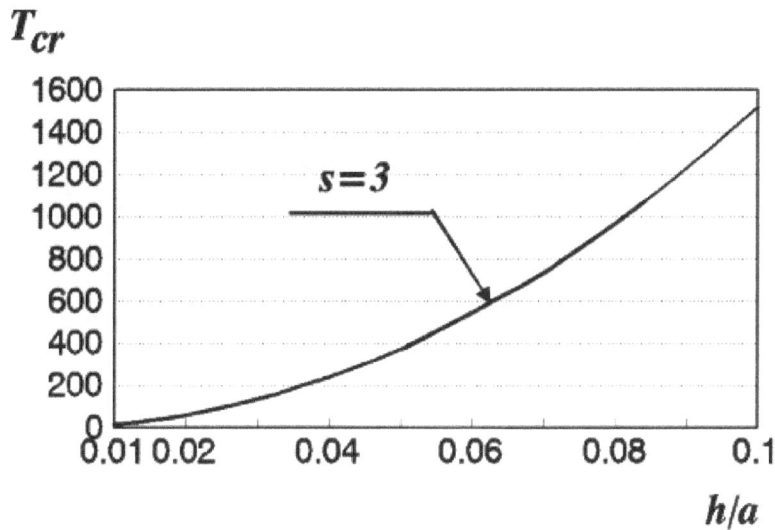

Figure 2.2-4: buckling temperature for clamped supported plate vs. ratio of h/a, for the case linear gradient through the thickness.

under uniform temperature rise versus h/a for clamped and simply supported edge conditions. Figure (2.2-2) shows the buckling temperature for the same plate under linear gradient through the thickness temperature variation. In both figures, the buckling temperature increases as h/a increase. Also, the clamped boundary condition provides more stable plate configuration.

Table (2.2-1) Thermal buckling of circular isotropic plate.

Type of load	Simply supported	Clamped Supported
uniform temperature rise	$\Delta T_{cr} = \frac{4.2}{12(1+\nu)}\left(\frac{h}{a}\right)^2\frac{1}{\alpha_T}$	$\Delta T_{cr} = \frac{14.68}{12(1+\nu)}\left(\frac{h}{a}\right)^2\frac{1}{\alpha_T}$
across thickness	$\Delta T_{cr} = \frac{8.4}{12(1+\nu)}\left(\frac{h}{a}\right)^2\frac{1}{\alpha_T}$	$\Delta T_{cr} = \frac{29.28}{12(1+\nu)}\left(\frac{h}{a}\right)^2\frac{1}{\alpha_T}$
along radius		$\Delta T_{cr} = \frac{19.87}{12(1+\nu)}\left(\frac{h}{a}\right)^2\frac{1}{\alpha_T}$

Now, consider a single layer orthotropic composite circular plate for the case $s = 3$, with the material properties; $E_{rr} = 2.1 GPa$, $E_{r\theta} = 0.7 GPa$, $E_{\theta\theta} = 19 GPa$, $\alpha_r = 7 \times 10^{-6}(1/^\circ C)$, and $\alpha_\theta = 23 \times 10^{-6}(1/^\circ C)$. The buckling temperature for three types of temperature variations are shown in Figs. (2.2-3), (2.2-4), and (2.2-5). In all cases the buckling temperature increases by the increase of h/a, as expected. The solid circular plate under radial temperature variation is more stable than the other two types of loading. Clamped boundary conditions is assumed for three types of loading shown in Figs. (2.2-3) to (2.2-5).

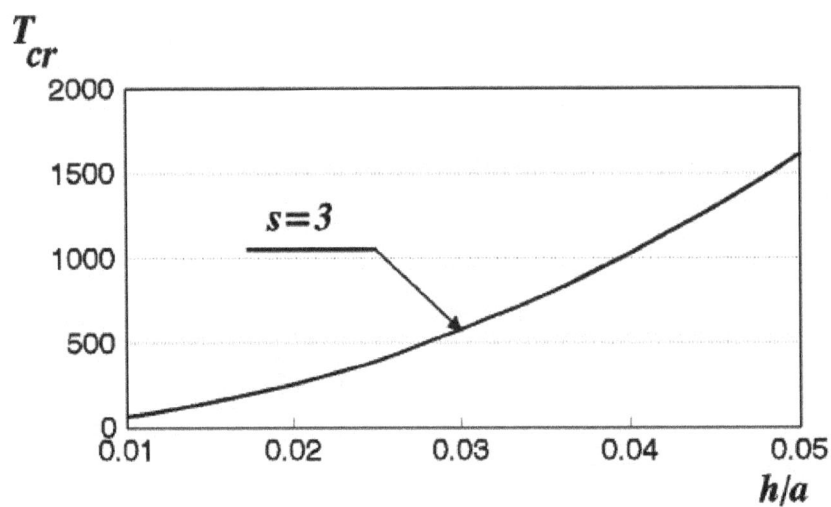

Figure 2.2-5: buckling temperature for clamped supported plate vs. ratio of h/a, for the case linear temperature variation along the radius.

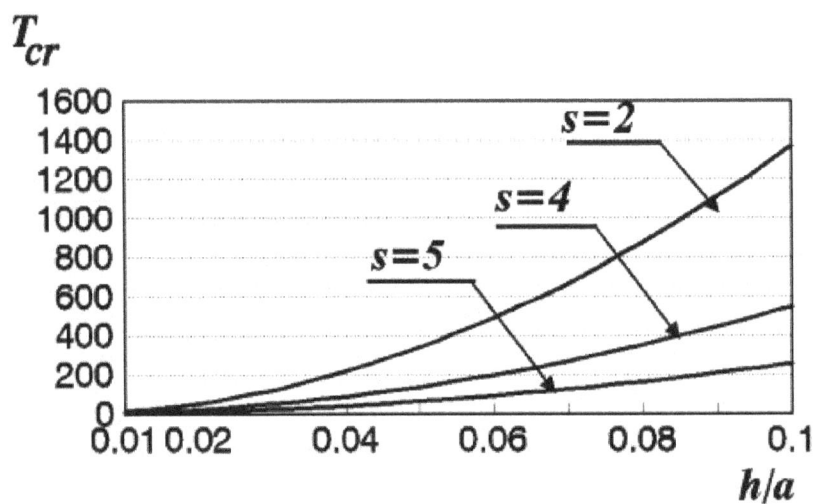

Figure 2.2-6: buckling temperature for clamped supported plate vs. ratio of h/a, for the case uniform temperature rise.

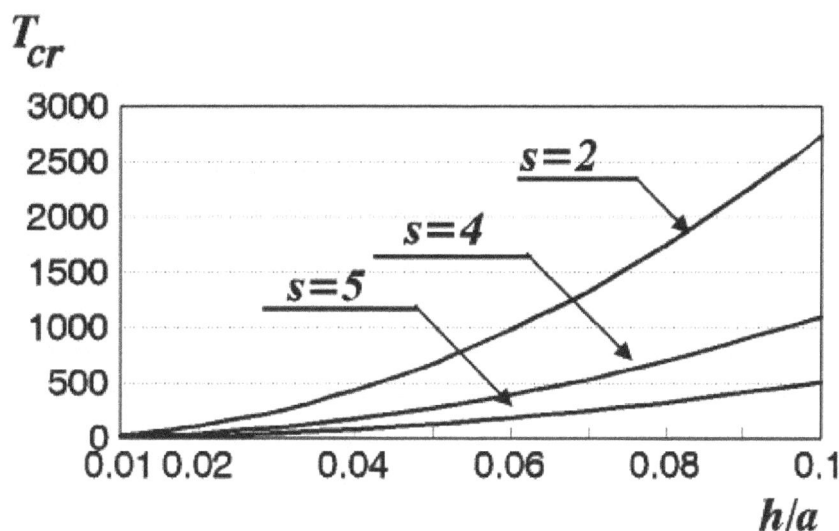

Figure 2.2-7: buckling temperature for clamped supported plate vs. ratio of h/a, for the case linear gradient through the thickness.

To study the effect of orthotropy ratio, the value of s is changed and Eq. (2.2-40) is solved for different values of s. As an example, the material constants for the case $s = 2$ are selected as $E_{rr} = 8GPa$, $E_{r\theta} = 2GPa$, $E_{\theta\theta} = 32GPa$, $\alpha_r = 6.6 \times 10^{-6}(1/^\circ C)$, $\alpha_\theta = 2.6 \times 10^{-6}(1/^\circ C)$, and for $s = 4$ are, $E_{rr} = 10GPa$, $E_{r\theta} = 3GPa$, $E_{\theta\theta} = 160GPa$, $\alpha_r = 30 \times 10^{-6}(1/^\circ C)$, $\alpha_\theta = 10^{-6}(1/^\circ C)$, and for $s = 5$ are, $s = 5$, $E_{rr} = 20GPa$, $E_{r\theta} = 77GPa$, $E_{\theta\theta} = 500GPa$, $\alpha_r = 45 \times 10^{-6}(1/^\circ C)$, $\alpha_\theta = 10^{-6}(1/^\circ C)$.

Figures (2.2-6) and (2.2-7) represent the buckling temperatures related to the uniform temperature rise and linear gradient through the thickness versus h/a for clamped boundary conditions and different values of s. It is observed that as the values of s are decreased, the buckling temperatures are increased.

Circular plate are widely used in structural design problems. When such a member is subjected to thermal environment, its thermal buckling capacity is important in the design stage. For a design engineer, the closed form solution for buckling temperature of such a member is essential as he may quickly check his design. In this section, the closed-form solutions for buckling temperature of circular plates under commonly practiced temperature distribution in design are given. More complicated temperature distribution in design may be expected in circular plates under working conditions, but the related buckling temperature may then be obtained through numerical methods. The aim of this paper has been to avoid such complicated analysis and rather provide simple closed-form solutions for commonly practiced circular plates design problems.

3   References

1 - Meissner, E., " Uber das Knicken kreisringformiger Scheiben ", Schweiz, Bauztg, Vol. 101, pp. 87-89, 1963.

2 - Yamaki, N., " Buckling of a Thin Annular Plate under Uniform Compression ", J. App. Mech. Vol. 25, pp. 267-273, 1958.

3 - Grigolyuk, E.I., " Some Problems of Stability of Circular Plates under Nonuniform Heating ", Inzh. Sb., Vol. 6, p. 73, 1950.

4 - Mansfield, E.H., " Bending, Buckling and Curling of a Heated Thin Plate ", Proc. Royal Soc. A 268, pp. 316-327, 1962.

5 - Klosner, J.M., and Forray, M.J., " Buckling of Simply Supported Plates under Arbitrary Symmetrical Temperature Distributions ", J. Aerospace. Sci. Vol. 25, pp. 181-184, 1958.

6 - Sarkar, S.K., " Buckling of a Heated Circular Plate ", Indian J. Mech. Math. Vol. 5, pp. 39-42, 1967.

7 - Das, A., " On Stresses and Displacement of Elastic Plates and Shells ", PhD Thesis, North Bengal University, Darjeeling, 1972.

8 - De, A., " Buckling of a Heated Circular Plate of Variable Thickness ", Indian J. Tech. Vol. 19, pp. 272-275, 1981.

9 - Hong, G.M., and Wang, C.M., " Elastic Buckling of Circular Plates Allowing for Prebuckling Deformation ", J. Engineering Mechanics, Vol. 119, No. 5, pp. 904-915, 1993.

10 - Stavsky, Y., " Thermoelastic Stability of Laminated Orthotropic Circular Plates ", Acta Mech. Vol. 22, pp. 31-51, 1975.

11 - Chang, J.S., and Shiao, F.J., " Thermal Buckling Analysis of Isotropic a Composite Plates With a Hole ", J. Thermal Stresses, Vol. 10, pp. 315-332, 1990.

12. Ye, J., " Axisymmetric Buckling of Homogeneous and Laminated Circular Plates ", J. Structural Engineering, Vol. 121, No. 8, pp. 1221-1224, 1995.

13 - Krizhevsky, G., and Stavsky, Y., " Refined Theory for Vibrations

and Buckling of Laminated Isotropic Annular Plates ", Int. J. Mech. Sci., Vol. 38, No. 5, pp. 539-555, 1996.

14 - Nowacki, W., *Thermoelasticity*, Pergamon, Oxford, 1962.

15 - Kovalenko, A.D., *Thermoelasticity; Basic Theory and Application*, Wolters-Noordhoff Pub., Groningen, The Netherlands, 1969.

16 - Brush, D.O., and Almorth, B.O., *Buckling of Bars, Plates and Shells*, McGraw-Hill, New York, 1975.

17 - Zakerzadeh, H., " Thermoelastic Stability of Isotropic Homogeneous Circular Plates ", MSc Thesis, Islamic Azad University, Tehran South Branch, 1999.

Chapter 3

Buckling of Cylindrical Shells

1   Isotropic Cylindrical Shells

1.1 Introduction

The theoretical developments of the elastic buckling of shells may be attributed to the works of Donnell [1-3] who formulated the critical buckling load of a short cylindrical shell, which are frequently used in industry. The prime importance of his formulation is simplicity and the closed form solutions that he obtained for the buckling loads of some loading conditions that frequently occur in practical design problems. These formulations include the effect of imperfection and present the analysis for short cylindrical shell under axial compression and external pressure on the basis that he ignores the shear force in circumferential direction and the rotations. These assumptions lead to results that are acceptable for short cylindrical shells, where the transverse shear force and rotations are small and can be ignored. The equilibrium and stability equations that he obtained was essentially based on force summation method and they have become the basis of many other developments for the shell buckling theory such as treatments presented by Donnell [4], Koiter [5,6], Flugge [7], Morley [8], and Brush and Almroth [9]. The analysis presented in these papers are restricted to the mechanical loads such as axial compression, lateral pressure, twisting moments, or combinations thereof. The method of solution is generally based on trigonometric approximation for the displacement components and substitution into the stability equations and setting the determinant of coefficients to zero. The pre-buckling stresses are the associated membrane stresses for each individual loading.

Thermal buckling of cylindrical shells based on the Donnell equations are given by Johns [10], Cheng and Cord [11], and for shell of revolution

by Bushnell [12]. Thermal buckling of short cylindrical shells fixed between two disks that are allowed to move in the direction of shell axis,
where the disks are kept at low temperatures and the shell at a high
temperature is considered in [13].

More complete treatment of shell buckling analysis include the effect
of shear force and rotations in the equilibrium and stability equation.
The inclusion of these terms in the governing equations of cylindrical
shell upgrade the accuracy but at the same time cause coupling between
the governing equations when written in terms of the displacement components, which in turn cause complications in analytical treatments. Eslami and Shariyat [14-15] used the improved equations to obtain the
elastic, plastic and creep buckling of thin cylindrical shells under differ
mechanical loading conditions. In this section the improved equilibrium
and stability equations are obtained and employed to compute the critical thermoelastic buckling loads of thin cylindrical shells under radial
thermal loading, axial temperature difference, and critical uniform final
temperature for simply supported shells. The initial imperfection analysis is included and the critical thermoelastic loads including imperfections
are included. The results are extended to both short and long circular
cylindrical shells.

1.2   Basic equations

A thin cylindrical shell of mean radius R and thickness h with length
L is considered. The normal and shear strains at a distance z from the
middle plane of the shell are [4]

$$\epsilon_x = \epsilon_{xm} + zk_x$$
$$\epsilon_\theta = \epsilon_{\theta m} + zk_\theta$$
$$\gamma_{x\theta} = \gamma_{x\theta m} + zk_{x\theta} \tag{3.1-1}$$

where the ϵ's are the normal stains, γ's are the shear strains, and k_{ij} are
the curvatures. The subscript m refers to the strain at the middle surface
of the shell. The indices x and θ refer to the axial and circumferential
directions, respectively. According to the Sanders assumption [16], the
general strain-displacement relations can be simplified to give the following term for the strains at the middle surface and the curvatures in terms
of displacement components

$$\epsilon_{xm} = u_{,x} + 0.5 w_{,x}$$
$$\epsilon_{\theta m} = (v_{,\theta} + w)/R + (v - w_{,\theta})^2 / 2R^2$$

$$\gamma_{x\theta m} = u_{,\theta}/R + v_{,x} + (-w_{,x}v + w_{,x}w_{,\theta})/R$$
$$k_x = -w_{,xx}$$
$$k_\theta = (v_{,\theta} - w_{\theta\theta})/R^2$$
$$k_{x\theta} = (v_{,x} - 2w_{,x\theta})/2R \qquad (3.1\text{-}2)$$

where u, v, and w are the displacement components and $(,)$ indicates a partial derivative. The Hooke's law in terms of forces and moments per unit length is

$$N_x = C(\epsilon_{xm} + \nu\epsilon_{\theta m}) - E\alpha T_0/(1-\nu)$$
$$N_\theta = C(\epsilon_{\theta m} + \nu\epsilon_{xm}) - E\alpha T_0/(1-\nu)$$
$$N_{x\theta} = N_{\theta x} = C(1-\nu)\gamma_{x\theta m}/2$$
$$M_x = D(k_x + \nu k_\theta) - E\alpha T_1/(1-\nu)$$
$$M_\theta = D(k_\theta + \nu k_x) - E\alpha T_1/(1-\nu)$$
$$M_{x\theta} = M_{x\theta} = D(1-\nu)k_{x\theta} \qquad (3.1\text{-}3)$$

where N_{ij} and M_{ij} are related to Q_{ij} through the shell thickness according to the first-order shell theory, E is the elastic modulus, ν is the Poisson's ratio, and α is the coefficient of thermal expansion. Also,

$$T_0 = \int_{-h/2}^{h/2} T dz \qquad\qquad T_1 = \int_{-h/2}^{h/2} Tz dz \qquad (3.1\text{-}4)$$

and $C = Eh/(1-\nu^2)$, $D = Eh^3/12(1-\nu^2)$. The total potential energy of the shell is the sum of membrane strain energy U_m, the bending strain energy U_b, and the thermal stain energy U_T, expressed as

$$U = U_m + U_b + U_T \qquad (3.1\text{-}5)$$

where

$$U_m = RC/2 \int\int [\epsilon_{xm}^2 + \epsilon_{\theta m}^2 + 2\nu\epsilon_{xm}\epsilon_{\theta m} + (1-\nu)\gamma_{x\theta m}^2/2] dx d\theta$$
$$U_b = RD/2 \int\int [k_x^\theta + k_\theta^2 + 2\nu k_x k_\theta + 2(1-\nu)k_{x\theta}^2] dx d\theta$$
$$U_T = -RE\alpha/(1-\nu) \int\int [(\epsilon_{xm} + \epsilon_{\theta m})T_0 + (k_x + k_\theta)T_1] dx d\theta$$
$$+RE\alpha^2/(1-\nu) \int\int\int T^2 dx d\theta dz \qquad (3.1\text{-}6)$$

Assuming that the cylindrical shell under thermal stress alone, the total potential energy is a function of the displacement components and their derivatives and can be written as

$$U = \int\int\int F(u, v, w, u_{,x}, u_{,\theta}, v_{,x}, v_{,\theta}, w_{,x}, w_{,\theta}, w_{,xx}, w_{,\theta\theta}, w_{,x\theta}) dx d\theta dz$$

Minimizing the functional of potential energy leads to the Euler equations

$$\frac{\partial F}{\partial u} - \frac{\partial}{\partial x}\frac{\partial F}{\partial u_{,x}} = 0$$

$$\frac{\partial F}{\partial v} - \frac{\partial}{\partial x}\frac{\partial F}{\partial v_{,x}} - \frac{\partial}{\partial \theta}\frac{\partial F}{\partial v_{,\theta}} = 0$$

$$\frac{\partial F}{\partial w} - \frac{\partial}{\partial x}\frac{\partial F}{\partial w_{,x}} - \frac{\partial}{\partial \theta}\frac{\partial F}{\partial w_{,\theta}} + \frac{\partial^2}{\partial x^2}\frac{\partial F}{\partial w_{,xx}}$$

$$+ \frac{\partial^2}{\partial x\theta}\frac{\partial F}{\partial w_{,x\theta}} + \frac{\partial^2}{\partial \theta^2}\frac{\partial F}{\partial w_{,\theta\theta}} = 0 \qquad (3.1\text{-}7)$$

Upon substitution from Eqs.(3.1-6) into (3.1-7) and using Eqs. (3.1-2) and (3.1-3), the equilibrium equations for general thin cylindrical shell are obtained as

$$RN_{x,x} + N_{x\theta,\theta} = 0$$

$$RN_{x\theta,x} + N_{\theta,\theta} + 1/RM_{\theta,\theta} + M_{x\theta,x} - (N_\theta\beta_\theta + N_{x\theta}\beta_x) = 0$$

$$RM_{x,xx} + 2RM_{x\theta,x\theta} + M_{\theta,\theta\theta} - RN_\theta \qquad (3.1\text{-}8)$$

$$-R[RN_x\beta_{x,x} + N_{x\theta}(R\beta_{\theta,x} + \beta_{x,\theta})]N_\theta\beta_{\theta,\theta} = 0$$

where

$$\beta_x = w_{,x}, \ \beta_\theta = (v - w_{,\theta})/R \qquad (3.1\text{-}9)$$

are the rotations in the x and θ-directions. It is to be noted that the first and the third equilibrium equations (3.1-8) are identical to the Donnell equations. The second equilibrium equations (3.1-8)has extra terms compared to the Donnell equations. Here we recalled that the transverse shear force in circumferential direction is

$$Q_\theta = M_{\theta,\theta}/R + M_{x\theta,x} \qquad (3.1\text{-}10)$$

Comparison shows that in Donnell equations for short cylindrical shells the shear force in circumferential direction and rotations β_x and β_θ are ignored. These approximations are justified for short cylinders, but for long cylinders these assumptions are no longer sufficient and the inclusion of these terms add to the accuracy of the results.

The stability equations of thin cylindrical shells can be derived by the variational formulations. If V is the total potential energy of the shell, the expansion of V about the equilibrium state into the Taylor series yields

$$\Delta V = \delta V + \tfrac{1}{2}\,\delta^2 V + \frac{1}{3!}\,\delta^3 V + ... \qquad (3.1\text{-}11)$$

The first variation δV is associated with the state of equilibrium. The stability of the original configuration of shell in the neighborhood of equilibrium state can be determined by the sign of the second variation $\delta^2 V$ as follows:

1 - The equilibrium is stable if $\delta^2 V > 0$ for all virtual displacements.
2 - The equilibrium is unstable if $\delta^2 V < 0$ for at least one admissible set of virtual displacements.

The condition $\delta^2 V = 0$ is used to derive the stability equations for many practical buckling problems as discussed by Langhaar [17].

Let us assume that u_i^* denote the displacement components of the equilibrium state and δu_i^* the virtual displacement corresponding to a neighboring state. Denoting by δ the variation with respect to $\hat{\delta} u_i^*$, the following rule known as the Treffetz rule, is stated for the determination of the lowest critical load.
The external load acting on the original configuration is considered to be the critical buckling load if the following variational equation is satisfied

$$\hat{\delta}(\delta^2 V) = 0 \tag{3.1-12}$$

This rule provides the governing equations that determines the lowest critical load.

Consider the state of stable equilibrium of a general cylindrical shell under thermal load that is designated by u_0, v_0 and w_0. The displacement components of the neighboring state are

$$
\begin{aligned}
u &= u_0 + u_1 \\
\nu &= \nu_0 + \nu_1 \\
w &= w_0 + w_1
\end{aligned}
\tag{3.1-13}
$$

similarly, the components of forces and moments related to the neighboring state are related to the state of equilibrium as

$$
\begin{array}{ll}
N_x = N_{x0} + \Delta N_x & M_x = M_{x0} + \Delta M_x \\
N_\theta = N_{\theta 0} + \Delta N_\theta & M_\theta = M_{\theta 0} + \Delta M_\theta \\
N_{x\theta} = N_{x\theta 0} + \Delta N_{x\theta} & M_{x\theta} = M_{x\theta 0} + \Delta M_{x\theta}
\end{array}
\tag{3.1-14}
$$

If N_{x1}, $N_{\theta 1}$, $N_{x\theta 1}$, ... express the linear portion of ΔN_x, ΔN_θ, $\Delta N_{x\theta}$, ..., in terms of u_1, v_1, and w_1 they become

$$
N_{x1} = C(\epsilon_{x1} + \nu \epsilon_{\theta 1}) \qquad\qquad M_{x1} = D(k_{x1} + \nu k_{\theta 1})
$$

$$N_{\theta 1} = C(\epsilon_{\theta 1} + \nu \epsilon_{x1}) \qquad\qquad M_{\theta 1} = D(k_{\theta 1} + \nu k_{x1})$$
$$N_{x\theta 1} = N_{\theta x1} = C(1-\nu)/2\gamma_{x\theta 1} \qquad M_{x\theta 1} = M_{\theta x1} = D(1-\nu)k_{x\theta 1}$$

$$(3.1\text{-}15)$$

and

$$N_{x0} = C(\epsilon_{x0} + \nu \epsilon_{\theta 0}) - E\alpha T_0/(1-\nu)$$
$$N_{\theta 0} = C(\epsilon_{\theta 0} + \nu \epsilon_{x0}) - E\alpha T_0/(1-\nu)$$
$$N_{x\theta 0} = C(1-\nu)/2\gamma_{x\theta 0}$$
$$M_{x0} = D(k_{x0} + \nu k_{\theta 0}) - E\alpha T_1/(1-\nu)$$
$$M_{\theta 0} = D(k_{\theta 0} + \nu k_{x0}) - E\alpha T_1/(1-\nu)$$
$$M_{x\theta 0} = D(1-\nu)k_{x\theta 0} \qquad\qquad (3.1\text{-}16)$$

Using Eqs. (3.1-13) for the displacement components of a neighboring state of stable equilibrium and employing and decomposing only the linear part of strain displacement relations, the linear strain relation for the equilibrium state (0) and the first variation (1) are obtained as

$$e_x = e_{x0} + e_{x1} \qquad\qquad \beta_x = \beta_{x0} + \beta_{x1}$$
$$e_\theta = e_{\theta 0} + e_{\theta 1} \qquad\qquad \beta_\theta = \beta_{\theta 0} + \beta_{\theta 1}$$
$$k_x = k_{x0} + k_{x1} \qquad\qquad k_\theta = k_{\theta 0} + k_{\theta 1}$$
$$e_{x\theta} = e_{x\theta 0} + e_{x\theta 1} \qquad\qquad k_{x\theta} = k_{x\theta 0} + k_{x\theta 1} \qquad (3.1\text{-}17)$$

Substituting into the functional of potential energy yields

$$V = V_0 + \Delta V_1$$
$$\frac{RC}{2} \int\int \left\{ \left[(e_{x0} + e_{x1}) + \tfrac{1}{2}(\beta_{x0} + \beta_{x1})^2 \right]^2 \right.$$
$$+ \left[(e_{\theta 0} + e_{\theta 1}) + \tfrac{1}{2}(\beta_{\theta 0} + \beta_{\theta 1}^2) \right]^2$$
$$+ 2\nu \left[(e_{x0} + e_{x1}) + \tfrac{1}{2}(\beta_{x0} + \beta_{x1})^2 \right]$$
$$\times \left[(e_{\theta 0} + e_{\theta 1}) + \tfrac{1}{2}(\beta_{\theta 0} + \beta_{\theta 1})^2 \right]$$
$$\left. + \frac{(1-\nu)}{2} \left[(e_{x\theta 0} + e_{x\theta 1}) + (\beta_{x0} + \beta_{x1})(\beta_{\theta 0} + \beta_{\theta 1}) \right]^2 \right\} dx d\theta$$
$$+ \frac{RD}{2} \int\int \left[(k_{x0} + k_{x1})^2 + (k_{\theta 0} + k_{\theta 1})^2 + 2(1-\nu)(k_{x\theta 0} + k_{x\theta 1}) \right] dx d\theta$$
$$\frac{RE\alpha}{(1-\nu)} \int\int \left\{ \left[(e_{x0} + e_{x1}) + \tfrac{1}{2}(\beta_{x0} + \beta_{x1})^2 + (e_{\theta 0} + e_{\theta 1}) \right. \right.$$

$$+ \tfrac{1}{2}\left(\beta_{\theta 0} + \beta_{\theta 1}\right)^2\Big] T_0 + (k_{x0} + k_{\theta 0} + k_{x1} + k_{\theta 1})T_1\Big\}$$

$$+ \frac{RE\alpha^2}{(1-\nu)} \int\int\int T^2 dx d\theta dz \tag{3.1-18}$$

Terms with subscript 0 are related to V_0 and those with subscript 1 are related to ΔV. Neglecting the prebuckling rotations β_x and β_θ, the second variation of the potential energy is obtained as

$$\tfrac{1}{2}\,\delta^2 V = \frac{RC}{2} \int\int [e_{x1}^2 + e_{\theta 1}^2 + 2\nu e_{x1}e_{\theta 1} + (1-\nu)/2e_{x\theta 1}^2]dx d\theta$$

$$+ \tfrac{1}{2}\,R \int\int [N_{x0}\beta_{x1}^2 + N_{\theta 0}\beta_{\theta 1}^2 + 2N_{x\theta 0}\beta_{x1}\beta_{\theta 1}]dx d\theta$$

$$+ \frac{RD}{2} \int\int [k_{x1}^2 + k_{\theta 1}^2 + 2\nu k_{x1}k_{\theta 1} + 2(1-\nu)k_{x\theta 1}^2]dx d\theta \tag{3.1-19}$$

where the prebuckling linearized forces are given in Eqs. (3.1-16). Substituting for the strains and curvatures from the linearized strain-displacement relations yield

$$\begin{array}{ll}
\epsilon_{x1} = e_{x1} = u_{1,x} & k_{x1} = -w_{1,xx} \\[4pt]
\epsilon_{\theta 1} = e_{\theta 1} = (v_{1,\theta} + w_1)/R & k_{\theta 1} = (v_{1,\theta} - w_{1,\theta\theta})/R^2 \\[4pt]
\gamma_{x\theta 1} = e_{x\theta 1} = v_{1,x} + u_{1,\theta}/R & k_{x\theta 1} = (v_{1,x} - 2w_{1,x\theta})/R \\[4pt]
\beta_{x1} = -w_{1,x} & \beta_{\theta 1} = (v_1 - w_{1,\theta})/R \tag{3.1-20}
\end{array}$$

Equations (3.1-20) are written in terms of the displacement components, which upon application of the Euler equations result into the following stability equations

$$RN_{x1,x} + N_{x\theta 1,\theta} = 0$$

$$RN_{x\theta 1,x}N_{\theta 1,\theta} + \frac{1}{R}M_{\theta 1,\theta} + M_{x\theta 1,x} - (n_{\theta 0}\beta_{\theta 1} + N_{\theta 0}\beta_{x1}) = 0$$

$$RM_{x1,xx} + 2M_{x\theta 1,x\theta} + M_{\theta 1,\theta\theta}/R - N_{\theta 1}$$

$$- [RN_{x0}\beta_{x1,x} + N_{x\theta 0}(R\beta_{\theta 1,x} + \beta_{x1,\theta}) + N_{\theta 0}\beta_{\theta 1,\theta}] = 0 \tag{3.1-21}$$

A comparison of these equations with the Donnell stability equations reveals that while the first and third equations are identical, the second equation include extra terms which are ignored in the Donnell equations as they are small for short cylinders. Inclusion of these terms remove the limitations imposed by Donnell equations in respect to the shell length. It should be noticed that the effect of these terms will improve the accuracy of the predicted design critical loads.

1.3   Thermal buckling, short cylinders

For short cylindrical shells the transverse shear force Q_θ and rotations β_x and β_θ are ignored and the equilibrium and stability equations reduce to the Donnell equations. Upon substitution for N_{ij} and M_{ij} their equivalences of strain from Eqs. (3.1-15) and finally in terms of displacements from Eqs. (3.1-20), uncoupled form of the Donnell equations are obtained as follows

$$\nabla^2 u_1 = -\frac{\nu}{R} w_{1,xxx} + \frac{1}{R^3} w_{1,x\theta\theta}$$

$$\nabla^2 v_1 = -\frac{1}{R^4} w_{1,\theta\theta\theta} - \frac{2+\nu}{R^2} w_{1,xx\theta}$$

$$D\nabla^8 w_1 + \frac{C(1-\nu^2)}{R^2} w_{1,xxxx}$$

$$-\nabla^2[N_{x0}w_{1,xx} + \frac{2}{R} N_{x\theta0}w_{1,x\theta} + \frac{1}{R} N_{\theta0}w_{1,\theta\theta}] = 0 \quad (3.1\text{-}22)$$

These equations are related to the thermal stresses through the prebuckling terms such as N_{x0} through equations (3.1-16). In the followings three types of thermal buckling loads are discussed and the buckling temperatures are calculated.

A. Critical uniform temperature rise

Consider a cylindrical shell of length L, radius R, and thickness h with both ends simply supported. The initial uniform temperature of the shell is assumed to be T_i. Under the simply supported boundary conditions, where the axial displacement is prevented, temperature can be uniformly raised to a final value T_f such that shell buckles. To find the critical $\Delta T = T_f - T_i$, the prebuckling thermal stresses are

$$N_{x0} = -\frac{E\alpha}{(1-\nu)} \int_{-h/2}^{+h/2} T dz = -\beta\Delta T \qquad (3.1\text{-}23)$$

where $\beta = E\alpha h/(1-\nu)$. For this type of loading $N_{\theta0} = N_{x\theta0} = 0$. From the edge conditions

$$w_1 = w_{1,xx} = 0 \qquad (3.1\text{-}24)$$

Assume the solution in the form

$$w_1 = C_1 \sin \bar{m}x \sin n\theta \qquad (3.1\text{-}25)$$

where $\bar{m} = m\pi R/L$ and C_1 is a constant coefficient. Substituting Eq. (3.1-23) into the third of Eqs. (3.1-22) yields

$$D\nabla^8 w_1 + \frac{C(1 - \nu^2)}{R^2} w_{1,xxxx} + \beta\Delta T\nabla^4 w_{1,xx} = 0 \qquad (3.1\text{-}26)$$

Substituting the solution w_1 from Eq. (3.1-25) yields

$$\Delta T = \frac{D(\bar{m}^2 + n^2/R^2)^2}{\bar{m}^2} + \frac{C\bar{m}^2(1 - \nu^2)}{R^2(\bar{m}^2 + n^2/R^2)^2} \qquad (3.1\text{-}27)$$

The critical temperature depends upon m and n. Denoting $\gamma = (\bar{m}^2 + n^2/R^2)^2/\bar{m}^2$, Eq. (3.1-27) reduces to

$$\beta\Delta T = D\gamma + C(1 - \nu^2)/R^2\gamma \qquad (3.1\text{-}28)$$

Minimizing ΔT with respect to γ gives

$$\gamma = [C(1 - \nu^2)/DR^2]^{1/2} \qquad (3.1\text{-}29)$$

where C and D are constants defined in Eq. (3.1-3). Substituting into Eq. (3.1-28) yields the critical temperature difference as

$$\Delta T_{crit} = \frac{Eh^2/R}{\beta[3(1 - \nu^2)]^{1/2}} = \frac{h}{R\alpha}\left[\frac{1 - \nu}{3(1 + \nu)}\right]^{1/2} \qquad (3.1\text{-}30)$$

For $\nu = 0.3$ we have

$$\Delta T_{crit} = 0.424/R\alpha \qquad (3.1\text{-}31)$$

This value for the critical temperature can be compared with the relation given by Johns [10]. Johns considered a cylindrical shell stiffened by solid rings under uniform temperature. The critical initial-final temperature that he obtained is

$$\Delta T_{crit} = Kh/R\alpha \qquad (3.1\text{-}32)$$

For simply supported cylindrical shells $K = 5.3$, as given by Johns [10] compared to 0.424 given by Eq. (3.1-31). While the same parameters are involved in both equations Eq. (3.1-32) predicts critical temperature of about 12 times higher. This factor is reasonable because of the circumferentially solid ringed constraint in Johns assumption. It is further noticed that the critical temperature for short cylindrical shells does not depend upon the shell length.

B. Critical radial temperature

Assume a linear temperature variation across the shell thickness as

$$T(z) = \Delta T(z + h/2)/h \qquad (3.1\text{-}33)$$

where z measures from the middle plane of shell. For simply supported edges the prebuckling axial force in shell is

$$N_{x0} = -\frac{E\alpha h\Delta T}{2(1-\nu)} \qquad (3.1\text{-}34)$$

With a similar method, the critical temperature difference between inside and outside surfaces is

$$\Delta T_{crit} = (T_i - T_o)_{crit} = [\frac{4(1-\nu)}{3(1+\nu)}]^{1/2}\frac{h}{R\alpha} \qquad (3.1\text{-}35)$$

For $\nu = 0.3$

$$\Delta T_{crit} = \frac{0.848h}{R\alpha} \qquad (3.1\text{-}36)$$

C. Critical axial temperature

Consider a cylindrical shell of length L under axial temperature difference and simply supported edges, where edges motion in axial direction is prevented. Assuming a linear temperature variation in axial direction x as

$$T(x) = \frac{\Delta T x}{L} \qquad (3.1\text{-}37)$$

where $\Delta T = T(L) - T(0)$. The prebuckling axial force is

$$N_{x0} = -\frac{E\alpha h}{(1-\nu)}\frac{x\Delta T}{L} = -\frac{\beta x\Delta T}{L} \qquad (3.1\text{-}38)$$

Substituting this into the third of the stability equation gives

$$D\nabla^8 w_1 + \frac{Eh}{R^2}w_{1,xxx} + \frac{\beta\Delta T x}{L}\nabla^4 w_{1,xx} = 0 \qquad (3.1\text{-}39)$$

Expanding $T(x)$ from Eq. (3.1-37) into a Fourier series, keeping only the first two terms, assuming a series solution for w_1 with two terms as

$$w_1 = \sum_{m=1}^{2} a_m \sin n\theta \sin \bar{m}x \qquad (3.1\text{-}40)$$

and substituting Eq. (3.1-40) into Eq. (3.1-39) results in two equations for a_1 and a_2 that should be solved simultaneously if Eq. (3.1-39) is to be satisfied. The determinant of the resulting two equations for a_1 and a_2 yields the following relation for the critical axial temperature difference

$$\Delta T_{crit} = \frac{\pm\sqrt{[1/4(k_{1n}p_{2n} + k_{2n}p_{1n})^2 - 0.22p_{1n}p_{2n}k_{1n}k_{2n}]}}{0.22\beta p_{1n}p_{2n}} \qquad (3.1\text{-}41)$$

where

$$k_{mn} = D(\bar{m}^2 + n^2/R^2)^4 + \frac{Eh\bar{m}^4}{R^2}$$

$$p_{mn} = (\bar{m}^2 + n^2/R^2)^2 \bar{m}^2$$

$$\bar{m} = \frac{m\pi R}{L} \qquad\qquad (3.1\text{-}42)$$

Equation (3.1-41) is used to obtain the critical buckling temperature. As an example, for a cylindrical shell with $R = 1000$ mm, $h = 5$ mm, $\alpha = 11.9 \times 10^{-6}$ $1/^\circ C$, and $\nu = 0.3$, the critical temperature for two values of L/R are given in Table (3.1-1). The value of m in Eq. (3.1-41) is either 1 or 2, and value of n, the associated circumferential waves at which the buckling appears, are also given in Table (3.1-1).

It is noticed that when L/R is increased, the number of circumferential waves n is decreased. Also, when L/R is increased, the critical axial temperature also is increased.

Table (3.1-1) Buckling temperature for axial load

L/R	n	ΔT_{crit}
0.5	11	270.48
10	3	320.79

1.4 Thermal buckling, long cylinders

For long cylindrical shells the improved equations, which include the transverse shear force Q_θ and rotations β_x and β_θ, must be used to obtain the critical load. In this case the governing equations in terms of the displacement components u_1, v_1, and w_1 are no longer separable and they have the form

$$Ru_{,xx} + \frac{(1-\nu)}{2R} u_{,\theta\theta} + \frac{(1+\nu)}{2} v_{1,x\theta} + \nu w_{1,x} = 0$$

$$\frac{(CR^2 + D)}{R^3} v_{1,\theta\theta} + \frac{(1-\nu)(CR^2 + D)}{2R} v_{1,xx}$$

$$+ \frac{C(1+\nu)}{2} u_{1,x\theta} - \frac{D}{R^3} w_{1,\theta\theta\theta} - \frac{D}{R} w_{1,xx\theta}$$

$$+ \frac{(C + N_{\theta 0})}{R} w_{1,\theta} - \frac{1}{R} v_1 N_{\theta 0} = 0$$

$$Dw_{1,xxxx} + \frac{2D}{R^2} w_{1,xx\theta\theta} + \frac{D}{R^4} w_{1,\theta\theta\theta\theta} - \frac{D}{R^2} v_{1,xx\theta}$$

$$+ \frac{1}{R^2} v_{1,\theta\theta\theta} + \frac{C}{R^2} (v_{1,\theta} + w_1 - \nu R u_{1,x})$$

$$+N_{x0}w_{1,xx} - \frac{1}{R^2}\,N_{\theta 0}w_{1,\theta\theta} + \frac{1}{R^2}\,N_{\theta 0}v_{1,\theta} = 0 \qquad (3.1\text{-}43)$$

With simply supported edges at $x = 0$ and $x = L$, the following set of approximate solutions are considered

$$u_1 = A \sin n\theta \cos \lambda x/R$$
$$v_1 = B \cos n\theta \sin \lambda x/R$$
$$w_1 = C \sin n\theta \sin \lambda x/R \qquad (3.1\text{-}44)$$

where $\lambda = m\pi R/L$. Three types of thermal loads considered.

A. Critical uniform temperature rise

In this case $N_{x\theta 0} = N_{\theta 0} = 0$ and $N_{x0} = E\alpha h\Delta T/(1 - \nu)$. Let us define

$$\Phi = N_{x0}/C \qquad (3.1\text{-}45)$$

Substituting Eqs. (3.1-44) and (3.1-45) into Eqs. (3.1-43), and setting the determinant of the coefficients equal to zero yields

$$\Phi = R/S \qquad (3.1\text{-}46)$$

where

$$R = (1 - \nu)\lambda^4 + \beta \Big[(n^2 + \lambda^2)^4 - 2\lambda^2 n^2(\lambda^2 + n^2)(2 + \nu)$$
$$+ (\lambda^2 + n^2)^2 + 2\lambda^2 n^2(3\nu + 1) - \nu^2\lambda^4 - 2n^6\Big]$$
$$S = \lambda^2 \Big[(1 + \beta)(\lambda^2 + n^2)^2 + \beta\lambda^2 n^2(1 + \nu)^2/4(1 - \nu)\Big] \quad (3.1\text{-}47)$$

For short cylinders the buckling appears in short wave lengths, and thus λ is large and β is small. When this assumption is applied to Eqs. (3.1-43) it reduces Eq. (3.1-46) to

$$\Phi = (1 - \nu^2)/x + \beta x \qquad (3.1\text{-}48)$$

where $x = (\lambda^2 + n^2)^2/\lambda^2$. The minimum of Φ with respect to x is obtained which finally results into the following expression for the critical uniform temperature rise

$$\Delta T_{crit} = \left[\frac{1 - \nu}{3(1 + \nu)}\right]^{1/2} \frac{h}{R\alpha} \qquad (3.1\text{-}49)$$

which for $\nu = 0.3$ reduces to

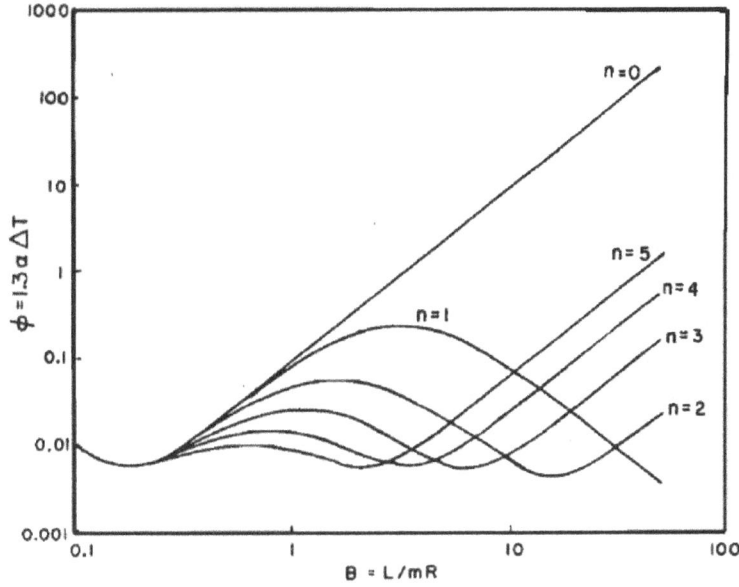

Figure 3.1-1: Thermal buckling under uniform temperature rise.

$$\Delta T_{crit} = 0.424h/R\alpha \qquad (3.1\text{-}50)$$

which is identical with Eq. (3.1-31) for short cylindrical shells. This result shows that the critical temperature for short cylinders does not depend upon the length.

For long cylinders the buckling appears with long wave lengths and thus λ^2 is small. Ignoring λ^2 and higher powers of λ in the numerator and denominator of Φ in Eqs. (3.1-46) and (3.1-47) yields

$$\Phi = (1 - \nu^2)\lambda^4/n^4 + \beta(n^2 - 1)^2/\lambda^2 \qquad (3.1\text{-}51)$$

The minimum value of Φ with respect to λ^2 is obtained for $n = 2$ and is expressed as

$$\Delta T_{crit} = \frac{3h}{4R\alpha} \sqrt{(1 - \nu)/3(1 + \nu)} \qquad (3.1\text{-}52)$$

which for $\nu = 0.3$ reduces to

$$\Delta T_{crit} = 0.318h/R\alpha \qquad (3.1\text{-}53)$$

In general, when λ is kept in the equation for Φ, by fixing the values of n the $\Phi - \lambda$ or $\Delta T_{crit} - \lambda$ relationship is solved and plotted numerically. In Fig. (3.1-1) the horizontal axis shows the value of $L/mR = \pi/\lambda$ and the vertical axis shows $\Phi = N_{x0}(1 - \nu^2)/Eh$. Since $N_{x0} = E\alpha h\Delta T/(1 - \nu)$, then $\Phi = (1 + \nu)\alpha T$; and for $\nu = 0.3$, the value of $\Phi = 1.3\alpha\Delta T$ is shown on the vertical axis of Fig. (3.1-1). The exact relationship between Φ and λ is obtained from Eqs. (3.1-46) and (3.1-47) and for $n = 0,1,2,3,4$, and 5 are plotted in Fig. (3.1-1), where each value of n is corresponding to its associated mode of buckling.

B. Critical radial temperature

The temperature distribution is assumed to be the same as Eq. (3.1-37), and the prebuckling axial force is the same as Eq. (3.1-34). A comparison of the prebuckling force given by Eq. (3.1-34) and $N_{x0} = Eh\alpha\Delta T/(1-\nu)$, obtained for the uniform temperature rise of long shells, reveals that ΔT_{crit} should be replaced by $\frac{1}{2}\Delta T_{crit}$ in Eq. (3.1-52). Thus the critical buckling temperature for this type of thermal load is

$$\Delta T_{crit} = \frac{3h}{4R\alpha} \sqrt{4(1 - \nu)/3(1 + \nu)} \tag{3.1-54}$$

For $\nu = 0.3$ Eq. (3.1-54) reduces to

$$\Delta T_{crit} = 0.6355h/R\alpha \tag{3.1-55}$$

Comparing Eq. (3.1-54) with Eq. (3.1-36) for short cylindrical shells reveals that the critical radial temperature of long cylindrical is equivalent to %75 of that for short cylindrical shells.

C. Critical axial temperature difference

The axial temperature distribution and the resulting axial thermal force are the same as Eqs. (3.1-37) and (3.1-38). However, instead of foregoing method, the Galerkin method may be employed. The displacement components for a cylindrical shell with simply supported ends are approximated by

$$u_1 = \sum_{m=1}^{\infty} A_{mn} \cos\left(m\pi x/L\right) \sin\left(n\theta\right)$$

$$v_1 = \sum_{m=1}^{\infty} B_{mn} \sin\left(m\pi x/L\right) \cos\left(n\theta\right)$$

$$w_1 = \sum_{m=1}^{\infty} C_{mn} \sin\left(m\pi x/L\right) \sin\left(n\theta\right) \tag{3.1-56}$$

Substituting these approximate solutions into the stability equations (3.1-43), for $m = 1,2$ and $n = 1,2$ the determinant of matrix of coefficients which, is 12x12, is set equal to zero. The resulting equation has four roots as

$$\frac{\Phi}{2} \left(\frac{\pi R}{L}\right)^2 = \gamma \left[\left(\frac{\pi R}{L}\right)^2 + 1\right]^2 + 1$$

$$\frac{\Phi}{2} \left(\frac{\pi R}{L}\right)^2 = \gamma \left[\left(\frac{\pi R}{L}\right)^2 + 4\right]^2 + 1$$

$$\frac{\Phi}{2} \left(\frac{\pi R}{L}\right)^2 = \gamma/4 \left[\left(\frac{2\pi R}{L}\right)^2 + 1\right]^2 + 1$$

$$\frac{\Phi}{2} \left(\frac{\pi R}{L}\right)^2 = \gamma/4 \left[\left(\frac{2\pi R}{L}\right)^2 + 4\right]^2 + \frac{1}{4} \qquad (3.1\text{-}57)$$

where $\Phi = (1 + \nu)\alpha\Delta T$ and $\gamma = h^2/12R^2$. Among these roots the state causing minimum value of Φ results into the critical temperature. For short cylindrical shells the value of $(\pi R/L)^2$ is large and causes the first of Eqs. (3.1-57) to provide the minimum Φ. Neglecting 1 compared to $(\pi R/L)^2$, we get

$$\Delta T_{crit}|_{short} = \pi^2 h^2/6(1 + \nu)\alpha L^2 \qquad (3.1\text{-}58)$$

This relation can be compared with Eq. (3.1-41) which is obtained for the same cylindrical shell but from a different approach.

For a very long cylindrical shell the third of Eqs. (3.1-57) constitute a criterion for critical temperature. It reduces to

$$\Phi = \frac{\gamma}{2} \left[\frac{2\pi R}{L} + \frac{L}{\pi R}\right]^2 + \frac{1}{4} \left[\frac{L}{\pi R}\right]^2 \qquad (3.1\text{-}59)$$

The critical temperature for $\nu = 0.3$ is then

$$\Delta T_{crit} = \frac{1}{\alpha} \left\{\frac{h^2}{31.2R^2} \left[\frac{2\pi R}{L} + \frac{L}{\pi R}\right]^2 + \frac{1}{5.2} \left(\frac{L}{\pi R}\right)^2\right\} \qquad (3.1\text{-}60)$$

For a very long cylindrical shell where $\pi R/L$ is small, the critical temperature is further simplified to the form

$$\Delta T_{crit} = (2\gamma + 1)/[5.2\alpha(\pi R/L)^2] \qquad (3.1\text{-}61)$$

Frequently, the thermal loading encountered in buckling design problems of cylindrical shells include simple temperature distributions along the shell. The radial temperature distribution between the inside and outside surfaces, uniform temperature rise, and the axial temperature distribution are commonly practiced in design problems. The simple design formula presented in this section should provide an effective means for the analysis of practical problems of thermal buckling of cylindrical shells.

2   Composite Cylindrical Shells

2.1   Introduction

The Donnell equilibrium and stability equations, which are the basis of many works in the area of shell buckling analysis, are based on the Love-Kirchhoff hypothesis and the Sanders nonlinear strain-displacement relations [9]. Mathematically, the assumptions used in the derivation of Donnell equations lead to uncoupled stability equations in terms of displacements and provide an easy means of analysis for predicting critical buckling loads of practical loading conditions. Physically, the assumptions used in this theory provide an acceptable approximation for the evaluation of buckling loads of short cylindrical shells.

A more complete treatment of shell buckling analysis include the effect of shear force and rotations in the equilibrium and stability equations. Eslami and Shariyat [14,15,18] used the improved equations to obtain the elastic, plastic, and creep buckling of thin cylindrical shells under different mechanical loading conditions. Also Eslami and et al. [19] obtained the thermoelastic buckling of thin cylindrical shells under a number of practical thermal loading based on the Donnell and improved Donnell equations.

Radhamohan and Venkataramana [20] studied the thermal buckling of composite cylindrical shells subjected to a uniform temperature rise. Herakovich and Johnson [21] studied the buckling of composite cylinders under combined compression and torsion using both theoretical and experimental results. Shaw and Simitses [22] studied the buckling of laminated cylinders in torsion. Buckling of laminated composite plates and shell panels is given by Leissa [23]. A review of various investigations on the analysis of laminated shells is given by Kapania [24]. Thangaratnam, et al. [25] analyzed the buckling of laminated composite cylindrical and conical shells under thermal load using the finite element method. Soldatos [26] studied the buckling problem of axially compressed circular and oval cylindrical shells of a regular antisymmetric cross-ply laminated arrangement based on linearized transverse shear deformable shell theory. Savoia and Reddy [27] studied the post-buckling problem of stiffened, cross-ply laminated circular cylinders under uniform axial compression. Effect of temperature on buckling and post buckling behavior of reinforced and unstiffened composite plates and cylindrical shells are considered by Birman and Bert [28].

In this section the Donnell and improved Donnell equations are employed to obtain the critical buckling loads of laminated circular cylindrical shells under mechanical and thermal loads. The shell material is considered specially orthotropic laminated composite [29].

2.2 Basic equations

A thin cylindrical shell of mean radius R and thickness h with length L is considered. The normal and shear strains at a point with distance z from the middle surface of the shell according to the Love-Kirchhoff assumptions are

$$\epsilon_x = \epsilon_{xm} + zk_x$$
$$\epsilon_\theta = \epsilon_{\theta m} + zk_\theta$$
$$\gamma_{x\theta} = \gamma_{x\theta m} + zk_{x\theta} \tag{3.2-1}$$

where subscript m refers to the strains on middle surface of the shell and k_x and k_θ are the bending strains in axial direction x, and circumferential direction θ, and $k_{x\theta}$ is the twist strain in $x\theta$ plane. According to the Sanders assumption [28], the general nonlinear strain-displacement relations can be simplified to give the following terms for the strains on the middle surface and the curvatures in terms of the displacement components u, v, and w in the axial, circumferential, and normal to the middle plane directions,

$$\epsilon_{xm} = u_{,x} + \frac{1}{2} w_{,x}^2$$
$$\epsilon_{\theta m} = \frac{1}{R} (v_{,\theta} + w) + \frac{1}{2R^2} (v - w_{,\theta})^2$$
$$\gamma_{x\theta m} = (\frac{1}{R} u_{,\theta} + v_{,x}) + \frac{1}{R} (-w_{,x}v + w_{,x}w_{,\theta})$$
$$k_x = -w_{,xx}$$
$$k_\theta = \frac{1}{R^2} (v_{,\theta} - w_{,\theta\theta})$$
$$k_{x\theta} = \frac{1}{2R} (v_{,x} - 2w_{,x\theta}) \tag{3.2-2}$$

where $()_{,i}$ $(i = x, \theta)$ indicates the partial derivative.

A symmetrical laminated cylindrical shell with all layers specially orthotropic is considered. The constitutive equations are written as follows [29]

$$\left\{ \begin{array}{c} N_x \\ N_\theta \\ N_{x\theta} \end{array} \right\} = \left[\begin{array}{ccc} A_{11} & A_{12} & 0 \\ A_{12} & A_{22} & 0 \\ 0 & 0 & A_{66} \end{array} \right] \left\{ \begin{array}{c} \epsilon_x \\ \epsilon_\theta \\ \gamma_{x\theta} \end{array} \right\} - \left[\begin{array}{ccc} T'_{11} & T'_{12} & 0 \\ T'_{12} & T'_{22} & 0 \\ 0 & 0 & T'_{66} \end{array} \right] \left\{ \begin{array}{c} \alpha_x \\ \alpha_\theta \\ \alpha_{x\theta} \end{array} \right\}$$

$$\left\{ \begin{array}{c} M_x \\ M_\theta \\ M_{x\theta} \end{array} \right\} = \left[\begin{array}{ccc} D_{11} & D_{12} & 0 \\ D_{12} & D_{22} & 0 \\ 0 & 0 & D_{66} \end{array} \right] \left\{ \begin{array}{c} k_x \\ k_\theta \\ k_{x\theta} \end{array} \right\} - \left[\begin{array}{ccc} T"_{11} & T"_{12} & 0 \\ T"_{12} & T"_{22} & 0 \\ 0 & 0 & T"_{66} \end{array} \right] \left\{ \begin{array}{c} \alpha_x \\ \alpha_\theta \\ \alpha_{x\theta} \end{array} \right\}$$

$$(3.2\text{-}3)$$

where the forces and moments per unit length of shell element in normal and shear directions, N_{ij} and M_{ij}, based on the first order shell theory, are related to the stress components σ_{ij} trough the following relations

$$N_{ij} = \int_{-h/2}^{h/2} \{\sigma_{ij}\}_k dz = \sum_{k=1}^{N} \int_{z_{k-1}}^{z_k} \{\sigma_{ij}\}_k dz$$

$$M_{ij} = \int_{-h/2}^{h/2} \{\sigma_{ij}\}_k z dz = \sum_{k=1}^{N} \int_{z_{k-1}}^{z_k} \{\sigma_{ij}\}_k z dz \qquad (3.2\text{-}4)$$

The constants A_{ij} and D_{ij} are membrane and bending stiffnesses, and T'_{ij} and $T"_{ij}$ are the thermal forces and moments, respectively, and are defined as

$$A_{ij} = \sum_{k=1}^{N} (\bar{Q}_{ij})_k (z_k - z_{k-1})$$

$$D_{ij} = \frac{1}{3} \sum_{k=1}^{N} (\bar{Q}_{ij})_k (z_k^3 - z_{k-1}^3)$$

$$T'_{ij} = \sum_{k=1}^{N} (\bar{Q}_{ij})_k \int_{z_{k-1}}^{z_k} T dz$$

$$T"_{ij} = \sum_{k=1}^{N} (\bar{Q}_{ij})_k \int_{z_{k-1}}^{z_k} T z dz \qquad (3.2\text{-}5)$$

where N is the number of layers, α_{ij} are the coefficient of thermal expansion, and T is the temperature. $(\bar{Q}_{ij})$ are the transformed reduced stiffnesses which relate stresses to strains in an orthotropic layer and are given in terms of the elastic modulus, Poisson's ratios, and the shear moduli [29].

The total potential energy function of the shell is the sum of the membrane, bending, and thermal strain energies, and the potential energy of the external forces. By substituting strains and curvatures from Eqs.

(3.2-1) and (3.2-2) into the expression of total potential energy function, with the aid of constitutive law (3.2-3), and applying the Euler equations, the general equilibrium equations are obtained as follows [9]

$$RN_{x,x} + N_{x\theta,\theta} = 0$$
$$RN_{x\theta,x} + N_{\theta,\theta} + \frac{1}{R}\,M_{\theta,\theta} + M_{x\theta,x} - (N_\theta\beta_\theta + N_{x\theta}\beta_x) = 0$$
$$R^2 M_{x,xx} + 2RM_{x\theta,x\theta} + M_{\theta,\theta\theta} - RN_\theta - R^2 N_x\beta_{x,x}$$
$$-RN_{x\theta}(R\beta_{\theta,x} + \beta_{x,\theta}) - RN_\theta\beta_{\theta,\theta} - R\beta_\theta(RN_{x\theta,x} + N_{\theta,\theta}) = F$$

$$(3.2\text{-}6)$$

where

$$\beta_x = -w_{,x} \qquad \beta_\theta = \frac{1}{R}\,(v - w_{,\theta}) \qquad\qquad (3.2\text{-}7)$$

are the rotations in x and θ directions. For axial load $F = -F_a$, for external pressure loading $F = -R^2 P_e$ and for thermal loading $F = 0$.

These equations can be compared with the Donnell equilibrium equations given as [9]

$$RN_{x,x} + N_{x\theta,\theta} = 0$$
$$RN_{x\theta,x} + N_{\theta,\theta} = 0$$
$$R^2 M_{x,xx} + 2RM_{x\theta,x\theta} + M_{\theta,\theta\theta} - RN_\theta - R^2 N_x\beta_{x,x}$$
$$-RN_{x\theta}(R\beta_{\theta,x} + \beta_{x,\theta}) - RN_\theta\beta_{\theta,\theta} = F$$

$$(3.2\text{-}8)$$

The second of Eqs. (3.2-6) includes β_x, β_θ, and $Q_\theta = 1/RM_{\theta,\theta} + M_{x\theta,x}$ in addition to the terms given in the Donnell equations.

The stability equations are derived by consideration of the second variation of potential energy function. For a thin cylindrical shell under general load for elastic and thermoelastic conditions, the stability equations are [9]

$$RN_{x1,x} + N_{x\theta1,\theta} = 0$$
$$RN_{x\theta1,x} + N_{\theta1,\theta} + \frac{1}{R}\,M_{\theta1,\theta} + M_{x\theta1,x} - (N_{\theta0}\beta_{\theta1} + N_{x\theta0}\beta_{x1}) = 0$$
$$RM_{x1,xx} + 2M_{x\theta1,x\theta} + \frac{1}{R}\,M_{\theta1,\theta\theta} - N_{\theta1} - RN_{x0}\beta_{x1,x}$$
$$-N_{x\theta0}(R\beta_{\theta1,x} + \beta_{x1,\theta}) - N_{\theta0}\beta_{\theta1,\theta} = 0$$

$$(3.2\text{-}9)$$

The subscript (1) refers to the state of stability and the subscript (0) refers to the state of equilibrium conditions.

The Donnell stability equations are

$$RN_{x1,x} + N_{x\theta1,\theta} = 0$$

$$RN_{x\theta1,x} + N_{\theta1,\theta} = 0$$

$$RM_{x1,xx} + 2M_{x\theta1,x\theta} + \frac{1}{R}\, M_{\theta1,\theta\theta} - N_{\theta1} - RN_{x0}\beta_{x1,x}$$

$$- N_{x\theta0}(R\beta_{\theta1,x} + \beta_{x1,\theta}) - N_{\theta0}\beta_{\theta1,\theta} = 0 \qquad (3.2\text{-}10)$$

Similar to the equilibrium equations, the first and third stability equations given in Eq. (3.2-9) and (3.2-10) are identical. In the second Donnell stability equation the terms β_{x1}, $\beta_{\theta1}$, and $Q_\theta = \frac{1}{R}\, M_{\theta1,\theta} + M_{x\theta1,x}$ are absent.

The Donnell equilibrium and stability equations are mathematically more convenient to deal with, as the stability equations in terms of displacement components can be decoupled. The Donnell approximation is more suitable for the short cylindrical shells. For the buckling analysis of longer shells, the effect of transverse shear force and rotations are significant and thus they can not be ignored in the stability equations. The introduction of these terms, however, provides a set of stability equations which are no longer separable for the displacement components and therefore their mathematical analysis is more complicated.

2.3   Buckling, short composite cylindrical shells

For the short cylindrical shells the transverse shear Q_θ and rotations β_x and β_θ are ignored and the equilibrium and stability equations reduce to the Donnell equations (3.2-8). The strain-displacement relations for the state of stability conditions are linearized and are [19]

$$\epsilon_{x1} = u_{1,x} \qquad \epsilon_{\theta1} = \frac{1}{R}\,(v_{1,\theta}, +w_1) \qquad \gamma_{x\theta1} = v_{1,x} + \frac{1}{R}\, u_{1,\theta}$$

$$k_{x1} = -w_{1,xx} \qquad k_{\theta1} = -\frac{1}{R^2}\, w_{1,\theta\theta} \qquad k_{x\theta1} = -\frac{1}{R}\, w_{1,x\theta}$$

$$(3.2\text{-}11)$$

The forces and moments of Eqs. (3.2-3) are divided into the forces and moments in the initial equilibrium state N_{ij0} and M_{ij0}, and forces and moments deviated from the equilibrium state associated with the stability condition as given below

$$N_{ij} = N_{ij0} + N_{ij1}$$
$$M_{ij} = M_{ij0} + M_{ij1} \qquad (3.2\text{-}12)$$

Each force or moment component is related to the state of strains and curvatures of equilibrium and stability conditions trough the following relations

$$\left\{ \begin{matrix} N_{x0} \\ N_{\theta 0} \\ N_{x\theta 0} \end{matrix} \right\} = \begin{bmatrix} A_{11} & A_{12} & 0 \\ A_{12} & A_{22} & 0 \\ 0 & 0 & A_{66} \end{bmatrix} \left\{ \begin{matrix} \epsilon_{x0} \\ \epsilon_{\theta 0} \\ \gamma_{x\theta 0} \end{matrix} \right\} - \begin{bmatrix} T'_{11} & T'_{12} & 0 \\ T'_{12} & T'_{22} & 0 \\ 0 & 0 & T'_{66} \end{bmatrix} \left\{ \begin{matrix} \alpha_x \\ \alpha_\theta \\ \alpha_{x\theta} \end{matrix} \right\}$$

$$\left\{ \begin{matrix} M_{x0} \\ M_{\theta 0} \\ M_{x\theta 0} \end{matrix} \right\} = \begin{bmatrix} D_{11} & D_{12} & 0 \\ D_{12} & D_{22} & 0 \\ 0 & 0 & D_{66} \end{bmatrix} \left\{ \begin{matrix} k_{x0} \\ k_{\theta 0} \\ k_{x\theta 0} \end{matrix} \right\} - \begin{bmatrix} T''_{11} & T''_{12} & 0 \\ T''_{12} & T''_{22} & 0 \\ 0 & 0 & T''_{66} \end{bmatrix} \left\{ \begin{matrix} \alpha_x \\ \alpha_\theta \\ \alpha_{x\theta} \end{matrix} \right\}$$

$$(3.2\text{-}13)$$

and the terms related to the stability conditions as

$$\left\{ \begin{matrix} N_{x1} \\ N_{\theta 1} \\ N_{x\theta 1} \end{matrix} \right\} = \begin{bmatrix} A_{11} & A_{12} & 0 \\ A_{12} & A_{22} & 0 \\ 0 & 0 & A_{66} \end{bmatrix} \left\{ \begin{matrix} \epsilon_{x1} \\ \epsilon_{\theta 1} \\ \gamma_{x\theta 1} \end{matrix} \right\}$$

$$\left\{ \begin{matrix} M_{x1} \\ M_{\theta 1} \\ M_{x\theta 1} \end{matrix} \right\} = \begin{bmatrix} D_{11} & D_{12} & 0 \\ D_{12} & D_{22} & 0 \\ 0 & 0 & D_{66} \end{bmatrix} \left\{ \begin{matrix} k_{x1} \\ k_{\theta 1} \\ k_{x\theta 1} \end{matrix} \right\}$$

$$(3.2\text{-}14)$$

By substitution of Eqs. (3.2-11) and (3.2-14) into the Donnell stability equations, the stability equations in terms of displacement components are obtained as follows

$$RA_{11}u_{1,xx} + A_{12}(v_{1,x\theta} + w_{1,x}) + A_{66}(\frac{1}{R}\,u_{1,\theta\theta} + v_{1,x\theta}) = 0$$

$$A_{12}u_{1,x\theta} + \frac{1}{R}\,A_{22}(v_{1,\theta\theta} + w_{1,\theta}) + RA_{66}(\frac{1}{R}\,u_{1,x\theta} + v_{1,xx}) = 0$$

$$RD_{11}w_{1,xxxx} + \frac{2}{R}\,(D_{12} + D_{66})w_{1,xx\theta\theta} + \frac{1}{R^3}\,D_{22}w_{1,\theta\theta\theta\theta} + A_{12}u_{1,x}$$

$$+\frac{1}{R}\,A_{22}(v_{1,\theta} + w_1) - RN_{x0}w_{1,xx} - \frac{1}{R}\,N_{\theta 0}w_{1,\theta\theta} - 2N_{x\theta 0}w_{1,x\theta} = 0$$

$$(3.2\text{-}15)$$

Considering simply supported edges at $x = 0$ and $x = L$, the edge conditions can be written as [9,19,20]

$$w_1 = w_{1,xx} = 0$$
$$u_{1,x} = 0$$
$$v_1 = 0 \qquad\qquad (3.2\text{-}16)$$

These boundary conditions are commonly used for simply supported cylindrical shells under buckling loads. These conditions are equivalent to the corresponding set of boundary conditions which, for classical shell theories, are classified as $S2$ simply supported edge [26].

The following set of approximate solutions are proposed

$$u_1 = A \sin n\theta \cos \frac{m\pi x}{L}$$
$$v_1 = B \cos n\theta \sin \frac{m\pi x}{L}$$
$$w_1 = C \sin n\theta \sin \frac{m\pi x}{L} \qquad\qquad (3.2\text{-}17)$$

where m and n are the number of half waves in longitudinal and circumferential directions, respectively. Also A, B and C are constant coefficients. Now, the following buckling loads are considered:

A. Composite cylindrical shells under external lateral pressure

Consider a composite cylindrical shell of length L radius R and thickness h with both ends simply supported. The cylinder has been subjected to uniform external lateral pressure P_e. The governing stability equations are Eqs. (3.2-15). For this type of loading $N_{x0} = N_{x\theta0} = 0$ and the only prebuckling force is obtained from equilibrium equations as

$$N_{\theta0} = -P_e R \qquad\qquad (3.2\text{-}18)$$

Substituting Eqs. (3.2-17) and (3.2-18) into Eqs. (3.2-15) and setting the determinant of the coefficients equal to zero yields

$$
\begin{aligned}
P_e = &\{(A_{11}\bar{m}^2 + A_{66}n^2)(A_{66}\bar{m}^2 + A_{22}n^2)[D_{11}\bar{m}^4 + 2(D_{12} + D_{66})\bar{m}^2 n^2 \\
&+ D_{22}n^4 + A_{22}R^2] + 2A_{12}A_{22}(A_{12} + A_{66})R^2\bar{m}^2 n^2 - A_{12}^2 R^2\bar{m}^2(A_{66}\bar{m}^2 \\
&+ A_{22}n^2) - A_{22}^2 R^2 n^2(A_{11}\bar{m}^2 + A_{66}n^2) - (A_{12} + A_{66})^2\bar{m}^2 n^2[D_{11}\bar{m}^4 \\
&+ 2(D_{12} + D_{66})\bar{m}^2 n^2 + D_{22}n^4 + A_{22}R^2]\}/\{[(A_{11}\bar{m}^2 \\
&+ A_{66}n^2)(A_{66}\bar{m}^2 + A_{22}n^2) - (A_{12} + A_{66})^2\bar{m}^2 n^2]R^3 n^2\} \qquad (3.2\text{-}19)
\end{aligned}
$$

where $\bar{m} = \frac{m\pi R}{L}$. The critical external pressure is obtained for the values of m and n which make the above expression a minimum. Since material constants of composites, such as elastic and shear moduli, are considerably lower compared to the metallic materials, the buckling waves appear at higher modes. Thus, m and n assume higher values to make the expression (3.2-19) a minimum.

B. Composite cylindrical shells under axial compression

The cylinder is subjected to axial compression F_a. For this type of loading $N_{\theta 0} = N_{x\theta 0} = 0$ and the prebuckling force is obtain from equilibrium equations as

$$N_{x0} = -\frac{F_a}{2\pi R} \tag{3.2-20}$$

Substituting Eqs. (3.2-17) and (3.2-20) in to Eqs. (3.2-15) and setting the determinant of the coefficients equal to zero yields

$$
\begin{aligned}
F_a = &\{(A_{11}\bar{m}^2 + A_{66}n^2)(A_{66}\bar{m}^2 + A_{22}n^2)[D_{11}\bar{m}^4 + 2(D_{12} + D_{66})\bar{m}^2n^2 \\
&+ D_{22}n^4 + A_{22}R^2] + 2A_{12}A_{22}(A_{12} + A_{66})R^2\bar{m}^2n^2 - A_{12}^2R^2\bar{m}^2(A_{66}\bar{m}^2 \\
&+ A_{22}n^2) - A_{22}^2R^2n^2(A_{11}\bar{m}^2 + A_{66}n^2) - (A_{12} + A_{66})^2\bar{m}^2n^2[D_{11}\bar{m}^4 \\
&+ 2(D_{12} + D_{66})\bar{m}^2n^2 + D_{22}n^4 + A_{22}R^2]\}/\{[(A_{11}\bar{m}^2 \\
&+ A_{66}n^2)(A_{66}\bar{m}^2 + A_{22}n^2) - (A_{12} + A_{66})^2\bar{m}^2n^2]\frac{R\bar{m}^2}{2\pi}\}
\end{aligned}
\tag{3.2-21}
$$

C. Composite cylindrical shells under uniform temperature rise

The initial uniform temperature of the shell is assumed to be T_i. Under simply supported conditions, temperature can be uniformly raised to a final value T_f such that shell buckles [19]. To find the critical $\Delta T = T_f - T_i$, the prebuckling thermal stresses are

$$
\begin{aligned}
N_{x0} &= -(A_{11}\alpha_x + A_{12}\alpha_\theta)\Delta T \\
N_{\theta 0} &= N_{x\theta 0} = 0
\end{aligned}
\tag{3.2-22}
$$

substituting Eqs. (3.2-17) and (3.2-22) into Eqs. (3.2-15) and setting the determinant of the coefficients equal to zero yields

$$
\begin{aligned}
\Delta T = &\{(A_{11}\bar{m}^2 + A_{66}n^2)(A_{66}\bar{m}^2 + A_{22}n^2)[D_{11}\bar{m}^4 + 2(D_{12} + D_{66})\bar{m}^2n^2 \\
&+ D_{22}n^4 + A_{22}R^2] + 2A_{12}A_{22}(A_{12} + A_{66})R^2\bar{m}^2n^2 - A_{12}^2R^2\bar{m}^2(A_{66}\bar{m}^2 \\
&+ A_{22}n^2) - A_{22}^2R^2n^2(A_{11}\bar{m}^2 + A_{66}n^2) - (A_{12} + A_{66})^2\bar{m}^2n^2[D_{11}\bar{m}^4
\end{aligned}
$$

$$+2(D_{12} + D_{66})\bar{m}^2 n^2 + D_{22}n^4 + A_{22}R^2]\}/\{[(A_{11}\bar{m}^2 + A_{66}n^2)(A_{66}\bar{m}^2$$
$$+A_{22}n^2) - (A_{12} + A_{66})^2\bar{m}^2 n^2]\,R^2(A_{11}\alpha_x + A_{12}\alpha_\theta)\bar{m}^2\}$$

$$(3.2\text{-}23)$$

D. Composite cylindrical shells under radial temperature difference

Assume a linear temperature variation across the shell thickness as [19]

$$T(z) = \frac{\Delta T}{h}\left(z + \frac{h}{2}\right) \qquad (3.2\text{-}24)$$

where z is measured from the middle surface of the shell. For simply supported edges the prebuckling forces are

$$N_{x0} = -(A_{11}\alpha_x + A_{12}\alpha_\theta)\frac{\Delta T}{2}$$
$$N_{\theta 0} = N_{x\theta 0} = 0 \qquad (3.2\text{-}25)$$

By substituting Eqs. (3.2-17) and (3.2-25) into Eqs. (3.2-15), ΔT for buckling of composite cylindrical shells under radial temperature difference is obtained. Calculated ΔT is twice of ΔT for buckling of cylinder under initial-final temperature which was found in Eq. (3.2-23).

$$\Delta T(\text{radial temperature}) = 2\Delta T(\text{initial-final temperature}) \qquad (3.2\text{-}26)$$

E. Composite cylindrical shells under axial temperature difference

Consider a cylindrical shell of length L under axial temperature difference and simply supported edges. Assume a linear temperature variation along the axial x direction [19]

$$T(x) = \Delta T\,\frac{x}{L} \qquad (3.2\text{-}27)$$

where $\Delta T = T(L) - T(0)$. The prebuckling forces are

$$N_{x0} = -(A_{11}\alpha_x + A_{12}\alpha_\theta)T(x)$$
$$N_{\theta 0} = N_{x\theta 0} = 0 \qquad (3.2\text{-}28)$$

The function $T(x)$ of Eq. (3.2-27) is expanded by Fourier series and only the first two terms are kept. Replacing $T(x)$ in Eqs. (3.2-28) by the expanded form, and substituting Eqs. (3.2-17) and (3.2-28) into Eqs. (3.2-15) and employing the Galerkin's method, results in three equations. Setting the determinant of the coefficients of the three resulting equations to zero yields the same equation obtained for cylinder under radial temperature difference, thus

$$\Delta T(\text{axial temperature}) = \Delta T(\text{radial temperature}) \qquad (3.2\text{-}29)$$

2.4 Buckling, long composite cylindrical shells

For the long cylindrical shell, the improved equations, i.e., Eqs. (3.2-9), which include the transverse shear Q_θ and rotations β_x and β_θ, will be used to obtain the critical load. The linear strain-displacement relations for this case reduce to

$$\epsilon_{x1} = u_{1,x} \qquad \epsilon_{\theta 1} = \frac{1}{R}\left(v_{1,\theta} + w_1\right) \qquad \gamma_{x\theta 1} = v_{1,x} + \frac{1}{R}u_{1,\theta}$$

$$k_{x1} = -w_{1,xx} \qquad k_{\theta 1} = \frac{1}{R^2}\left(v_{1,\theta} - w_{1,\theta\theta}\right) \qquad k_{x\theta 1} = \frac{1}{2R}\left(v_{1,x} - 2w_{1,x\theta}\right)$$

$$(3.2\text{-}30)$$

By substitution of Eqs. (3.2-30) and (3.2-14) into the improved stability equations, stability equations in terms of displacement components are obtained as follows:

$$RA_{11}u_{1,xx} + A_{12}(v_{1,x\theta} + w_{1,x}) + A_{66}(\frac{1}{R}u_{1,\theta\theta} + v_{1,x\theta}) = 0$$

$$A_{12}u_{1,x\theta} + \frac{1}{R}A_{22}(v_{1,\theta\theta} + w_{1,\theta}) + RA_{66}(\frac{1}{R}u_{1,x\theta} + v_{1,xx})$$

$$-\frac{1}{R}D_{12}w_{1,xx\theta} + \frac{1}{R^3}D_{22}(v_{1,\theta\theta} - w_{1,\theta\theta\theta}) + \frac{1}{2R}D_{66}(v_{1,xx}$$

$$-2w_{1,xx\theta} - \frac{1}{R}N_{\theta 0}(v_1 - w_{1,\theta}) + N_{x\theta 0}w_{1,x} = 0$$

$$RD_{11}w_{1,xxxx} - \frac{1}{R}D_{12}(v_{1,xx\theta} - 2w_{1,xx\theta\theta}) - \frac{1}{R^3}D_{22}(v_{1,\theta\theta\theta} - w_{1,\theta\theta\theta\theta})$$

$$-\frac{1}{R}D_{66}(v_{1,xx\theta} - 2w_{1,xx\theta\theta}) + A_{12}u_{1,x} + \frac{1}{R}A_{22}(v_{1,\theta} + w_1) - RN_{x0}w_{1,xx}$$

$$+\frac{1}{R}\,N_{\theta 0}(v_{1,\theta}-w_{1,\theta\theta})+N_{x\theta 0}(v_{1,\theta}-2w_{1,x\theta})=0$$

(3.2-31)

Considering simply supported edges at $x = 0$ and $x = L$, the edge conditions are as in Eq. (3.2-16), and a set of approximate solutions as in Eqs. (3.2-17) are considered. The following buckling loads are considered:

A. Composite cylindrical shells under external lateral pressure

For this type of loading $N_{x0} = N_{x\theta 0} = 0$ and the prebuckling force is obtain from Eq. (3.2-18). Substituting Eqs. (3.2-17) and (3.2-18) in to the Eqs. (3.2-31) and setting the determinant of the coefficients equal to zero yields

$$
\begin{aligned}
P_e = &\{(A_{11}\bar{m}^2 + A_{66}n^2)[(A_{66}+\frac{1}{2R^2}\,D_{66})\bar{m}^2+(A_{22}+\frac{1}{R^2}\,D_{22})n^2] \\
&\times[D_{11}\bar{m}^4+2(D_{12}+D_{66})\bar{m}^2n^2+D_{22}n^4+A_{22}R^2]+2A_{12}(A_{12}\\
&+A_{66})n\bar{m}^2[(D_{12}+D_{66})n\bar{m}^2+D_{22}n^3+A_{22}R^2n]-A_{12}^2\bar{m}^2[(A_{66}R^2\\
&+\tfrac{1}{2}\,D_{66})\bar{m}^2+(A_{22}R^2+D_{22})n^2]-(A_{11}\bar{m}^2+A_{66}n^2)[(D_{12}+D_{66})\bar{m}^2n\\
&+D_{22}n^3+A_{22}R^2n][\frac{1}{R^2}\,(D_{12}+D_{66})\bar{m}^2n+\frac{1}{R^2}\,D_{22}n^3\\
&+A_{22}n]-(A_{12}+A_{66})^2\bar{m}^2n^2[D_{11}\bar{m}^4+2(D_{12}+D_{66})\bar{m}^2n^2+D_{22}n^4\\
&+A_{22}R^2]\}/\{R^3n^2(A_{11}\bar{m}^2+A_{66}n^2)[(A_{66}-\frac{1}{R^2}\,D_{12}-\frac{1}{2R^2}\,D_{66})\bar{m}^2\\
&+A_{22}(n^2-1)]-R^3\bar{m}^2n^2(A_{12}+A_{66})[(n^2-1)A_{12}+n^2A_{66}]\}
\end{aligned}
$$

(3.2-32)

B. Composite cylindrical shells under axial compression

For this type of loading $N_{\theta 0} = N_{x\theta 0} = 0$ and the prebuckling force is obtained from Eq. (3.2-20). Substituting Eqs. (3.2-20) and (3.2-17) into Eqs. (3.2-31) and setting determinant of the coefficients equal to zero yields

$$
\begin{aligned}
F_a = &\{(A_{11}\bar{m}^2 + A_{66}n^2)[(A_{66}+\frac{1}{2R^2}\,D_{66})\bar{m}^2+(A_{22}+\frac{1}{R^2}\,D_{22})n^2] \\
&\times[D_{11}\bar{m}^4+2(D_{12}+D_{66})\bar{m}^2n^2+D_{22}n^4+A_{22}R^2]+2A_{12}(A_{12}\\
&+A_{66})n\bar{m}^2[(D_{12}+D_{66})n\bar{m}^2+D_{22}n^3+A_{22}R^2n]-A_{12}^2\bar{m}^2[(A_{66}R^2\\
&+\tfrac{1}{2}\,D_{66})\bar{m}^2+(A_{22}R^2+D_{22})n^2]-(A_{11}\bar{m}^2+A_{66}n^2)[(D_{12}+D_{66})\bar{m}^2n
\end{aligned}
$$

$$+D_{22}n^3 + A_{22}R^2 n][\frac{1}{R^2}(D_{12}+D_{66})\bar{m}^2 n + \frac{1}{R^2}D_{22}n^3$$

$$+A_{22}n] - (A_{12}+A_{66})^2 \bar{m}^2 n^2 [D_{11}\bar{m}^4 + 2(D_{12}+D_{66})\bar{m}^2 n^2 + D_{22}n^4$$

$$+A_{22}R^2]\}/\{\frac{R\bar{m}^2}{2\pi}\{(A_{11}\bar{m}^2+A_{66}n^2)[(A_{66}+\frac{1}{2R^2}D_{66})\bar{m}^2$$

$$+(A_{22}+\frac{1}{R^2}D_{22})n^2] - \bar{m}^2 n^2 (A_{12}+A_{66})^2]\}\}$$

$$(3.2\text{-}33)$$

C. Composite cylindrical shells under uniform temperature rise

For this type of loading, the prebuckling thermal stresses are as in Eqs. (3.2-22). Substituting Eqs. (3.2-22) and (3.2-17) into Eqs. (3.2-31) and setting determinant of the coefficient equal to zero yields

$$\Delta T = \{(A_{11}\bar{m}^2 + A_{66}n^2)[(A_{66}+\frac{1}{2R^2}D_{66})\bar{m}^2 + (A_{22}+\frac{1}{R^2}D_{22})n^2]$$

$$\times [D_{11}\bar{m}^4 + 2(D_{12}+D_{66})\bar{m}^2 n^2 + D_{22}n^4 + A_{22}R^2] + 2A_{12}(A_{12}$$

$$+A_{66})n\bar{m}^2[(D_{12}+D_{66})n\bar{m}^2 + D_{22}n^3 + A_{22}R^2 n] - A_{12}^2 \bar{m}^2[(A_{66}R^2$$

$$+\frac{1}{2}D_{66})\bar{m}^2 + (A_{22}R^2+D_{22})n^2] - (A_{11}\bar{m}^2+A_{66}n^2)[(D_{12}+D_{66})\bar{m}^2 n$$

$$+D_{22}n^3 + A_{22}R^2 n][\frac{1}{R^2}(D_{12}+D_{66})\bar{m}^2 n + \frac{1}{R^2}D_{22}n^3$$

$$+A_{22}n] - (A_{12}+A_{66})^2 \bar{m}^2 n^2 [D_{11}\bar{m}^4 + 2(D_{12}+D_{66})\bar{m}^2 n^2 + D_{22}n^4$$

$$+A_{22}R^2]\}/\{R^2\bar{m}^2(A_{11}\alpha_x + A_{12}\alpha_\theta)\{(A_{11}\bar{m}^2+A_{66}n^2)[(A_{66}$$

$$+\frac{1}{2R^2}D_{66})\bar{m}^2 + (A_{22}+\frac{1}{R^2}D_{22})n^2] - \bar{m}^2 n^2 (A_{12}+A_{66})^2]\}\}$$

$$(3.2\text{-}34)$$

D. Composite cylindrical shells under radial temperature difference

For this type of loading, the prebuckling forces are as in Eqs. (3.2-25). Comparing the prebuckling forces with the prebuckling thermal stresses given by Eqs. (3.2-22) for initial-final temperature, it is easily verified that ΔT in stability equations can be replaced by $\Delta T/2$, and conclude that

$$\Delta T(\text{radial temperature}) = 2\Delta T(\text{initial - final temperature}) \quad (3.2\text{-}35)$$

E. Composite cylindrical shells under axial temperature differ-

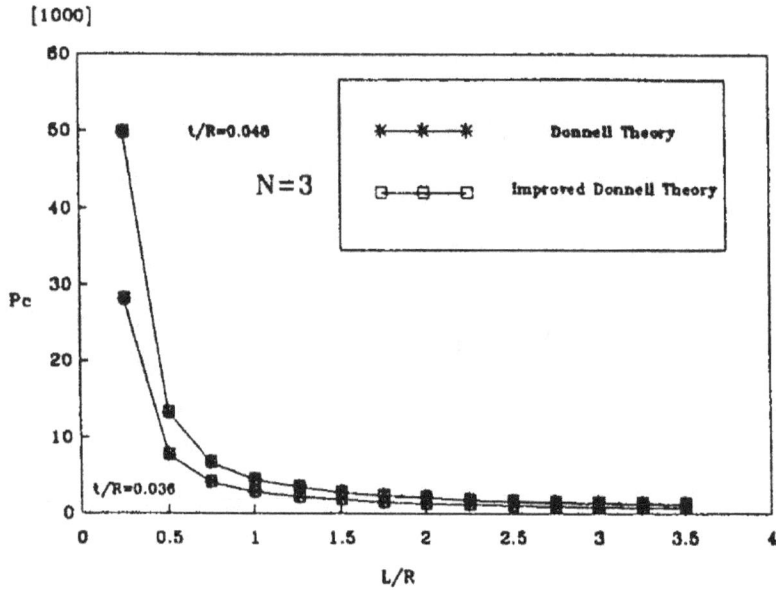

Figure 3.2-1: variation of the dimensionless critical load parameters $P_c = P/A_{11}$ versus the variation of the dimensionless length parameter L/R.

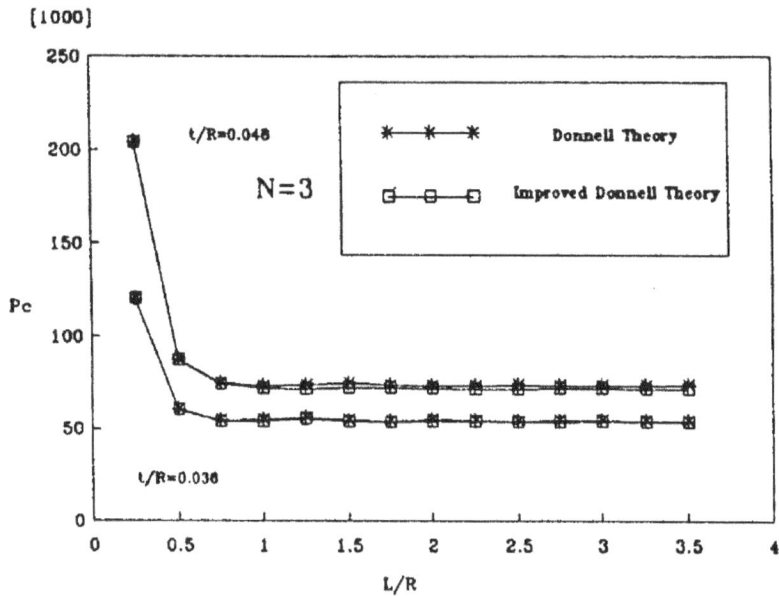

Figure 3.2-2: variation of the dimensionless critical load parameters $P_c = P/A_{11}$ versus the variation of the dimensionless length parameter L/R.

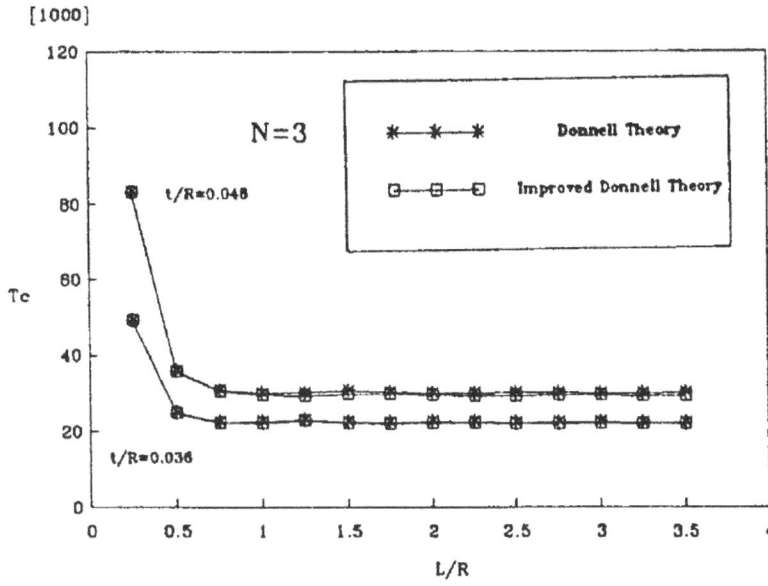

Figure 3.2-3: variation of the dimensionless critical temperature $T_c = T\alpha_\theta$ versus the variation of the dimensionless length parameter L/R.

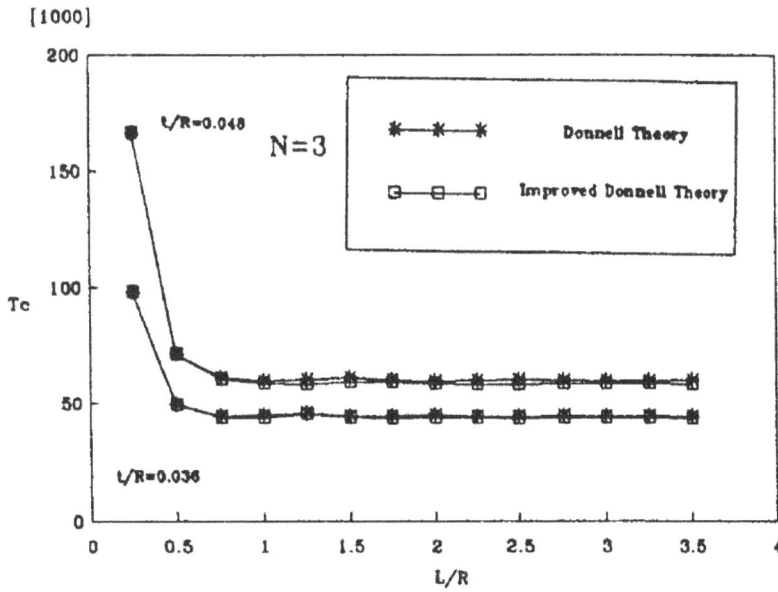

Figure 3.2-4: variation of the dimensionless critical temperature $T_c = T\alpha_\theta$ versus the variation of the dimensionless length parameter L/R.

ence

For this type of loading, the prebuckling forces are given by Eqs. (3.2-28). The function $T(x)$ of Eq. (3.2-27) is expanded into a Fourier series and only the first two terms are kept, as in the case of short cylinders. Substituting Eqs. (3.2-28) and (3.2-17) in to Eqs. (3.2-31) and employing the Galerkin's method and following the same procedure we obtain

$$\Delta T(\text{axial temperature}) = \Delta T(\text{radial temperature}) \qquad (3.2\text{-}36)$$

2.5   Results and discussion

The critical load values are calculated numerically and are plotted. Figs. (3.2-1) to (3.2-4) show the variation of the dimensionless critical load parameters $P_c = P/A_{11}$ and $T_c = T\alpha_\theta$ versus the variation of the dimensionless length parameter L/R. Two arbitrary dimensionless values of thickness ratio t/R are shown in each figure. The cross-ply circular cylindrical shell material is considered. The cylinder is made up of three angle-ply sequence being (90°, 0°, 90°) layers of equal thickness. All layers are constructed of the same E-glass epoxy material, with the following properties:

$$E_1 = 5.38 \times 10^7 \; Pa, \quad E_2 = 1.79 \times 10^7 \; Pa, \quad G_{12} = 8620000 \; Pa$$
$$\nu_{12} = 0.25, \quad \alpha_x = 0.0000063/ \; ^oC, \quad \alpha_\theta = 0.0000205/ \; ^oC \quad (3.2\text{-}37)$$

As expected, with increasing the dimensionless length parameter L/R, dimensionless critical load parameter decreases. Also with increasing the dimensionless thickness ratio t/R, the dimensionless critical load parameter increases, specially for short cylinders.

Figures (3.2-5) to (3.2-8) are drawn for a cross-ply circular cylindrical shell with five angle ply sequence being (0°, 90°, 0°, 90°, 0°) layers of equal thickness. The layer thickness and the material constants of each layer are the same as the previous example. It is observed that the dimensionless buckling load parameter increases with the increase of the number of layers.

It is noticed that for long composite cylindrical shells, the difference between the critical load values based on the Donnell and improved Donnell theories are not negligible. Independent of the kind of loading, the critical load values based on the improved Donnell theory were always lower than the corresponding values based on the Donnell theory. For $N = 3$, $L/R = 3.5$, and $t/R = 0.036$ the relative difference between the critical buckling loads predicted by the two theories is about 3.74% for

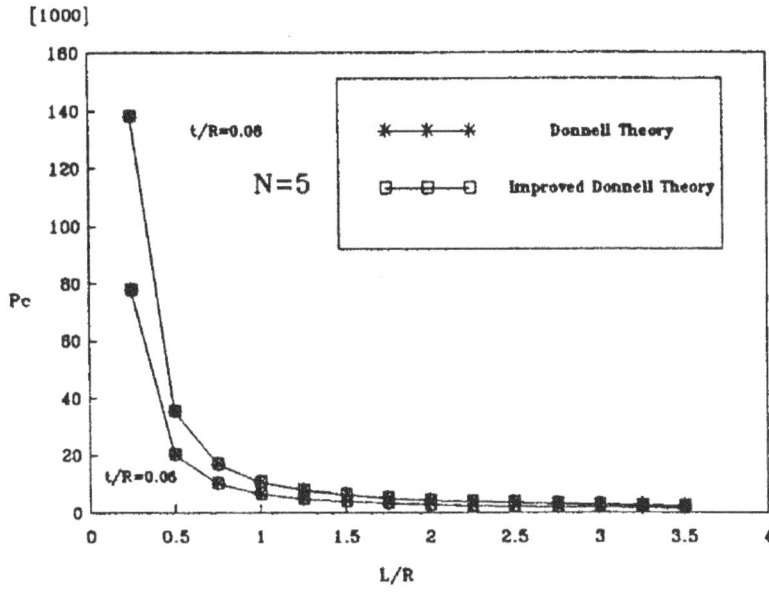

Figure 3.2-5: variation of the dimensionless critical load parameters $P_c = P/A_{11}$ versus the variation of the dimensionless length parameter L/R.

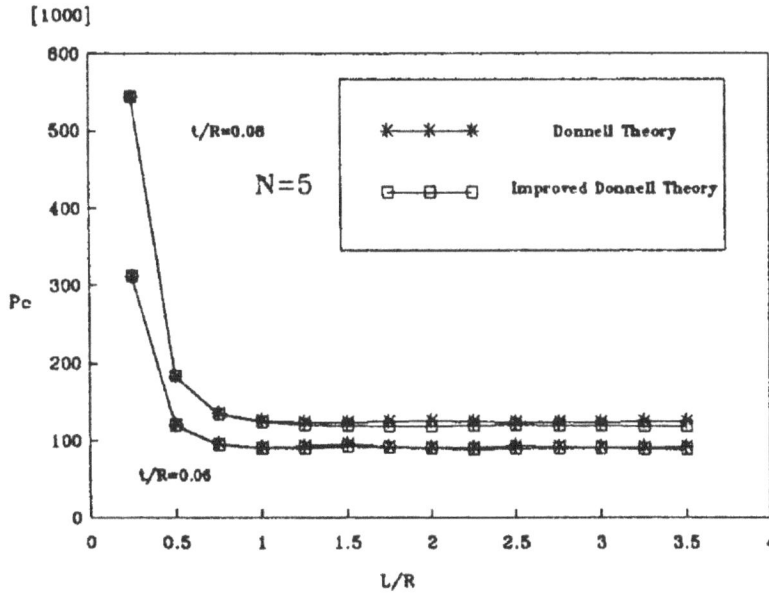

Figure 3.2-6: variation of the dimensionless critical load parameters $P_c = P/A_{11}$ versus the variation of the dimensionless length parameter L/R.

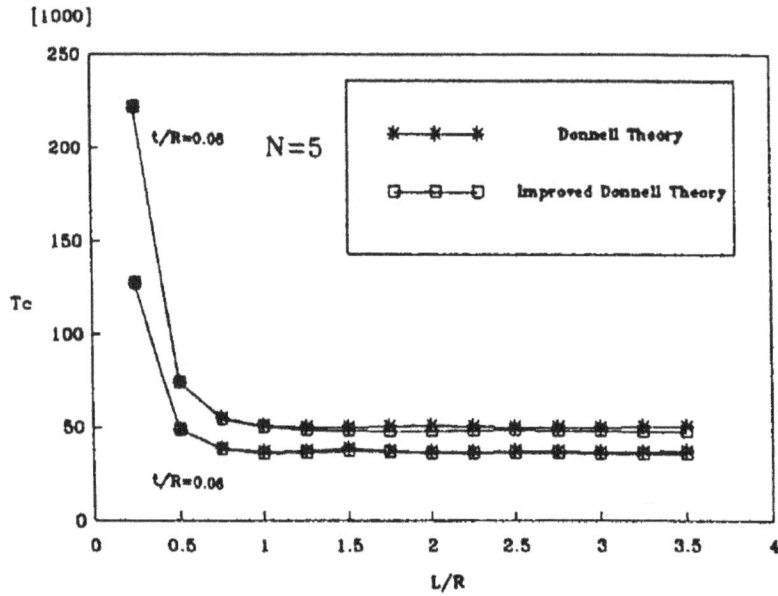

Figure 3.2-7: variation of the dimensionless critical temperature $T_c = T\alpha_\theta$ versus the variation of the dimensionless length parameter L/R.

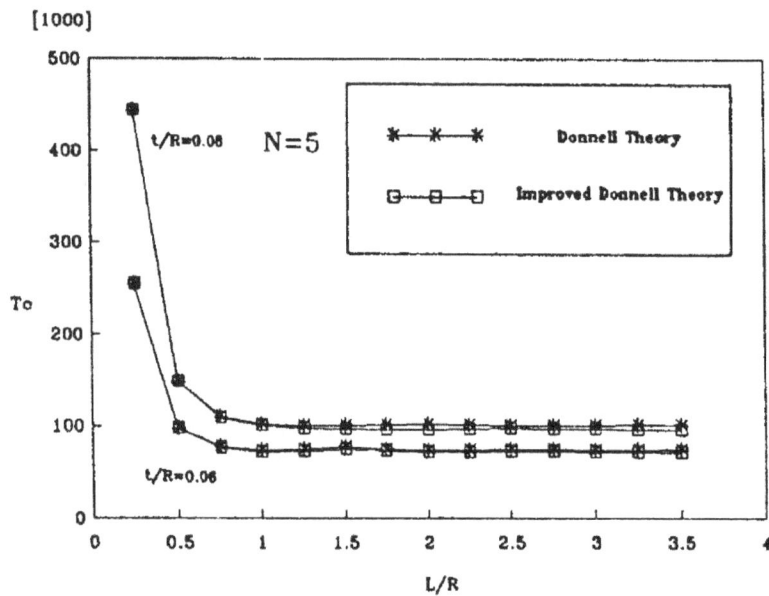

Figure 3.2-8: variation of the dimensionless critical temperature $T_c = T\alpha_\theta$ versus the variation of the dimensionless length parameter L/R.

cylinder under external lateral pressure, and 1.87% for other kinds of loading. These differences are about 4.54% for a cylinder under external lateral pressure, and 2.76% for other kinds of loading for $t/R = 0.048$. The difference between the two theories remains almost constant as the shell length increases. This conclusion, where the difference between the two theories remains almost constant for long cylindrical shells, is also obtained and discussed by Soldatos [26].

The buckling analysis of cross-ply laminated circular cylindrical shells under mechanical and thermal loads based on the Donnell and improved Donnell equations are investigated. It is concluded that:
1. The value of the critical buckling load decreases by increasing the length of cylinder.
2. The value of the critical buckling load increases by increasing the thickness of cylinder.
3. For long cylinders, the critical load values based on the Donnell and improved Donnell theories show noticeable differences.

3 Ring and Stringer Stiffened Shells

3.1 Introduction

The theoretical developments of the elastic buckling of shells is mostly based on the works of Donnell [4], who formulated the critical buckling load of short thin cylindrical shells which are frequently used in industry. The prime importance of his formulation is simplicity and the closed form solutions that he obtained for cylindrical shells at some loading conditions that are frequently occurred in practical design problems. The basic assumptions used in Donnell equations are that he ignored the shear force in circumferential directions and the rotations β_x and β_θ. These assumptions leads to results that are acceptable for short cylindrical shells.

A more complete treatment of shell buckling analysis include the effect of these terms in equilibrium and stability equations. The inclusion of these terms increase the accuracy of predicted buckling load and remove the limitations regarding the shell length, but cause coupling between the governing equations written in terms of displacement components and results complications in analytical treatments.

Eslami and Shariyat [14,15,18] used the improved equations to obtain the elastic, plastic and creep buckling of thin cylindrical shells under different mechanical loading conditions. The thermoelastic buckling of thin cylindrical and conical shells based on improved equations are also presented by Eslami et al. [19,30]. The thermoelastic buckling properties of the composite thin cylindrical shells based on Donnell and improved stability equations are investigated by Eslami and Javaheri [31]. The effect of eccentricity of stiffeners on the stability of cylindrical shells under hydrostatic pressure is discussed by Brush and Singer [32].

In this section the Donnell and improved equilibrium and stability equations are employed to compute the critical thermoelastic buckling load of thin ring and stringer stiffened cylindrical shells under radial thermal loading, axial temperature difference, and critical uniform initial-final temperature for simply supported shells. The results are extended to both short and long circular cylindrical shells.

3.2   Basic equations

A thin cylindrical shell of mean radius R and thickness h with length L is considered. The shell is assumed to be stiffened by circumferential rings and longitudinal stringers, as shown in Fig. (3.3-1). The normal and shear strain at a distance z from the middle plane of the shell are [9]

$$\epsilon_x = \epsilon_{xm} + zk_x$$
$$\epsilon_\theta = \epsilon_{\theta m} + zk_\theta$$
$$\gamma_{x\theta} = \gamma_{x\theta m} + zk_{x\theta} \qquad (3.3\text{-}1)$$

where the $\epsilon's$ is the normal strain, $\gamma's$ is the shear strain, and $k's$ is the curvature. The subscript m refers to the strain at the middle surface of the shell. The indices x and θ refer to the axial and circumferential directions, respectively. According to the Sanders assumption [9], the general strain-displacement relations can be simplified to give the following terms for the strains at the middle surface and the curvature in terms of displacement components

$$\epsilon_{xm} = u_{,x} + \frac{1}{2}w_{,x}^2$$
$$\epsilon_{\theta m} = (v_{,\theta} - w)/R + (v + w_{,\theta})^2/2R^2$$
$$\gamma_{x\theta m} = u_{,\theta}/R + (w_{,x}v + w_{,x}w_{,\theta})/R$$
$$k_{,x} = w_{,xx}$$
$$k_\theta = (v_{,\theta} + w_{,\theta\theta})/R^2$$
$$k_{,x\theta} = (v_{,x} + 2w_{,x\theta})/2R \qquad (3.3\text{-}2)$$

where u, v, and w are the displacement components and (,) indicates a partial derivative. The Hooke's law in terms of forces and moments per unit length for a stiffened cylindrical shell is

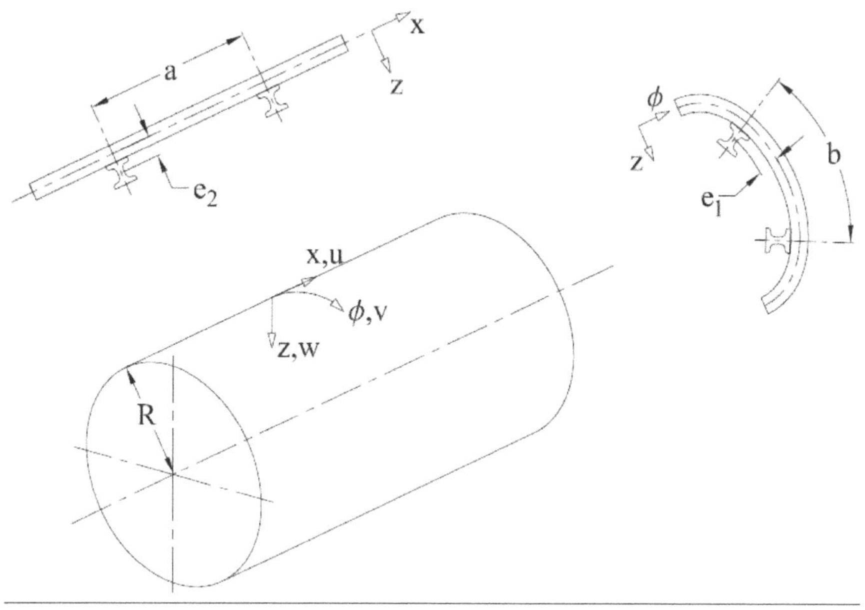

Figure 3.3-1: Ring and stringer stiffened cylindrical shell.

$$N_x = C[\epsilon_{xm}(1 + \mu_1) + \nu\epsilon_{xm} - RX_1k_x] - \frac{E\alpha T_0}{1 - \nu} - E_1\alpha_1 T_{s0}$$

$$N_\theta = C[\epsilon_{\theta m}(1 + \mu_2) + \nu\epsilon_{\theta m} - RX_2k_\theta] - \frac{E\alpha T_0}{1 - \nu} - E_2\alpha_2 T_{r0}$$

$$N_{x\theta} = Ch\gamma_{x\theta m}$$

$$M_x = -D[k_x(1 + \eta_{01}) + \nu k_x - \xi_1\epsilon_{xm}/R] - \frac{E\alpha T_1}{1 - \nu} - E_1\alpha_1 T_{s0}$$

$$M_\theta = -D[k_\theta(1 + \eta_{02}) + \nu k_\theta - \xi_2\epsilon_{\theta m}/R] - \frac{E\alpha T_1}{1 - \nu} - E_2\alpha_2 T_{r0}$$

$$M_{x\theta} = D[(1 - \nu) + \eta_{t1}]k_{x\theta}$$

$$M_{\theta x} = D[(1 - \nu) + \eta_{t2}]k_{\theta x} \tag{3.3-3}$$

where N_{ij} and M_{ij} are related to σ_{ij} through the shell thickness according to the first order shell theory, E is the elastic modulus, ν is the Poisson's ratio, and α is the coefficient of thermal expansion. The subscripts 1 and 2 stand for the properties of the stringer and ring, respectively. Also, the properties of ring and stringer stiffened cylindrical shells are as follows

$$\mu_1 = \frac{(1 - \nu^2)A_1 E_1}{bhE} \qquad \mu_2 = \frac{(1 - \nu^2)A_2 E_2}{bhE}$$

$$X_1 = \frac{(1 - \nu^2)A_1 E_1 e_1}{bhER} \qquad X_2 = \frac{(1 - \nu^2)A_2 E_2 e_2}{bhER}$$

$$\eta_{01} = \frac{E_1 I_{01}}{bD} \qquad\qquad\qquad \eta_{02} = \frac{E_2 I_{02}}{bD}$$

$$\eta_{t1} = \frac{G_1 I_{t1}}{bD} \qquad\qquad\qquad \eta_{t2} = \frac{G_2 I_{t2}}{bD}$$

$$\xi_1 = \frac{E_1 A_1 e_1 R}{bD} \qquad\qquad\qquad \xi_2 = \frac{E_2 A_2 e_2 R}{bD}$$

$$T_0 = \int_{shell} T dz \qquad T_{s0} = \int_{stringer} T_s dz \qquad T_{r0} = \int_{ring} T_r dz$$

$$T_1 = \int_{shell} T z dz \qquad T_{s1} = \int_{stringer} T_s z dz \qquad T_{r1} = \int_{ring} T_r z dz$$

$$C = \frac{Eh}{(1-\nu^2)} \qquad D = \frac{Eh^3}{12(1-\nu^2)} \qquad\qquad\qquad (3.3\text{-}4)$$

In these expressions A_1 and A_2 are the cross sectional areas of stringer and ring, a is the distance between rings and b is the distance between stringers, e_1 and e_2 are the distance between centroid of stiffener cross section and the middle surface of shell, see Fig. (3.3-1), I_{11} and I_{22} are the moments of inertia of stiffener cross section about its centroidal axis, I_{01} and I_{02} are the moments of inertia of stiffener cross section about the middle surface of the shell, and G_1 and G_2 are the shear modulus of the stringers and rings, respectively.

The total potential energy of the shell is the sum of membrane strain energy U_m the bending strain energy U_b , and the thermal strain energy U_T expressed as

$$U = U_m + U_b + U_T \qquad\qquad (3.3\text{-}5)$$

where

$$U_m = \frac{RC}{2} \int \int [\epsilon_{xm}^2 + \epsilon_{\theta m}^2 + 2\nu\epsilon_{xm}\epsilon_{\theta m} + \frac{1-\nu}{2}\gamma_{x\theta m}] dxd\theta +$$

$$+ \frac{RE_s}{2} \int \int [\epsilon_{xm}^2 \frac{A_s}{b} dxd\theta + \frac{RE_r}{2} \int \int [\epsilon_{\theta m}^2 \frac{A_r}{a} dxd\theta$$

$$U_b = \frac{RD}{2} \int \int [k_x^2 + k_\theta^2 + 2\nu k_x k_\theta + 2(1-\nu)k_{x\theta}^2] dxd\theta$$

$$+ \frac{RE_s}{2} \int \int \{\frac{k_x^2}{3}[(\frac{A_s}{b} + \frac{h}{2})^2 - \frac{h^3}{8}] - k_x\epsilon_{xm}(\frac{A_s^2}{b^2} + \frac{hA_s}{b})$$

$$+ \frac{4G_s}{3E_s} k_{x\theta}^2 [(\frac{A_s}{b} + \frac{h}{2})^3 - \frac{h^3}{8}]\} dxd\theta + \frac{RE_r}{2} \int \int \{\frac{k_\theta^2}{3}[(\frac{A_r}{a} + \frac{h}{2})^3 - \frac{h^3}{8}]$$

$$- k_\theta\epsilon_{\theta m}(\frac{A_r^2}{a^2} + \frac{ha_r}{a}) + \frac{4G_r}{3E_r} k_{x\theta}^2 [(\frac{A_r}{a} + \frac{h}{2})^3 - \frac{h^3}{8}]\} dxd\theta$$

$$U_T = \frac{ER\alpha^3}{1-\nu} \int \int \int T^2 dxd\theta dz[(\epsilon_{xm}\epsilon_{\theta m})T_0 - (k_x + k_\theta)T_1] dxd\theta$$

$$+\frac{E_s R \alpha_s^2}{2} \int \int \int T_s^2 dx d\theta dz - E_s \alpha_s R \int \int (T_{s0}\epsilon_{xm} - T_{s1}k_x)dxd\theta$$

$$+\frac{E_r R \alpha_r^2}{2} \int \int \int T_r^2 dx d\theta dz - E_r \alpha_r R \int \int (T_{r0}\epsilon_{\theta m} - T_{r1}k_\theta)dxd\theta$$

$$(3.3\text{-}6)$$

Assuming that the cylindrical shell is under thermal stress alone, the total potential energy is a function of the displacement components and their derivatives and can be written as

$$U = \int \int \int F(u, v, w, u_{,x}, u_{,\theta}, v_{,x}, v_{,\theta}, w_{,x}, w_{,\theta}, w_{,xx}, w_{,\theta\theta}, w_{,x\theta})dxd\theta dz$$

Minimizing the functional of potential energy leads to the Euler equations

$$\frac{\partial F}{\partial u} - \frac{\partial}{\partial x}\frac{\partial F}{\partial u_{,x}} - \frac{\partial}{\partial \theta}\frac{\partial F}{\partial u_{,\theta}} = 0$$

$$\frac{\partial F}{\partial v} - \frac{\partial}{\partial x}\frac{\partial F}{\partial v_{,x}} - \frac{\partial}{\partial \theta}\frac{\partial F}{\partial v_{,\theta}} = 0$$

$$\frac{\partial F}{\partial w} - \frac{\partial}{\partial x}\frac{\partial F}{\partial w_{,x}} - \frac{\partial}{\partial \theta}\frac{\partial F}{\partial w_{,\theta}} + \frac{\partial^2}{\partial x^2}\frac{\partial F}{\partial w_{,xx}} + \frac{\partial^2}{\partial \theta^2}\frac{\partial F}{\partial w_{,\theta\theta}} + \frac{\partial^2}{\partial x\partial\theta}\frac{\partial F}{\partial w_{,x\theta}} = 0$$

$$(3.3\text{-}7)$$

Upon substitution from Eqs. (3.3-6) into (3.3-7) and using Eqs. (3.3-2) and (3.3-3), the equilibrium equations for general thin cylindrical shell are obtained as

$$RN_{x,x} + N_{x\theta,\theta} = 0$$

$$RN_{x\theta,x} + N_{\theta,\theta} + \frac{1}{R}M_{\theta,\theta} + M_{x\theta,x} - (N_\theta\beta_\theta + N_{x\theta}\beta_x) = 0$$

$$R^2 M_{x,xx} + 2RM_{x\theta,x\theta} + M_{\theta,\theta\theta} - RN_\theta - R[RN_x\beta_{x,x} + N_{x\theta}(R\beta_{\theta,x} + \beta_{x,\theta})$$

$$+N_\theta\beta_{\theta,\theta}] = 0 \qquad (3.3\text{-}8)$$

where

$$\beta_x = w_{,x} \qquad\qquad \beta_\theta = (v + w_{,\theta})/R \qquad (3.3\text{-}9)$$

are the rotations in the x and θ-directions. Equations (3.3-8) can be compared with the Donnell equations for short cylinders as [9]

$$RN_{x,x} + N_{x\theta,\theta} = 0$$

$$RN_{x\theta,x} + N_{\theta,\theta} = 0$$

$$R^2 M_{x,xx} + 2RM_{x\theta,x\theta} + M_{\theta,\theta\theta} - RN_\theta - R[RN_x\beta_{x,x} + N_{x\theta}(R\beta_{\theta,x} + \beta_{x,\theta})$$

$$+N_\theta\beta_{\theta,\theta}] = 0 \qquad (3.3\text{-}10)$$

The first and third equations of Eqs. (3.3-8) and (3.3-10) are identical, but the second of Eqs. (3.3-8) is different. It is recalled that the shear force in circumferential direction is

$$Q_\theta = \frac{1}{R} M_{\theta,\theta} + M_{x\theta,x} \qquad (3.3\text{-}11)$$

Therefore, comparison reveals that in Donnell equations for short cylinders, the shear force in circumferential direction as well as angles of rotation β_x and β_θ are ignored. These approximations are justified for the short cylinders, but for the long cylindrical shells these assumptions are no longer valid and more accurate equilibrium equations must include these terms.

The stability equations of thin cylindrical shells can be derived by the variational formulations. If V is the total potential energy of the shell, the expansion of V about the equilibrium state into the Taylor series yields

$$\Delta V = \delta V + \frac{1}{2}\delta^2 V + \frac{1}{3!}\delta^3 V + ... \qquad (3.3\text{-}12)$$

The first variation is associated with the state of equilibrium. The stability of the original configuration of the shell in the neighborhood of equilibrium can be determined by the sign of second variation as follows

(1) The equilibrium is stable if $\delta^2 V > 0$ for all virtual displacements.

(2) The equilibrium is unstable if $\delta^2 V < 0$ for at least one admissible set of virtual displacements.

The condition $\delta^2 V = 0$ is used to derive the stability equations for many practical buckling problems as discussed by Langhaar [17].

Let us assume that u_i^* denotes the displacement components of the equilibrium state and δu_i^* the virtual displacement corresponding to a neighboring state. Denoting by $\bar{\delta}$ the variation with respect to δu_i^*. the following rule, known as the Trefftz rule, is stated for the determination of the lowest critical load. The external load acting on the original configuration is considered to be the critical buckling load if the following variational equation is satisfied.

$$\bar{\delta}(\delta^2 V) = 0 \qquad (3.3\text{-}13)$$

This rule provides the governing equations that determine the lowest critical load.

Consider the state of stable equilibrium of a general cylindrical shell under thermal load to be designated by u_0, v_0, and w_0. The displacement components of the neighboring state are

$$u = u_0 + u_1$$

$$v = v_0 + v_1$$
$$w = w_0 + w_1 \tag{3.3-14}$$

Similarly, the components of forces and moments related to the neighboring state are related to the state of equilibrium according to the relations

$$N_x = N_{x0} + \Delta N_x \qquad\qquad M_x = M_{x0} + \Delta M_x$$
$$N_\theta = N_{\theta 0} + \Delta N_\theta \qquad\qquad M_\theta = M_{\theta 0} + \Delta M_\theta$$
$$N_{x\theta} = N_{x\theta 0} + \Delta N_{x\theta} \qquad\qquad M_{x\theta} = M_{x\theta 0} + \Delta M_{x\theta} \tag{3.3-15}$$

If $N_{x1}, N_{\theta 1}, N_{x\theta 1}, \ldots$ express the linear portion of $\Delta N_{x1}, \Delta N_{\theta 1}, \Delta N_{x\theta 1}, \ldots$ in terms of $u_1, v_1,$ and w_1 they become

$$N_{x1} = C[\epsilon_{x1}(1 + \mu_1) + \nu\epsilon_{\theta 1} - RX_1 k_{x1}]$$
$$N_{\theta 1} = C[\epsilon_{\theta 1}(1 + \mu_2) + \nu\epsilon_{x1} - RX_2 k_{\theta 1}]$$
$$N_{x\theta 1} = N_{\theta x1} = Gh\gamma_{x\theta 1}$$
$$M_{x1} = -D[k_{x1}(1 + \eta_{01}) + \nu k_{\theta 1} - \frac{\xi_1}{R}\epsilon_{x1}]$$
$$M_{\theta 1} = -D[k_{\theta 1}(1 + \eta_{02}) + \nu k_{x1} - \frac{\xi_2}{R}\epsilon_{\theta 1}]$$
$$M_{x\theta 1} = D[(1 - \nu) + \eta_{t1}]k_{x\theta 1}$$
$$M_{\theta x1} = D[(1 - \nu) + \eta_{t2}]k_{x\theta 1} \tag{3.3-16}$$

and

$$N_{x0} = C[\epsilon_{x0}(1 + \mu_1) + \nu\epsilon_{\theta 0} - RX_1 k_{x0}] - \frac{E\alpha T_0}{1 - \nu} - E_s\alpha_s T_{s0}$$
$$N_{\theta 0} = C[\epsilon_{\theta 0}(1 + \mu_2) + \nu\epsilon_{x0} - RX_2 k_{\theta 0}] - \frac{E\alpha T_0}{1 - \nu} - E_r\alpha_r T_{r0}$$
$$N_{x\theta 0} = N_{\theta x0} = Gh\gamma_{x\theta 0}$$
$$M_{x0} = -D[k_{x0}(1 + \eta_{01}) + \nu k_{\theta 0} - \frac{\xi_1}{R}\epsilon_{x0}] - \frac{E\alpha T_1}{1 - \nu} - E_s\alpha_s T_{s1}$$
$$M_{\theta 0} = -D[k_{\theta 0}(1 + \eta_{02}) + \nu k_{x0} - \frac{\xi_2}{R}\epsilon_{\theta 0}] - \frac{E\alpha T_1}{1 - \nu} - E_r\alpha_r T_{r1}$$
$$M_{x\theta 0} = D[(1 - \nu) + \eta_{t1}]k_{x\theta 0}$$
$$M_{\theta x0} = D[(1 - \nu) + \eta_{t2}]k_{x\theta 0} \tag{3.3-17}$$

Using Eqs. (3.3-14) for the displacement components of a neighboring state of stable equilibrium and employing the decomposed linear part of strain displacement relations, the linear strain relations for the equilibrium state (0) and the first variation (1) are obtained as

$$e_x = e_{x0} + e_{x1} \qquad\qquad \beta_\theta = \beta_{\theta 0} + \beta_{\theta 1}$$

$$e_\theta = e_{\theta 0} + e_{\theta 1} \qquad\qquad k_x = k_{x0} + k_{x1}$$

$$e_{x\theta} = e_{x\theta 0} + e_{x\theta 1} \qquad\qquad k_\theta = k_{\theta 0} + k_{\theta 1}$$

$$\beta_x = \beta_{x0} + \beta_{x1} \qquad\qquad k_{x\theta} = k_{x\theta 0} + k_{x\theta 1} \qquad (3.3\text{-}18)$$

Substitution into the functional of potential energy yields

$$V = V_0 + \Delta V$$
$$= \frac{RC}{2} \int\int \{[(e_{x0} + e_{x1}) + \frac{1}{2}(\beta_{x0} + \beta_{x1})^2]^2 + [(e_{\theta 0} + e_{\theta 1}) + \frac{1}{2}(\beta_{\theta 0} + \beta_{\theta 1})^2]^2$$
$$+ \{[(e_{x0} + e_{x1}) + \frac{1}{2}(\beta_{x0} + \beta_{x1})^2][(e_{\theta 0} + e_{\theta 1}) + \frac{1}{2}(\beta_{\theta 0} + \beta_{\theta 1})^2]$$
$$+ \frac{1-\nu}{2}[(e_{x\theta 0} + e_{x\theta 1}) + (\beta_{x0} + \beta x1)(\beta_{\theta 0} + \beta\theta 1)]^2\} dx d\theta$$
$$+ \frac{RD}{2}\int\int\{[(k_{x0} + k_{x1})^2 + (k_{\theta 0} + k_{\theta 1})^2] + 2(1-\nu)(k_{x\theta 0} + k_{x\theta 1})^2\} dx d\theta$$
$$- \frac{RE\alpha}{1-\nu}\int\int\{[(e_{x0} + e_{x1}) + \frac{1}{2}(\beta_{x0} + \beta_{x1})^2 + (e_{\theta 0} + e_{\theta 1}) + \frac{1}{2}(\beta_{\theta 0} + \beta_{\theta 1})^2]T_0$$
$$+ (k_{x0} + k_{x1} + k_{\theta 0} + k_{\theta 1})T_1\} dx d\theta + \frac{RE\alpha^2}{1-\nu}\int\int\int T^2 dx d\theta dz$$
$$\frac{RE_s}{2}\int\int\{[(e_{x0} + e_{x1}) + \frac{1}{2}(\beta_{x0} + \beta_{x1})^2]^2 \frac{A_s}{b}[(\frac{A_s^2}{b^2} + \frac{hA_s}{b})(k_{x0} + k_{x1})]$$
$$+ \frac{(k_{x0} + k_{x1})^2}{3}[(\frac{A_s}{b} + \frac{h}{2})^3 - \frac{h^3}{8}] + \alpha_s^2 T_{s0}^2 - 2\alpha_s T_{s0}[(e_{x0} + e_{x1}) + \frac{1}{2}(\beta_{x0} + \beta_{x1})^2]$$
$$+ 2\alpha_s T_{s1}(k_{x0} + k_{x1}) + \frac{4G_s}{3E_s}[(\frac{A_s}{b} + \frac{h}{2})^3 - \frac{h^3}{8}](k_{x\theta 0} + k_{x\theta 1})^2\} dx d\theta$$
$$+ \frac{RE_r}{2}\int\int\{[(e_{\theta 0} + e_{\theta 1}) + \frac{1}{2}(\beta_{\theta 0} + \beta_{\theta 1})^2]^2 \frac{A_r}{a}[(\frac{A_r^2}{a^2} + \frac{hA_r}{a})(k_{\theta 0} + k_{\theta 1})]$$
$$+ \frac{(k_{\theta 0} + k_{\theta 1})^2}{3}[(\frac{A_r}{a} + \frac{h}{2})^3 - \frac{h^3}{8}] + \alpha_r^2 T_{r0}^2 - 2\alpha_r T_{r0}[(e_{\theta 0} + e_{\theta 1}) + \frac{1}{2}(\beta_{\theta 0} + \beta_{\theta 1})^2]$$
$$+ 2\alpha_r T_{r1}(k_{\theta 0} + k_{\theta 1}) + \frac{4G_r}{3E_r}[(\frac{A_r}{a} + \frac{h}{2})^3 - \frac{h^3}{8}](k_{x\theta 0} + k_{x\theta 1})^2\} dx d\theta \qquad (3.3\text{-}19)$$

Terms with subscript o are related to V_0 and those with subscript 1 are related to ΔV. Neglecting the prebuckling rotations β_x and β_θ, the second variation of the potential energy is obtained as

$$\frac{1}{2}\delta^2 V = \frac{RC}{2}\int\int\{(e_{x1}^2 + e_{x0}\beta_{x1}^2 + e_{\theta 1}^2 + e_{\theta 0}\beta_{\theta 1}^2) + 2\nu(e_{x1}e_{\theta 1} + \frac{1}{2}\beta_{\theta 1}^2 e_{x0} + \frac{1}{2}\beta_{x1}^2 e_{\theta 0})$$
$$+ \frac{1-\nu}{2}(e_{x\theta 1}^2 + e_{x\theta 0}\beta_{x1}\beta_{\theta 1}0\} dx d\theta + \frac{RD}{2}\int\int\{(k_{x1}^2 + k_{\theta 1}^2 + 2\nu k_{x1}k_{\theta 1} + 4k_{x\theta 1}^2\} dx d\theta$$
$$- \frac{RE\alpha}{2(1-\nu)}\int\int(\beta_{x1}^2 + \beta_{\theta 1}^2)T_0 dx d\theta$$
$$+ \frac{RE_s}{2}\int\int\{[e_{x1}^2 + e_{x0}\beta_{x1}^2](\frac{A_s}{b}) - [k_{x1}e_{x1} + \frac{1}{2}k_{x0}\beta_{x1}^2](\frac{A_s}{b})(2e_s)$$

$$+\frac{k_{x1}^2}{3}[(\frac{A_s}{b}+\frac{h}{2})^3-\frac{h^3}{8}]-\alpha_s T_{s0}\beta_{x1}^2+\frac{4G_s}{3E_s}[(\frac{A_s}{b}+\frac{h}{2})^3-\frac{h^3}{2}]k_{x\theta1})^2\}dxd\theta$$

$$+\frac{RE_r}{2}\int\int\{[e_{\theta1}^2+e_{\theta0}\beta_{\theta1}^2](\frac{A_r}{a})-[k_{\theta1}e_{\theta1}+\frac{1}{2}k_{\theta0}\beta_{\theta1}^2](\frac{A_r}{a})(2e_r)$$

$$+\frac{k_{\theta1}^2}{3}[(\frac{A_r}{a}+\frac{h}{2})^3-\frac{h^3}{8}]-\alpha_r T_{r0}\beta_{\theta1}^2+\frac{4G_r}{3E_r}[(\frac{A_r}{a}+\frac{h}{2})^3-\frac{h^3}{8}]k_{x\theta1})^2\}dxd\theta \quad (3.3\text{-}20)$$

where the prebuckling linearized forces are given in Eqs. (3.3-17). Substituting for the strains and curvatures from the linearized strain-displacement relations yield

$$\epsilon_{x1}=e_{x1}=u_{1,x} \qquad\qquad k_{x1}=w_{1,xx}$$
$$\epsilon_{\theta1}=e_{\theta1}=(v_{1,\theta}-w_1)/R \qquad k_{\theta1}=(v_{1,\theta}+w_{1,\theta\theta})/R^2$$
$$\gamma_{x\theta1}=e_{x\theta1}=(v_{1,x}-u_{1,\theta})/R \qquad k_{x\theta1}=(v_{1,x}+2w_{1,x\theta})/2R$$
$$\beta_{x1}=w_{1,x} \qquad\qquad \beta_{\theta1}=(v_1+w_{1,\theta})/R \quad (3.3\text{-}21)$$

Equations (3.3-21) are written in terms of the displacement components, which upon application of Euler equations result in the stability equations

$$RN_{x1,x}+N_{x\theta1,\theta}=0$$
$$RN_{x\theta1,x}+N_{\theta1,\theta}+\frac{1}{R}M_{\theta1,\theta}+M_{x\theta1,x}-(N_{\theta0}\beta_{\theta1}+N_{x\theta0}\beta_{x1})=0$$
$$R^2M_{x1,xx}+2RM_{x\theta1,x\theta}+M_{\theta1,\theta\theta}-RN_{\theta1}-R[RN_{x0}\beta_{x1,x}$$
$$+N_{x\theta0}(R\beta_{\theta1,x}+\beta_{x1,\theta})+N_{\theta0}\beta_{\theta1,\theta}]=0 \quad (3.3\text{-}22)$$

Equations (3.3-22) can be compared with Donnell equations for short cylinders as [9]

$$RN_{x1,x}+N_{x\theta1,\theta}=0$$
$$RN_{x\theta1,x}+N_{\theta1,\theta}=0$$
$$R^2M_{x1,xx}+2RM_{x\theta1,x\theta}+M_{\theta1,\theta\theta}-RN_{\theta1}-R[RN_{x0}\beta_{x1,x}$$
$$+N_{x\theta0}(R\beta_{\theta1,x}+\beta_{x1,\theta})+N_{\theta0}\beta_{\theta1,\theta}]=0 \quad (3.3\text{-}23)$$

A comparison of Eqs. (3.3-22) and (3.3-23) reveals that while the first and third equations are identical, the second equation of (3.3-22) includes extra terms that are ignored in the Donnell equations because they are small for short cylinders. Including these terms removes the length limitations imposed by Donnell equations. It should be noted that these terms improve the accuracy of the predicted design critical loads.

3.3   Thermal buckling, Donnell equations

For short cylindrical shells the transverse shear force Q_θ and rotations β_x and β_θ are ignored and the equilibrium and stability equations reduce to the Donnell equations. Upon substitution for N_{ij} and M_{ij} their equivalence of strain from Eqs. (3.3-16) and finally in terms of displacements from Eqs. (3.3-21), the uncoupled forms of the Donnell equations are obtained as

$$R^2(1+\mu_1)u_{1,xx} + \frac{R(1+\nu)}{2}v_{1,\theta x} + \frac{1-\nu}{2}u_{1,\theta\theta} - R\nu w_{1,x} - X_1 R^3 w_{1,xxx} = 0$$

$$\frac{R(1+\nu)}{2}u_{1,\theta x} + \frac{R^2(1-\nu)}{2}v_{1,xx} + (1+\mu_2)v_{1,\theta\theta} - X_2 w_{1,\theta\theta\theta}$$

$$-(1+\mu_2)w_{1,\theta} = 0$$

$$R^4(1+\eta_{01})w_{1,xxxx} + R^2(2+\eta_{t1}+\eta_{t2})w_{1,xx\theta\theta} + (1+\eta_{02})w_{1,\theta\theta\theta\theta} +$$

$$[12(\frac{R}{h})^2 X_2 + \xi_2]w_{1,\theta\theta} - \xi_2 v_{1,\theta\theta\theta} - 12(\frac{R}{h})^2(1+\mu_2)(v_{1,\theta} - w_1)$$

$$-\xi_1 R^3 u_{1,xxx} - 12R(\frac{R}{h})^2 \nu u_{1,x} - 12\frac{R^2}{c}(\frac{R}{h})^2 N_{x0}w_{1,xx}$$

$$-24\frac{R}{c}(\frac{R}{h})^2 N_{x\theta 0}w_{1,x\theta} - 12\frac{1}{c}(\frac{R}{h})^2 N_{\theta 0}w_{1,\theta\theta} = 0 \qquad (3.3\text{-}24)$$

These equations are related to the thermal forces through the prebuckling terms such as N_x through Eqs. (3.3-17). In the next section three cases of thermal buckling are discussed and the critical temperatures are calculated. In all cases the Galerkin method implied. The displacement components for a cylindrical shell with simply supported ends are approximated by

$$u_1 = \sum_{m=1}^{\infty}\sum_{n=1}^{\infty} A_{mn}\cos m\pi x/L \sin n\theta$$

$$v_1 = \sum_{m=1}^{\infty}\sum_{n=1}^{\infty} B_{mn}\sin m\pi x/L \cos n\theta$$

$$w_1 = \sum_{m=1}^{\infty}\sum_{n=1}^{\infty} C_{mn}\sin m\pi x/L \sin n\theta \qquad (3.3\text{-}25)$$

Substituting these approximate solutions into the stability equations (3.3-24), for M=1,2 and n=1,2 the determinant of the matrix of coefficients, which is 12x12, is set equal to zero for each case and critical Temperature is found.

A. Critical uniform temperature rise

Consider a cylindrical shell of length L, radius R, and thickness h with both ends simply supported. The initial uniform temperature of the shell is assumed to be T_i. If the shell is simply supported and the axial displacement is prevented, the temperature can be uniformly raised to a final value T_f such that the shell buckles. For simplicity it is assumed that the material of shell and stiffener are identical. So

$$E_r \alpha_r = E_s \alpha_s = E\alpha \qquad (3.3\text{-}26)$$

And from Eqs. (3.3-4) T_0, T_{s0}, and T_{r0} are as follows

$$T_0 = h\Delta T$$
$$T_{s0} = \frac{A_s}{b}\Delta T$$
$$T_{r0} = \frac{A_r}{a}\Delta T \qquad (3.3\text{-}27)$$

which $\Delta T = T_f - T_i$. The prebuckling thermal forces are

$$N_{x0} = -\frac{E\alpha h\Delta T}{1-\nu} - E\alpha\frac{A_s}{b}\Delta T$$
$$N_{\theta 0} = -\frac{E\alpha h\Delta T}{1-\nu} - E\alpha\frac{A_r}{a}\Delta T \qquad (3.3\text{-}28)$$

For this type of loading $N_{x\theta} = 0$. After substituting Eqs. (3.3-25) and (3.3-28) in Eqs. (3.3-24), the determinant of the matrix of coefficient is set equal to zero which the resulting equation has four roots. Among these roots the state that causes the minimum value of $\alpha\Delta T$ results in the critical temperature. For symmetrical buckling $(n = 0)$, and when the following condition between the dimensions of rings and stringers are met

$$x_1\xi_1 < (1 + \eta_{01})(1 + \mu_1) \qquad (3.3\text{-}29)$$

the critical initial-final temperature which cause thermal buckling is

$$\Delta T|_{crit} = \frac{kh}{R\alpha} \qquad (3.3\text{-}30)$$

where the expression for k is given for the following cases:
1 - Cylindrical shell with rings and stringers

$$k = \frac{\{4RD\sqrt{3[\nu^2 - (1+\mu_1)(1+\mu_2)][x_1\xi_1 - (1+\eta_{01})(1+\mu_1)]} - D\mu h(\xi_1 + 12R^2 x_1/h^2)\}}{REh^2(1+\mu_1)[h/(1-\mu) + A_s/b]}$$
$$(3.3\text{-}31)$$

2 - Cylindrical shell with stringers

$$k = \frac{\{4RD\sqrt{3[\nu^2 - (1+\mu_1)]}[x_1\xi_1 - (1+\eta_{01})(1+\mu_1)] - D\mu h(\xi_1 + 12R^2 x_1/h^2)\}}{REh^2(1+\mu_1)[h/(1-\mu) + A_s/b]} \qquad (3.3\text{-}32)$$

3 - Cylindrical shell with rings

$$k = \sqrt{\frac{1 + \mu_2 - \nu^2}{3(1+\nu)^2}} \qquad (3.3\text{-}33)$$

4 - Cylindrical shell with no rings and stringers

$$k = \sqrt{\frac{1 - \nu}{3(1+\nu)}} \qquad (3.3\text{-}34)$$

The results of this case is identical with the results of [19]. For $\nu = 0.3$, the critical initial-final buckling temperature reduces to

$$\Delta T|_{crit} = \frac{0.424h}{R\alpha} \qquad (3.3\text{-}35)$$

B. Critical radial temperature

Assume linear temperature variation across the shell thickness as

$$T(z) = \Delta T(z + h/2)/h \qquad (3.3\text{-}36)$$

where z measures from the middle plane of the shell. In this case T_{r0} and T_{s0} are the same as before, and $T_0 = \frac{h\Delta T}{2}$. For simply supported edges the prebuckling thermal forces are

$$N_{x0} = -\frac{E\alpha h\Delta T}{2(1-\nu)} - E\alpha \frac{A_s}{b}\Delta T$$

$$N_{\theta 0} = -\frac{E\alpha h\Delta T}{2(1-\nu)} - E\alpha \frac{A_r}{a}\Delta T \qquad (3.3\text{-}37)$$

With a similar method, the critical buckling temperature difference between inside and outside surface is found. For symmetrical buckling ($n = 0$), and assuming identical materials for shell, rings, and stringers, the critical radial temperature difference that cause thermal buckling is

$$\Delta T|_{crit} = \frac{kh}{R\alpha} \qquad (3.3\text{-}38)$$

where the expression for k is given for the following cases:

1 - Cylindrical shell with rings and stringers

$$k = \frac{\{4\sqrt{3[\nu^2 - (1+\mu_1)(1+\mu_2)]}[x_1\xi_1 - (1+\eta_{01})(1+\mu_1)] - 2\mu\xi_1(h/R)\}}{12(1-\nu^2)(1+\mu_1)[0.5(1-\nu) + A_s/bh]}$$

(3.3-39)

2 - Cylindrical shell with stringers

$$k = \frac{\{4\sqrt{3[\nu^2 - (1+\mu_1)]}[x_1\xi_1 - (1+\eta_{01})(1+\mu_1)] - 2\nu\xi_1(h/R)\}}{12(1-\nu^2)(1+\mu_1)[0.5(1-\nu) + A_s/b]}$$

(3.3-40)

3 - Cylindrical shell with rings

$$k = 2\sqrt{\frac{1 + \mu_1 - \nu^2}{3(1+\nu)^2}}$$

(3.3-41)

4 - Cylindrical shell with no rings and stringers

$$k = 2\sqrt{\frac{1-\nu}{3(1+\nu)}}$$

(3.3-42)

The results of this case is identical with the results of [19]. For $\nu = 0.3$, the critical initial-final buckling temperature reduces to

$$\Delta T|_{crit} = \frac{0.848h}{R\alpha}$$

(3.3-43)

C. Critical axial temperature

Consider a cylindrical shell of length L under axial temperature difference and with simply supported edges, where the motion of the edges in the axial direction is prevented. Assume a linear temperature variation in the axial direction x

$$T(x) = \Delta T \frac{x}{L}$$

(3.3-44)

where $\Delta T = T(L)_T(0)$. For calculating prebuckling thermal forces Fourier series of $T(x)$ are used and only 3 terms is considered

$$T(x) = \Delta T(\frac{L}{2} - \frac{4}{\pi^2}\cos\frac{\pi x}{L} - \frac{4}{9\pi^2}\cos\frac{3\pi x}{L})$$

(3.3-45)

The prebuckling thermal forces are

$$N_{x0} = N_{\theta 0} = -E\alpha h\Delta T(\frac{L}{2} - \frac{4}{\pi^2}\cos\frac{\pi x}{L} - \frac{4}{9\pi^2}\cos\frac{3\pi x}{L})(\frac{h}{1-\nu} + \frac{A_s}{b})$$

(3.3-46)

By considering Galerkin method, which similar before cases, the critical axial temperature is found.

3.4 Thermal buckling - improved equations

For the long cylindrical shell the improved equations, which include the transverse shear force Q_θ and rotations β_x and $\beta_{|theta}$, provide more accurate results for the critical buckling loads. In this case the governing equations in terms of displacement components u_1, v_1 and w_1 are no longer separable and are coupled. Using the improved stability equations and substituting for the forces and moments their respective values in terms of the displacement components result

$$R^2(1+\mu_1)u_{1,xx} + \frac{R(1+\nu)}{2}v_{1,\theta x} + \frac{1-\nu}{2}u_{1,\theta\theta} - R\nu w_{1,x} - X_1 R^3 w_{1,xxx} = 0$$

$$\frac{R(1+\nu)}{2}u_{1,\theta x} + \{\frac{R^2(1-\nu)}{2} + \frac{D}{c}[(1-\nu) + \frac{\eta_{t1}+\eta_{t2}}{2}]\}v_{1,xx}$$

$$+[(1+\mu_2) - X_2 - \frac{\xi D}{R^2 c} + \frac{D}{R^2 c}(1+\eta_{02})]v_{1,\theta\theta} + [-(1+\mu_2)$$

$$+\frac{D\xi_2}{R^2 c}]w_{1,\theta} - [X_2 - \frac{D}{R^2 c}(1+\eta_{02})]w_{1,\theta\theta\theta}$$

$$+\frac{2D}{c}[(1-\nu) + \frac{\eta_{t1}+\eta_{t2}}{2}]w_{1,xx\theta} + \frac{D\nu}{c}w_{1,xx\theta} - \frac{RN_{\theta0}}{c}\beta_{\theta1} - \frac{RN_{x\theta0}}{c}\beta_{x1} = 0$$

$$R^4(1+\eta_{01})w_{1,xxxx} + \{1 + \frac{\eta_{t1}+\eta_{t2}}{2}\}v_{1,\theta xx} + 2R^2[\nu - 2 + \frac{\eta_{t1}+\eta_{t2}}{2}\}w_{1,\theta\theta xx}$$

$$-\frac{\xi}{R^3}u_{1,xxx} + (1+\eta_{02})w_{1,\theta\theta\theta\theta} +$$

$$[12(\frac{R}{h})^2 X_2 + \xi_2]w_{1,\theta\theta} - \xi_2 v_{1,\theta\theta\theta} - 12(\frac{R}{h})^2(1+\mu_2)(v_{1,\theta} - w_1)$$

$$-\xi_1 R^3 u_{1,xxx} - 12R(\frac{R}{h})^2\nu u_{1,x} - 12\frac{R^2}{c}(\frac{R}{h})^2 N_{x0}w_{1,xx}$$

$$-24\frac{R}{c}(\frac{R}{h})^2 N_{x\theta0}w_{1,x\theta} - 12\frac{1}{c}(\frac{R}{h})^2 N_{\theta0}w_{1,\theta\theta} = 0 \qquad (3.3\text{-}47)$$

A. Critical uniform temperature rise

In this case Eqs. (3.3-25) and (3.3-28). Substituting in to Eqs. (3.3-34) and with a similar method, the critical uniform temperature rise is obtained.

B. Critical radial temperature

The temperature distribution is assumed to be specified by Eq. (3.3-29), and the resulting prebuckling axial force is specified by Eqs. (3.3-30). By considering Eqs. (3.3-30) and (3.3-34), the critical temperature difference between inside and outside is obtained.

C. Critical axial temperature difference

The axial temperature distribution and the resulting axial thermal force are the same as specified by Eqs. (3.3-31) and (3.3-33). With a similar method, the critical axial temperature is calculated.

4 Imperfection Analysis

4.1 Introduction

Thermal buckling analysis of perfect cylindrical shells of isotropic and homogeneous materials and cylindrical shells of composite materials based on Donnell and improved Donnell stability equations are studied by Eslami et al. [19,33]. A number of thermal buckling problems of practical importance, such as uniform temperature rise, radial, and axial temperature differences are discussed in the papers. It is to be emphasized here that while the Donnell equilibrium equation is basically developed on Love-Kirchhoff hypothesis and the Sanders nonlinear strain-displacement relations, the improved equilibrium equation considers the circumferential shear force and rotations β_x and β_θ in addition to the Donnell equations. It is shown by Eslami that resulting buckling loads predicted for long cylindrical shells are improved when the improved stability equations are used.

A more rigorous and complete treatment of the equilibrium and stability equations for mechanical buckling analysis of cylindrical shells is given by Soldatos [26,34]. His formulations are basically derived for composite shells, where the effects of transverse shear stresses are significant.

However, the more precise and exact mathematical modeling for buckling analysis of shells will require more sophisticated mathematical tools, such as numerical solutions. In this case, the closed form solution, which is of practical importance, is not possible to obtain. Eslami and Shariyat considered the flexural theory and with the full Green nonlinear strain-displacement relation, instead of the simplified Sanders assumption, formulated the dynamic mechanical and thermal buckling of imperfect cylindrical shells [35]. The higher order shear deformation theory including the normal stress was used and the mixed formulation was established to simplify the approach of both kinematical and forced boundary conditions. The technique was then improved by the same authors to incorporate the layer-wise theory of composite imperfect shells under mechanical and thermal loads [36]. The work was extended by the same authors to exact three dimensional analysis of circular cylindrical shells based on the equilibrium equation and the full nonlinear Green strain-displacement relation [37]. Due to the mathematical complexity of the resulting models, a numerical method was established to solve the problem, and no closed form solutions were obtained. The Donnell and improved Donnell stability equations are employed to present a closed form solution for elasto-plastic and creep buckling of cylindrical shells under mechanical loads at elevated temperature [18].

The present section is devoted to thermal buckling analysis of imperfect circular cylindrical shells of isotropic materials. The Donnell and improved Donnell stability equations are considered and two models for imperfection, the Wan-Donnell and Koiter models, are adopted. The thermal loads include the uniform temperature rise, linear radial, and linear axial temperature differences.

4.2   Basic equations

Consider a thin cylindrical shell of mean radius R and thickness h with length L. The normal and shear strains at distance z from the shell middle surface are [9].

$$
\begin{aligned}
\varepsilon_x &= \varepsilon_{xm} + zk_x \\
\varepsilon_\theta &= \varepsilon_{\theta m} + zk_\theta \\
\gamma_{x\theta} &= \gamma_{x\theta m} + zk_{x\theta}
\end{aligned}
\tag{3.4-1}
$$

where ε_{ij} are the normal strains, γ_{ij} are the shear strains, and k_{ij} are the curvatures. The subscript m refers to the strain at the middle surface of the shell. The indices x and θ refer to the axial and circumferential directions, respectively. The general strain displacement relations may be simplified according to the Sanders assumption as follows [9].

$$
\begin{aligned}
\varepsilon_{xm} &= u_{,x} + \tfrac{1}{2}w_{,x}^2 \\
\varepsilon_{\theta m} &= (v_{,\theta} + w)/R + (v - w_{,\theta})^2/2R^2 \\
\gamma_{x\theta m} &= u_{,\theta}/R + v_{,x} + w_{,x}(w_{,\theta} - v)/R \\
k_x &= -w_{,xx} \\
k_\theta &= (v_{,\theta} - w_{,\theta\theta})/R^2 \\
k_{x\theta} &= (v_{,x} - 2w_{,x\theta})/2R
\end{aligned}
\tag{3.4-2}
$$

where u, v, and w are the axial, circumferential, and lateral deflections of the shell, respectively, and the index $(,)$ indicates partial derivative.

The functional of total potential energy, including the membrane, bending, and thermal strain energies are written and the Euler equation is applied to the functional to obtain its stationary value, which corresponds to the Donnell nonlinear equilibrium equation

$$
\begin{aligned}
RN_{x,x} + N_{x\theta,\theta} &= 0 \\
RN_{x\theta,x} + N_{\theta,\theta} &= 0 \\
RM_{x,xx} + 2M_{x\theta,x\theta} + M_{\theta,\theta\theta}/R - N_\theta - [RN_x\beta_{x,x} + N_{x\theta}(R\beta_{\theta,x} \\
+ \beta_{x,\theta}) + N_\theta\beta_{\theta,\theta}] &= 0
\end{aligned}
\tag{3.4-3}
$$

Here, N_{ij} are the forces per unit length of shell, M_{ij} are the moments per unit length of shell, $\beta_x = -w_{,x}$ and $\beta_\theta = (v - w_{,\theta})/R$ are the rotations along the x and θ-directions, respectively. The definitions of forces N_{ij} and moments M_{ij} for the Donnell theory follows the Love-Kirchhoff hypothesis and are based on the first order shell theory [9].

The stability equations are obtained by consideration of the second variation of the total functional of strain energy. The displacement components are related to the terms representing the stable equilibrium and the terms of neighboring state. Accordingly, the forces N_{ij} and the moments M_{ij} are divided into two terms representing the stable equilibrium and the neighboring state, and through the linear strain-displacement relations the expression for total potential function is obtained. This expression through the Taylor expansion, results in the sum of first and second variations of the total potential energy function. The proper stationary value of the second variation of the total potential function, through the Euler equations, is the stability equations [9,17].

By the assumptions of this paper, the stability equations for thin cylindrical shells under thermoelastic loading condition are [19].

$$
\begin{aligned}
RN_{x1,x} + N_{x\theta1,\theta} &= 0 \\
RN_{x\theta1,x} + N_{\theta1,\theta} &= 0 \\
RM_{x1,xx} + 2M_{x\theta1,\theta} + M_{\theta1,\theta\theta}/R - N_{\theta1} - RN_{x0}\beta_{x1,x} - RN_{x\theta0}\beta_{\theta1,x} \\
- N_{x\theta0}\beta_{x1,\theta} - N_{\theta0}\beta_{\theta1,\theta} &= 0
\end{aligned}
\tag{3.4-4}
$$

In Eqs. (3.4-4), terms with the subscript (0) are related to the state of equilibrium and terms with the subscript (1) are those characterizing the state of stability. It is further noticed that while the equilibrium equations are nonlinear, the stability equations are linear. The terms with the subscript (0) are the solution of the equilibrium equation for the given load.

The forces and moments related to the state of equilibrium are given in terms of the strains and temperature through the Hooke's law as [19].

$$N_{x_0} = C(\varepsilon_{x_0} + \nu\varepsilon_{\theta_0}) - E\alpha T_0/(1-\nu)$$
$$N_{\theta_0} = C(\varepsilon_{\theta_0} + \nu\varepsilon_{x_0}) - E\alpha T_0/(1-\nu)$$
$$N_{x\theta_0} = \frac{C(1-\nu)}{2}\gamma_{x\theta_0}$$
$$M_{x_0} = D(k_{x_0} + \nu k_{\theta_0}) - E\alpha T_1/(1-\nu)$$
$$M_{\theta_0} = D(k_{\theta_0} + \nu k_{x_0}) - E\alpha T_1/(1-\nu)$$
$$M_{x\theta_0} = D(1-\nu)k_{x\theta_0}$$

$$(3.4\text{-}5)$$

where $C = Eh/(1-\nu^2)$, $D = Eh^3/12(1-\nu^2)$, and

$$T_0 = \int_{-h/2}^{h/2} T\,dz \qquad T_1 = \int_{-h/2}^{h/2} Tz\,dz \qquad (3.4\text{-}6)$$

The forces and moments related to the state of stability are

$$N_{x1} = C(\varepsilon_{x1} + \nu\varepsilon_{\theta1})$$
$$N_{\theta1} = C(\varepsilon_{\theta1} + \nu\varepsilon_{x1})$$
$$N_{x\theta1} = N_{\theta x1} = \frac{C(1-\nu)}{2}\gamma_{x\theta1}$$
$$M_{x1} = D(k_{x1} + \nu k_{\theta1})$$
$$M_{\theta1} = D(k_{\theta1} + \nu k_{x1})$$
$$M_{x\theta1} = M_{\theta x1} = D(1-\nu)k_{x\theta1}$$

$$(3.4\text{-}7)$$

4.3   Thermal buckling load, Wan-Donnell model

The Wan-Donnell model for the radial imperfection is

$$w^* = \frac{K-1}{2}w \qquad (3.4\text{-}8)$$

where the coefficient K is a constant value $0 \leq K \leq 1$. The value of $K = 1$ represents a perfect shell. The imperfection w^* is thus defined as a function of w, the lateral deflection of the cylindrical shell. Considering an axisymmetric imperfection, due to the dependency of w to w^*, the lateral deflection w must be assumed to be axisymmetric. This assumption results in an axisymmetric buckling mode of the cylindrical shell.

Considering a change of variable $y = R\theta$, the total strain-displacement relations (3.4-2) for the axisymmetric shell deformation, where $w_{,\theta} = 0$, are reduced to

$$\varepsilon_{xm}^t = u_{,x} + \tfrac{1}{2}w_{,x}^2$$
$$\varepsilon_{ym}^t = (w + Rv_{,y})/R + v^2/2R^2 \qquad (3.4\text{-}9)$$
$$\gamma_{xym}^t = v_{,x} + u_{,y} - w_{,x}v/R$$

The initial strains due to the initial axisymmetric imperfection w^* are

$$\varepsilon_{xm}^* = \frac{1}{2}w_{,x}^{*2} \qquad \varepsilon_{ym}^* = \frac{w^*}{R} \qquad \gamma_{xym}^* = 0 \qquad (3.4\text{-}10)$$

The net strains are obtained by replacing w by $w + w^*$ in Eqs. (3.4-9) and subtracting Eqs. (3.4-10) to yield

$$\begin{aligned}
\varepsilon_{xm} &= u_{,x} + \tfrac{1}{2}Kw_{,x}^2 \\
\varepsilon_{ym} &= \tfrac{w}{R} + v_{,y} \\
\gamma_{xym} &= v_{,x} + u_{,y}
\end{aligned} \qquad (3.4\text{-}11)$$

where in the derivation of Eqs. (3.4-11) the term v^2 is assumed to be small compared to the other terms and is thus neglected.

The equilibrium equations (3.4-3) may be written in terms of the displacement components. Considering the above assumptions for the axisymmetric deformation, the third equation reduces to

$$Dw_{,xxxx} + \frac{N_y}{R} - N_x w_{,xx} = 0 \qquad (3.4\text{-}12)$$

Substituting

$$w = w_0 + w^* \qquad (3.4\text{-}13)$$

into Eqs. (3.4-12), yields

$$Dw_{0,xxxx} + \frac{1}{R}N_{y0} - N_{x0}(w_{0,xx} + w_{,xx}^*) = 0 \qquad (3.4\text{-}14)$$

Considering a cylindrical shell of simply supported edge conditions and under axisymmetric thermal loading, $N_{y0} = N_{xy0} = 0$, and Eq. (3.4-14) reduces to

$$Dw_{0,xxxx} - N_{x0}w_{0,xx} = N_{x0}w_{,xx}^* \qquad (3.4\text{-}15)$$

This equation describes the prebuckling lateral displacement of the shell and is used in the literature to obtain the prebuckling lateral deflection of the shell for imperfection analysis of cylindrical shells based on Wan-Donnell model.

To obtain the stability equations based on the Wan-Donnell imperfection assumption, the force summation method may be used. The deflection w and forces N_x and N_y are related to their values at stable condition, shown by the subscript (0), and the neighboring state shown by the subscript (1), as given

$$\begin{aligned} w &= w_0 + w^* + w_1 \\ N_x &= N_{x0} + N_{x1} \\ N_y &= N_{y0} + N_{y1} \end{aligned}$$

(3.4-16)

Substituting Eqs. (3.4-16) into (3.4-12), ignoring the small terms of higher order than one, and replacing N_{x1} and N_{y1} by the Airy stress function $\Phi_{,yy}$ and $\Phi_{,xx}$, respectively, yield

$$Dw_{1,xxxx} + \frac{1}{R}\Phi_{,xx} - (N_{x0}w_{1,xx} + \Phi_{,yy}w_{0,xx} + \Phi_{,yy}w^*_{,xx}) = 0 \quad (3.4\text{-}17)$$

Equation (3.4-17) is called the Wan-Donnell stability equation of imperfect cylindrical shell. This equation includes two dependent functions w and Φ.

To obtain a second equation relating the dependent functions w and Φ, the compatibility equation may be used, as follows

$$\varepsilon_{mx,yy} + \varepsilon_{my,xx} = \gamma_{mxy,xy} \tag{3.4-18}$$

Substituting for the strains from Eqs. (3.4-2) and using the Hooke's Law, the compatibility equation in terms of the Airy stress function reduces to the following equation

$$\nabla^4 \Phi = \frac{Eh}{R} w_{1,xx} \tag{3.4-19}$$

Equations (3.4-15), (3.4-17), and (3.4-19) are the basic equations used to obtain the critical buckling loads of imperfect cylindrical shells. Based on these equations thermal buckling loads of imperfect cylindrical shells of isotropic and homogeneous material are determined.

A. Uniform temperature rise

A thin cylindrical shell of thickness h and length L is considered. The ends are assumed to be simply supported, where the boundary conditions at $x = 0$ and $x = L$ are

$$w = w_{,xx} = 0 \tag{3.4-20}$$

The shell is initially at uniform temperature T_i and is raised to a uniform temperature T_f, such that the temperature rise is $\Delta T = T_f - T_i$. The axial force developed in the shell due to the temperature rise ΔT is

$$N_{x0} = \frac{-E\alpha h}{1-\nu}\Delta T = \beta\Delta T \qquad (3.4\text{-}21)$$

where $\beta = -E\alpha h/(1-\nu)$. The lateral imperfection of the shell may be assumed $w^* = \frac{(K-1)\zeta}{2h}\sin\frac{m\pi x}{L}$, where ζ is the imperfection amplitude. Substituting N_{x0} from Eq. (3.4-21) into Eq. (3.4-15) and assuming a solution in the form $w_0 = A\sin(m\pi x/L)$, which satisfies the simply supported edge conditions at $x = 0$ and L, the constant A is obtained and the final approximate solution of Eq. (3.4-14) is

$$w_0 = \frac{-\beta\Delta T(K-1)\zeta}{2D\frac{m^2\pi^2}{L^2} + 2\beta\Delta T}\sin\frac{m\pi x}{L} \qquad (3.4\text{-}22)$$

Substituting w_0 from Eq. (3.4-22) into Eqs. (3.4-17) and (3.4-19) yields

$$S\overline{w}_{1,xxxx} + \frac{1}{hR}\psi_{,xx} - \left[\frac{\beta'\Delta T}{h^2}\overline{w}_{1,xx} + (A' - K_1)\psi_{,yy}\frac{m^2\pi^2}{L^2}\sin\frac{m\pi x}{L}\right] = 0$$
$$(3.4\text{-}23)$$

$$\nabla^4\psi = \frac{1}{hR}\overline{w}_{1,xx} \qquad (3.4\text{-}24)$$

where
$$\overline{w}_1 = \frac{w_1}{h} \qquad S = \frac{1}{12(1-\nu^2)} \qquad A' = \frac{A}{h}$$

$$(3.4\text{-}25)$$

$$\psi = \frac{\Phi}{Eh^3} \qquad \beta' = \frac{\beta}{Eh} \qquad K_1 = \frac{(K-1)\zeta}{2h}$$

To solve the system of Eqs. (3.4-23) and (3.4-24), with the consideration of the boundary conditions, the approximate solutions are assumed

$$\begin{array}{ll}\overline{w}_1 = a\sin\frac{m\pi x}{L} & \\ \psi = b\sin\frac{m\pi x}{L}\cos\frac{y}{R} & \quad 0 \le x \le L \ , \ \ 0 \le y \le 2\pi R\end{array} \qquad (3.4\text{-}26)$$

where a and b are constant coefficients and $y = R\theta$. The system of Eqs. (3.4-23) and (3,4-24) are made orthogonal with respect to the approximate solutions (3.4-26) according to the Galerkin method

$$\int_x \int_y (S\overline{w}_{1,xxxx} + \tfrac{1}{hR}\psi_{,xx} - \left[\tfrac{\beta'\Delta T}{h^2}\overline{w}_{1,xx} + (A' - K_1)\psi_{,yy}\tfrac{m^2\pi^2}{L^2} \sin\tfrac{m\pi x}{L} \right]) \times$$
$$\sin\tfrac{m\pi x}{L} \cos\tfrac{y}{R}\,dxdy = 0 \qquad (3.4\text{-}27)$$
$$\int_x \int_y (\nabla^4\psi - \tfrac{1}{hR}\overline{w}_{1,xx}) \sin\tfrac{m\pi x}{L}\,dxdy = 0$$

The determinant of the system of Eqs. (3.4-27) for the coefficients a and b is calculated and solved for ΔT to give

$$\beta\Delta T = -(\frac{D}{2}\lambda^2 + \frac{3RD\lambda^3 L}{16H}) \qquad (3.4\text{-}28)$$

where $\lambda = \frac{m\pi x}{L}$ and $H = \frac{(K-1)\zeta}{2}$. For short cylindrical shells, the buckling waves appear in short wave lengths, or large λ. Therefore, in the derivation of Eq. (3.4-28) λ is ignored compared to λ^2 and λ^3. The lowest critical thermal load is associated with $m = 1$, the first mode of axial buckling. The critical buckling thermal load associated with uniform temperature rise is then

$$\boxed{\Delta T_{crit} = \left(\frac{h}{R}\right)^2 \frac{1}{\alpha} \left(\frac{0.079}{B^2} + \frac{0.093R}{HB^2}\right)} \qquad (3.4\text{-}29)$$

where $B = \frac{L}{2R}$ and $\nu = 0.3$

B. Radial temperature difference

Consider a linear radial temperature distribution across the shell thickness as

$$T(z) = \frac{\Delta T}{h}(z + \frac{h}{2}) \qquad -\frac{h}{2} \le z \le \frac{h}{2} \qquad (3.4\text{-}30)$$

The prebuckling axial load developed in the shell due to this temperature distribution is

$$N_{x0} = \frac{-E\alpha h}{1 - \nu}\frac{\Delta T}{2} = \beta\frac{\Delta T}{2} \qquad (3.4\text{-}31)$$

Comparing Eqs. (3.4-21) and (3.4-31), the critical buckling thermal load associated with the radial temperature rise for $\nu = 0.3$ is

$$\Delta T_{crit} = (\tfrac{h}{R})^2 \tfrac{1}{\alpha} (\tfrac{0.158}{B^2} + \tfrac{0.186R}{HB^2})$$

(3.4-32)

C. Axial temperature difference

Consider a linear axial temperature distribution across the shell length as

$$T(x) = \frac{\Delta T x}{L} \qquad\qquad 0 \leq x \leq L$$

(3.4-33)

the prebuckling axial load developed in the shell due to this temperature distribution is

$$N_{x0} = \frac{-E\alpha h}{1-\nu} \Delta T \frac{x}{L} = \beta \Delta T \frac{x}{L}$$

(3.4-34)

With the same method described in part (I-A) and for $\nu = 0.3$, the critical prebuckling thermal load associated with the axial temperature rise is

$$\Delta T_{crit} = (\tfrac{h}{R})^2 \tfrac{1}{\alpha} (\tfrac{0.158}{B^2} + \tfrac{0.186R}{HB^2})$$

(3.4-35)

The results of the critical buckling radial and axial temperature differences are identical.

4.4   Thermal buckling load, Koiter model

The Koiter model for the imperfection of cylindrical shell is axisymmetric and is assumed as [9]

$$w^* = -\mu h \cos \frac{m\pi x}{L} \qquad , \qquad -\frac{L}{2} \leq x \leq +\frac{L}{2}$$

(3.4-36)

In this relation μh represents the imperfection amplitude. In comparison with the Wan-Donnell imperfection model which does not consider any definite form for w^*, as w^* is assumed to be a function of w, the Koiter model considers a definite form for w^* as given by Eq. (3.4-36). Since w

is independent of w^* in the Koiter imperfection model, w is not assumed axisymmetric and may have any general form. Assuming a change of variable $y = R\theta$, Eqs. (3.4-2) for the total strains are written as

$$\begin{aligned}
\varepsilon^t_{xm} &= u_{,x} + \tfrac{1}{2}w^2_{,x} \\
\varepsilon^t_{ym} &= (w + Rv_{,y})/R + (v - Rw_{,y})^2/2R^2 \\
\gamma^t_{xym} &= v_{,x} + u_{,y} + w_{,x}(w_{,y} - \tfrac{v}{R})
\end{aligned} \qquad (3.4\text{-}37)$$

The initial strains due to the initial imperfection w^* are

$$\begin{aligned}
\varepsilon^*_{xm} &= \tfrac{1}{2}w^{*2}_{,x} \\
\varepsilon^*_{ym} &= \tfrac{w^*}{R} + \tfrac{1}{2}w^{*2}_{,y} \\
\gamma^*_{xym} &= w^*_{,x}w^*_{,y}
\end{aligned} \qquad (3.4\text{-}38)$$

The net strains are obtained by replacing $(w + w^*)$ for w in Eqs. (3.4-37) and subtracting the resulting relations from Eqs. (3.4-38) gives

$$\begin{aligned}
\varepsilon_{xm} &= u_{,x} + \tfrac{1}{2}w^2_{,x} + w_{,x}w^*_{,x} \\
\varepsilon_{ym} &= v_{,y} + \tfrac{w}{R} + \tfrac{1}{2}w^{*2}_{,y} \\
\gamma_{xym} &= u_{,y} + v_{,x} + w^*_{,x}w_{,y} + w_{,x}w_{,y}
\end{aligned} \qquad (3.4\text{-}39)$$

In these equations the term v^2 is assumed to be small compared to the other terms and is neglected. Equations (3.4-39) may be compared with Eqs. (3.4-11) which were obtained for the Wan-Donnell model.

The third of equilibrium equations (3.4-3) in terms of the displacement components are

$$D\nabla^4 w + \frac{N_y}{R} - (N_x w_{,xx} + N_y w_{,yy} + 2N_{xy}w_{,xy}) = 0 \qquad (3.4\text{-}40)$$

Substituting $w = w_0 + w^*$ in Eq. (3.4-40) gives

$$D\nabla^4 w_0 + \frac{N_{y0}}{R} - [N_{x0}(w_0 + w^*)_{,xx} + N_{y0}w_{0,yy} + 2N_{xy0}w_{0,xy}] = 0$$

$$(3.4\text{-}41)$$

Koiter considered the following assumptions for the prebuckling displacements

$$u_0 = u_0(x) \ , \quad v_0 = 0 \ , \quad w_0 = w_0(x) \qquad (3.4\text{-}42)$$

Considering a cylindrical shell of simply supported edge conditions and under axisymmetric thermal loading, $N_{y_0} = N_{xy_0} = 0$, Eq. (3.4-41) reduces to

$$Dw_{0,xxxx} - N_{x_0}(w_0 + w^*)_{,xx} = 0 \qquad (3.4\text{-}43)$$

To obtain the stability equation, Eqs. (3.4-16) are subtracted into Eq. (3.4-40). Neglecting the small terms of higher order than one, and replacing N_{x1} and N_{y1} by the Airy stress function $\Phi_{,yy}$ and $\Phi_{,xx}$, respectively, yields

$$D\nabla^4 w_1 + \frac{1}{R}\Phi_{,xx} - (N_{x_0}w_{1,xx} + \Phi_{,yy}w_{0,xx} + \Phi_{,yy}w^*_{,xx}) = 0 \quad (3.4\text{-}44)$$

Equation (3.4-44) is the stability equation based on the Koiter model and is a function of w_1 and Φ.

To obtain a second equation relating w_1 and Φ, the compatibility equation is used. Similar procedure used to obtain Eq. (3.4-19), results into the following compatibility equation based on the Koiter model

$$\nabla^4\Phi = Eh\left(\frac{w_{1,xx}}{R} - w_{0,xx}w_{1,yy} - w^*_{,xx}w_{1,yy}\right) \qquad (3.4\text{-}45)$$

Equations (3.4-43), (3.4-44) and (3.4-45) are used in combination to obtain the thermal buckling loads of the imperfect cylindrical shell.

A. Uniform temperature rise

A thin cylindrical shell of thickness h and length L is considered. Assuming the coordinates system at the center of the shell, the simply supported edge conditions are

$$w = w_{,xx} = 0 \qquad\qquad x = \pm\frac{L}{2} \qquad (3.4\text{-}46)$$

A uniform temperature rise of $\Delta T = T_f - T_i$, produce an axial thermal force in the shell as given by Eq. (3.4-21). Substituting N_{x_0} from Eq. (3.4-21) into Eq. (3.4-43) and assuming a solution in the form $w_0 = A\cos\frac{m\pi x}{L}$, the constant A is obtained and the final approximate solution is

$$w_0 = \frac{\beta\Delta T\mu h}{D\frac{m^2\pi^2}{L^2} + \beta\Delta T}\cos\frac{m\pi x}{L} \qquad (3.4\text{-}47)$$

Substituting Eq. (3.4-47) into Eqs. (3.4-44) and (3.4-45) give

$$S\nabla^4\overline{w}_1 + \frac{1}{hR}\psi_{,xx} - \left[\frac{\beta'\Delta T}{h^2}\overline{w}_{1,xx} + (\mu h - A)\psi_{,yy}\frac{m^2\pi^2}{L^2}\cos\frac{m\pi}{L}x\right] = 0$$

$$\tag{3.4-48}$$

$$\nabla^4\psi - (\frac{1}{hR}\overline{w}_{1,xx} - \overline{w}_{0,xx}\overline{w}_{1,yy} - \overline{w}^*_{,xx}\overline{w}_{1,yy}) = 0 \tag{3.4-49}$$

where $\beta', S, \overline{w}_1$ and ψ are defined in Eq. (3.4-25). To solve the system of Eqs. (3.4-48) and (3.4-49), with the consideration of the boundary conditions (3.4-46), the approximate solutions are assumed

$$\overline{w}_1 = a\cos\frac{m\pi x}{L}\cos\frac{ny}{R}$$
$$-\frac{L}{2} \le x \le +\frac{L}{2} \ , \ \ 0 \le y \le 2\pi R \tag{3.4-50}$$
$$\psi = b\cos\frac{m\pi x}{L}\cos\frac{ny}{R}$$

where a and b are constant coefficients. Using the Galerkin method, we have

$$\int_x\int_y\left\{S\nabla^4\overline{w}_1 + \frac{1}{hR}\psi_{,xx} - \left[\frac{\beta'\Delta T}{h^2}\overline{w}_{1,xx} + (\mu h - A)\psi_{,yy}\frac{m^2\pi^2}{L^2}\cos\frac{m\pi x}{L}\right]\right\}$$
$$\times \cos\frac{m\pi x}{L}\cos\frac{ny}{R}dxdy = 0 \tag{3.4-51}$$

$$\int_x\int_y\left\{\nabla^4\psi - \left[\frac{1}{hR}\overline{w}_{1,xx} - \overline{w}_{0,xx}\overline{w}_{1,yy} - \overline{w}^*_{,xx}\overline{w}_{1,yy}\right]\right\}\cos\frac{m\pi x}{L}\cos\frac{ny}{R}dxdy = 0$$

$$\tag{3.4-52}$$

For the axial buckling with odd wave lengths, the system of Eqs. (3.4-51) and (3.4-52) are

$$a\left(S\frac{m^4\pi^4}{L^4} + S\frac{n^4}{R^4} + 2S\frac{m^2\pi^2n^2}{L^2R^2} + \beta'\Delta T + \frac{m^2\pi^2}{h^2L^2}\right)\frac{\pi RL}{2} \tag{3.4-53}$$

$$+b((\mu h - A)\frac{4m\pi^2n^2}{3LR} - \frac{m^2\pi^3}{2hL}) = 0$$

$$a\left(\frac{m^2\pi^2}{hRL^2} + A\frac{4m\pi^2n^2}{3hLR} - \mu\frac{4m\pi^2n^2}{3LR}\right) + b\left(\frac{m^4\pi^4}{L^4} + \frac{n^4}{R^4} + \frac{2m^2\pi^2n^2}{L^2R^2}\right)\frac{\pi RL}{2} = 0$$

$$\tag{3.4-54}$$

We let the determinant of the constant coefficients a and b to be zero, and obtain the buckling load as the lowest ΔT satisfying the equation obtained from the determinant of Eqs. (3.4-53) and (3.4-54). The results for $m = 1$ and different n-values are obtained and the curve associated with a value of n which provide minimum ΔT is selected as the critical buckling load.

B. Radial temperature difference

A linear radial temperature across the shell thickness is assumed as

$$T(z) = \frac{\Delta T}{h}(z + \frac{h}{2}) \tag{3.4-55}$$

The resulting axial thermal force is given by Eq. (3.4-31). Following the same procedure described in A - above, the critical thermal buckling load is obtained. It is found that the same results and curves given in A - above may be used by replacing ΔT by $\frac{\Delta T}{2}$.

C. Axial temperature difference

Considering a linear axial temperature along the shell length as

$$T(x) = \frac{\Delta T}{L}(x + \frac{L}{2}) \tag{3.4-56}$$

The axial thermal force developed in the shell is

$$N_{x_0} = \frac{\beta \Delta T}{L}(x + \frac{L}{2}) \tag{3.4-57}$$

Substituting Eq. (3.4-57) into Eq. (3.4-43), the approximate solution is obtained as

$$w_0 = \frac{\beta \Delta T \mu h}{2D\frac{m^2 \pi^2}{L^2} + \beta \Delta T} \cos \frac{m\pi x}{L} \tag{3.4-58}$$

Substituting Eqs. (3.4-57) and (3.4-58) into Eqs. (3.4-44) and (3.4-45) and applying the Galerkin method a set of two equations are obtained. Setting the determinant of constant coefficients a and b equal to zero, for $m = 1$ and different n-values the lowest ΔT associated with the thermal buckling load is obtained. It is found that the same results and curves obtained in A-above may be used by replacing ΔT by $\frac{\Delta T}{2}$.

4.5   Results and discussion

Thermal buckling analysis of imperfect short cylindrical shells based on Wan-Donnell model with simply supported edge conditions for the three types of thermal loading result into closed form solution. The thermal buckling loads are directly proportional to h/R and R and inversely proportional to the parameters α , B, and H. For both Wan-Donnell and the Koiter imperfection models, the thermal buckling load of the cylindrical shell for the axial and gradient through the thickness temperature difference are identical.

Figures (3.4-1) through (3.4-4) show the critical buckling temperature of a short thin cylindrical shell under uniform temperature rise based on the Wan-Donnell imperfection model in terms of the parameters H, B, h/R, and H/R. Figure (3.4-1) is a three dimensional plot of the uniform buckling temperature rise versus the parameters H and h/R. It is seen that the buckling temperature rise decreases by the increase of H and increases by the increase of h/R. In this figure $\alpha = 11.7 \times 10^{-6}1/^{\circ}C$ and $B = 2$ are considered. Figure (3.4-2) shows the same thermal buckling load versus the parameters H and B. The buckling temperature rise decreases with the increase of B. The numerical values used for this plot are $\alpha = 11.7 \times 10^{-6}1/^{\circ}C$ and $h/R = 0.002$. Figure (3.4-3) is the plot of buckling temperature versus H/R for different values of B. As B increases, the buckling temperature decreases. The variation of the buckling temperature versus H/R for different values of h/R are shown in Fig. (3.4-4). For a fixed value of H/R the curve associated with the smaller value of h/R gives the buckling temperature. Figure (3.4-3) is plotted for $\alpha = 11.7 \times 10^{-6}1/^{\circ}C$, $h/R = 0.002$, and $R = 1\ m$, and Fig. (3.4-4) is plotted for the same value of α, $R = 2\ m$ and $B = 2$.

The buckling temperature associated with the gradient through the thickness and the axial temperature differences, from Eq. (3.4-32) and (3.4-35), are obtained using Figs. (3.4-1) through (3.4-4) with the values of $\Delta T_{axial} = \Delta T_{radial} = 2\Delta T_{uniform}$.

The results of thermal buckling of a thin cylindrical shell based on the Koiter imperfection model is shown in Fig. (3.4-5). The buckling load related to the uniform temperature rise versus the imperfection amplitude μh for different values of n is shown in this figure. The lowest curve for ΔT is associated with $n = 1$. This figure is plotted for $L/2R = 2$, $R = 1m$, $\alpha = 11.7 \times 10^{-6}1/^{\circ}C$, and $h/R = 0.002$. Similar to the Wan-Donnell model, the buckling temperature difference is decreased with the increase of H. The buckling temperature associated with the gradient through the thickness and the axial temperature difference, from Eqs. (3.4-53) and (3.4-54) are obtained using Fig. (3.4-5) with the value of $\Delta T_{axial} = \Delta T_{radial} = 2\Delta T_{uniform}$

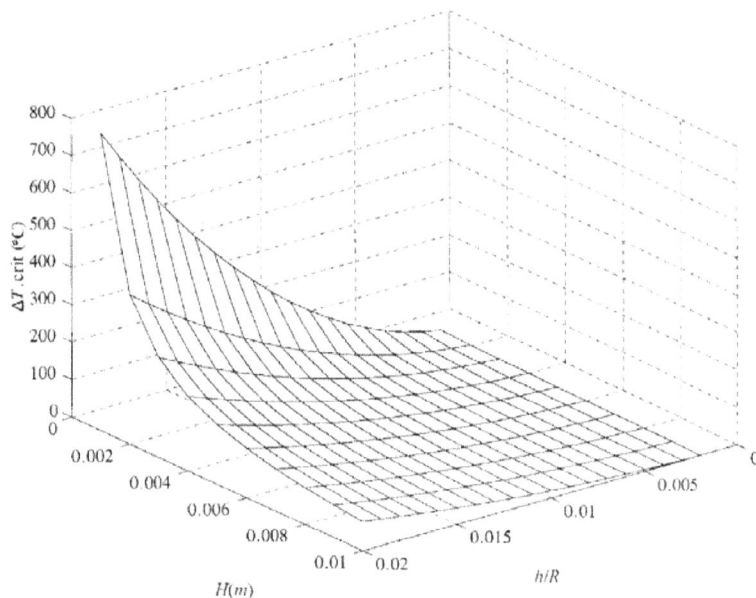

Figure 3.4-1: Variation of the critical uniform temperature rise based on the Wan-Donnell model versus H and h/r, $(H = (K-1)\xi/2)$.

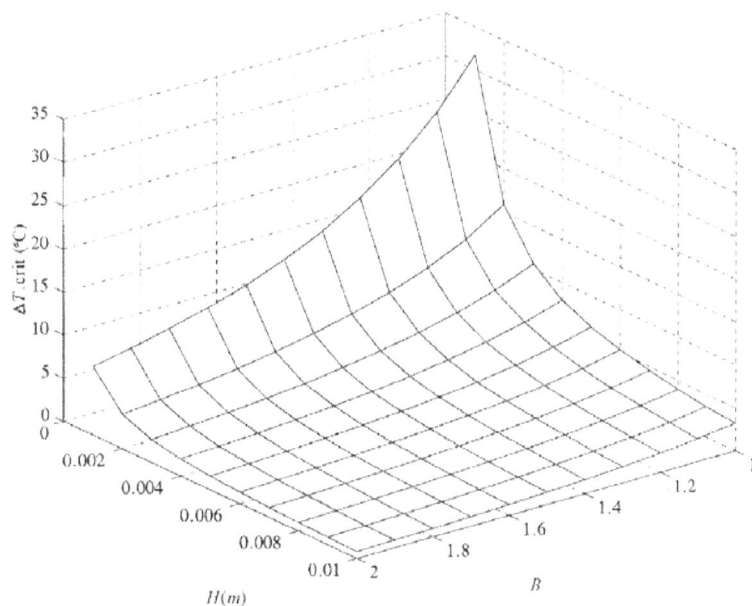

Figure 3.4-2: Variation of the critical uniform temperature rise based on the Wan-Donnell model versus H and B, $H = (K-1)\xi/2$, $B = L/2R$.

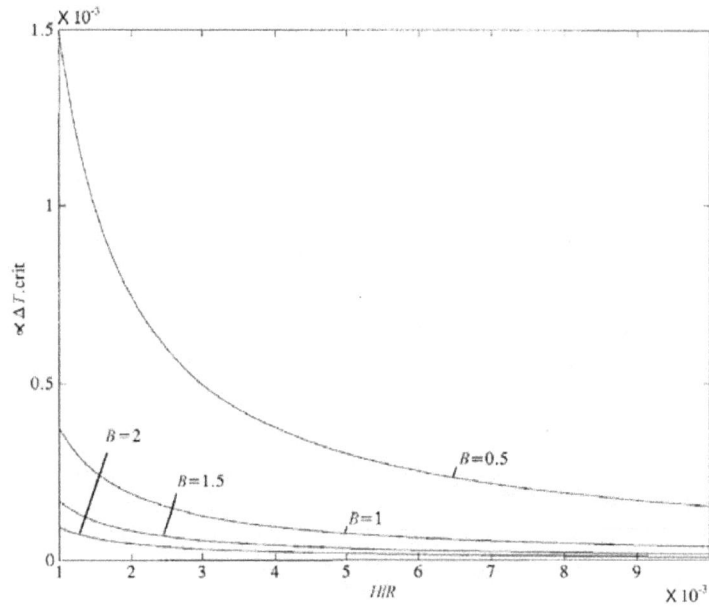

Figure 3.4-3: Variation of the critical uniform temperature rise based on the Wan-Donnell model versus H/R for different values of B, $H = (K-1)\xi/2$, $B = L/2R$.

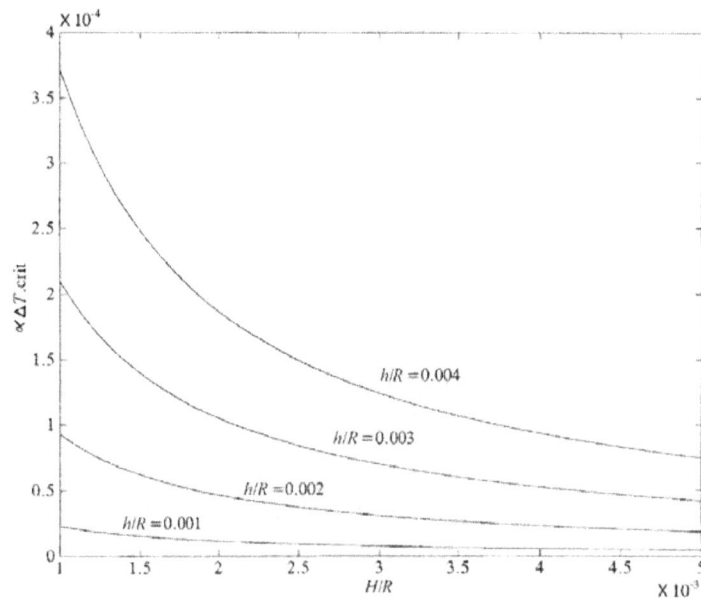

Figure 3.4-4: Variation of the critical uniform temperature rise based on the Wan-Donnell model versus H/R for different values of h/R, $H = (K-1)\xi/2$.

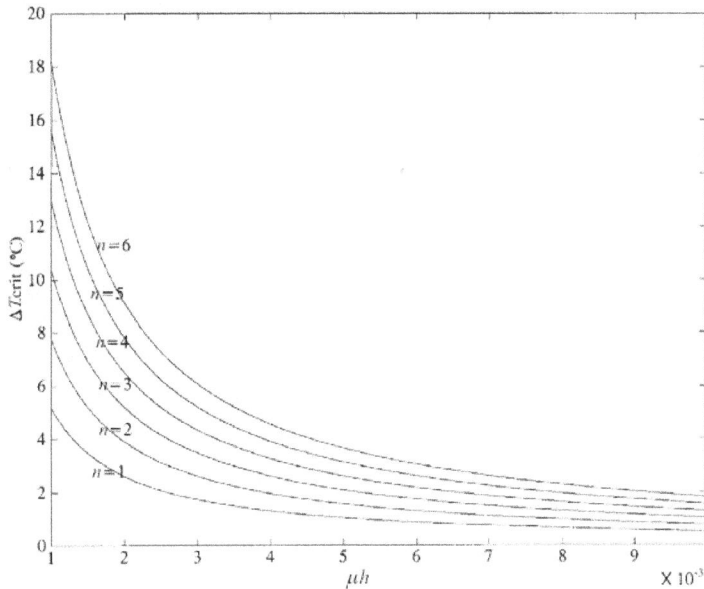

Figure 3.4-5: Variation of the critical uniform temperature rise based on the Koiter model versus μh for different values of n.

5 References

1 - Donnell, L.H., " Effect of Imperfection on Buckling of Thick Cylinders and Columns under Axial Compression ", Trans. ASME, J. Appl. Mech., Vol. 17, 1950.

2 - Donnell, L.H., " Effect of Imperfection on Buckling of Thin Cylinders under External Pressure ", Trans. ASME, J. Appl. Mech., Vol. 23, 1956.

3 - Donnell, L.H., " Effect of Imperfection on Buckling of Thin Cylinders with Fixed Edges under External Pressure ", Trans. ASME, J. Appl. Mech., Vol. 28, 1958.

4 - Donnell, L.H., *Beams, Plates, and Shells*, McGraw-Hill, New York, 1976.

5 - Koiter, W.T., *Theoretical and Applied Mechanics*, North Holland Pub. Co., New York, 1977.

6 - Koiter, W.T., " The Intrinsic Equations of Shell Theory with Some Applications ", in Nemat-Nasser (ed.), Mechanics Today, Vol. 5, Pergamon Press, Oxford, 1980.

7 - Flugge, W., *Stresses in Shells*, Springer Verlag, Berlin, 1973.

8 - Morley, L.S.D., " An Inmprovement on Donnell Approximation for Thin Walled Circular Cylinders ", J. Mech. Appl. Math., Vol. 12, PP. 89, 1959.

9 - Brush, D.O., and Almroth, B.O., *Buckling of Beams, Plates and Shells*, McGraw-Hill, New York, 1975.

10 - Johns, D.J., " Local Circumferential Buckling of Thin Circular Cylindrical Shells ", Collected Papers on Instability of Shell Structures, TN-D-1510, NASA, pp. 267-276, 1962.

11 - Chang, L.K., and Cord, M.F., " Thermal Buckling of Stiffened Cylindrical Shells ", Proc. AIAA/ASME 11th. Structures, Structural Dynamics, and Material Conf., pp. 260-272, April 1970.

12 - Bushnell, D., " Analysis of Ring-Stiffened Shells of Revolution Under Combined Thermal and Mechanical Loadings ", AIAA J., Vol. 9, No. 3, pp. 401-410, 1971.

13 - Lukasiewicz, S., " Thermal Stresses in Shells ", in Hetnarski, R.B. (ed.), *Thermal Stresses III*, North Holland, Amsterdam, 1980.

14 - Eslami, M.R., and Shariyat, M., " Elastic-Plastic Buckling of Cylindrical Shells ", Proc. ASME-ESDA Conf., Istanbul, Turkey, 1992.

15 - Eslami, M.R., and Shariyat, M., " Variational Approach to Elastic Plastic Buckling of Cylindrical Shells ", Proc. 7th. Int. Conf. on Pressure Vessel Tech., Vol. 1, Design, Analysis, Materials, May 31-June 4, Germany, pp. 241-253, 1992.

16 - Sanders, J.L., " Nonlinear Theories for Thin Shells ", Q. Appl. Mech., 1963.

17 - Langhaar, H.L., *Energy Methods in Applied Mechanics*, John Wiley, New York, 1962.

18 - Eslami, M.R., and Shariyat, M., " Elastic-Plastic and Creep Analysis of Imperfect Cylinders Under Mechanical and Thermal Loadings ", Trans. ASME, J. Pressure Vessel Tech., Vol. 118, No. 11, Nov. 1996.

19 - Eslami, M.R., Ziaii, A.R., and Ghorbanpour, A. " Thermoelastic Buckling of Thin Cylindrical Shells Based on Improved Stability Equations ", J. Thermal stresses, Vol. 19, No. 4, pp. 299-315, 1996.

20 - Radhamohan, S.K., and Venkataramana, J., " Thermal Buckling of Orthotropic Cylindrical Shells ", AIAA Journal, Vol. 13, No. 3, 1975.

21 - Herakovich, C.T., and Johnson, E.R., " Buckling of Composite Cylinders under Combined Compression and Torsion-Theoretical / Experimental Correlation ", Test Method for Fibrous Composites, ASTM, Tech. Pub. 734, pp. 341 - 360, 1981.

22 - Shaw, D., and Simitses, G.J., " Instability of Laminated Cylinders in Torsion ", Trans. ASME, J. Appl. Mech., Vol. 51, 1984.

23 - Leissa, A.W., " Buckling of Laminated Composite Plates and Shell Panels ", Air Force Wright Aeronautical Lab., AFWAL - TR - 85 - 3069, 1985.

24 - Kapania, R.K., " A Review on the Analysis of Laminated Shells ", Trans. ASME, J. Pressure Vessel Tech., Vol. 111, 1989.

25 - Kari Thangaratnam, R., Palaninathan, R., and Ramachandran, J., " Thermal Buckling of Laminated Composite Shells ", AIAA Journal, Vol. 28, No. 5, 1990.

26 - Soldatos, K.P., " Nonlinear Analysis of Transverse Shear Deformable Laminated Composite Cylindrical Shells. Part 22: Buckling of Axially Compressed Cross-Ply Circular and Oval Cylinders ", Trans. ASME, J. Pressure Vessel Tech., Vol. 114, 1992.

27 - Savoia, M., and Reddy, J.N., " Post-Buckling Behavior of Stiffened Cross-Ply Cylindrical Shells ", Trans. ASME, J. Appl. Mech., Vol. 61, 1994.

28 - Birman, V., and Bert, C.W., " Buckling and post-Buckling of Composite Plates and Shells Subjected to Elevated Temperature ", Trans. ASME, J. Appl. Mech., Vol. 60, 1993.

29 - Jones, R.M., *Mechanics of Composite Materials* , McGraw-Hill, New York, 1975.

30 - Eslami, M.R., and Rafeeyan, M., " Thermal and Mechanical Buck-

ling of Conical Shells", Proc. Joint ASME-ICPVT-8 Conference, Montreal, Canada, July 1996.

31 - Eslami, M.R., Javaheri, M.R., Thermal and Mechanical Buckling of Composite Cylindrical Shells, J. Thermal Stresses, Vol. 22, No. 6, pp. 527-545, Aug. 1999.

32 - Brush, M., and Singer, J. " Effect of Eccentricity of Stiffeners on the General Instability of Stiffened Cylindrical Shells under Hydrostatic Pressure", Journal of Mech. Eng., Vol. 5, No.1, pp. 23-27, 1963.

33 - Eslami, M.R., and Javaheri, M.R., " Thermal and Mechanical Buckling of Composite Cylindrical Shells", J. Thermal Stresses, Vol. 22, Aug. 1999.

34 - Soldatos, K.P., " Nonlinear Analysis of Transverse Shear Deformable Laminated Composite Cylindrical Shells-Part II: Buckling of Axially Compressed Cross- Ply Circular and Oval Cylinders ", ASME Journal of Pressure Vessel Technology, Vol. 144, pp. 110-114, Feb. 1992.

35 - Eslami, M.R., and Shariyat, M., " A High-Order Theory for Dynamic Buckling and Postbuckling Analysis of Laminated Cylindrical Shell", ASME Journal of Pressure Vessel Technology, Vol. 121, pp. 94-102, Feb. 1999.

36 - Eslami, M.R., and Shariyat, M., " Layerwise Theory for Dynamic Buckling and Postbuckling of Laminated Composite Cylindrical Shells ", J. AIAA, Vol. 36, pp. 1874 - 1882, Sep. 1998.

37 - Eslami, M.R., and Shariyat, M., " Dynamic Buckling and Postbuckling of Imperfect Orthotropic Cylindrical Shells under Mechanical and Thermal Loads, Based on the Three-Dimensional Theory of Elasticity ", ASME of Journal Applied Mechanics, Vol. 66, pp. 476-484, June 1999.

Chapter 4

Buckling of Spherical Shells

1 Thermal Buckling of Spherical Shells

1.1 Introduction

Spherical shells are used as vessels or as vessel's end closures. The stability of spherical shells under applied loads is an important control in the design stage. Numerous papers have been written on the buckling analysis of spherical shells under mechanical loads. Morris [1] and Kaplan [2] have given reviews of these papers. Zolley's linear buckling analysis of a complete sphere under uniform internal pressure is described in Timoshenko and Gere [3]. Elastic postbuckling behavior of spherical shells under external pressure was rather strikingly introduced by Karman and Tsien [4]. Budiansky [5] studied the buckling of clamped shallow spherical shells. He used numerical method to solve the shell stability equation. His work involved a highly nonlinear axisymmetric prebuckling solution followed by a linear eigenvalue problem. Weinitschke [6] analyzed the asymmetric buckling of thin shallow spherical shells. He used a power series method to solve the asymmetrical stability equations and obtained extensive numerical results. Huang [7] studied unsymmetrical buckling of thin shallow spherical shells. Before him, it had been found that there are discrepancies between the experimental observations of the buckling pressure and the theoretical predictions based on the axisymmetric buckling theory. The buckling pressures obtained by this theory are too high as compared to the experimental results. He found that initial imperfections of the shell and unsymmetrical buckling are the sources of this discrepancy. His work reduced the gap between experiment and theory and was in serious disagreement with Weinitschke's work. However, some test data obtained using very accurately manufactured shells are in reasonable agreement with the theoretical predictions [8]. Penning and

Thurston [9] studied the initial postbuckling analysis of shallow spherical shells under concentrated load. Their analysis showed that the shell retains its load carrying capacity as it makes the transition to asymmetric behavior. For the case of a concentrated load at the apex, Bushnell [10] analyzed buckling and Fitch [11] studied the postbuckling as well as the buckling behavior. They found that if the shell thickness is less than a certain critical value, asymmetric bifurcation will occur at a smaller load than that required for axisymmetric snap buckling. Also, Fitch and Budiansky [12] obtained solutions for the case in which a clamped spherical cap is uniformly loaded over a circular region concentrated at the apex, but not extending over the entire cap. They found that, as the area of the loaded region increases the deformation mode corresponding to loss of stability changes from asymmetric bifurcation to axisymmetric snap-through and then back to asymmetric bifurcation. Akkas [13] used the freezing in time technique to study the asymmetric buckling behavior of spherical caps under uniform step pressures.

In some cases shells are exposed to thermal loading. Thermal buckling of shells and plates have received less attention in the literature compared to the mechanical buckling [14-26]. It is the aim of this section to study thermal stability of thin spherical shells. Here, the thermal buckling of thin spherical shells based on the deep and shallow shell theories with the assumptions of isotropic material and geometrically perfect shell is considered. The Sanders and D.M.V. nonlinear strain-displacement relations are used. The thermal buckling loads of spherical shells under uniform temperature change and radial temperature difference are calculated. The results are compared with the conventional finite element programs and other known data in the literature.

1.2   Basic equations

A simply supported thin wall spherical shell as shown in Fig. (4.1-1) is considered. The variables ϕ and θ represent the latitude and circumferential coordinates, respectively, and the shell middle surface is used as the reference surface. The radius of the shell middle surface is denoted by R and the shell thickness is denoted by h. Points not on the middle surface are located by an additional radial coordinate z which is the distance of the point on shell wall from the corresponding point on the middle surface. The sign of z is taken to be positive outward. The coordinates ϕ, θ and z form an orthogonal coordinate system.

The normal and shear strains at a point with distance z from the middle plane of the spherical shell may be expressed by Eq. (4.1-1).

$$\varepsilon_\phi = \varepsilon_{\phi m} + z k_\phi$$

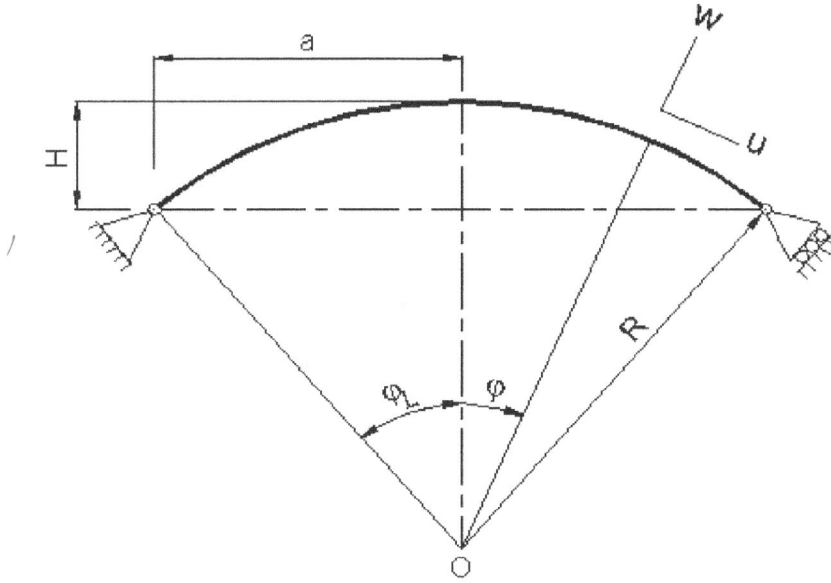

Figure 4.1-1: Geometry of a simply supported thin spherical shell.

$$\varepsilon_\theta = \varepsilon_{\theta m} + z k_\theta \tag{4.1-1}$$
$$\varepsilon_{\phi\theta} = \varepsilon_{\phi\theta m} + 2 z k_{\phi\theta}$$

where the subscript m refers to extensional and shearing strain components of the shell middle surface, k_ϕ and k_θ are flexural curvature changes of the surface in ϕ and θ-directions, respectively, and $k_{\phi\theta}$ is flexural twist of the surface. These strains, curvatures, and twist depend to middle surface displacement components u, v, and w , the displacement components in ϕ ,θ, and z-directions, via nonlinear kinematic relations. The deep shell kinematic relations for thin spherical shells in moderately large range of deflection are given by Sanders as [27]

$$\varepsilon_{\phi m} = \frac{1}{R}(u_{,\phi} + w) + \frac{1}{2}\beta_\phi^2 \quad , \quad \beta_\phi = \frac{1}{R}(u - w_{,\phi})$$

$$\varepsilon_{\theta m} = \frac{1}{R\sin\phi}(\cos\phi\, u + v_{,\theta} + \sin\phi\, w) + \frac{1}{2}\beta_\theta^2 \quad , \quad \beta_\theta = \frac{1}{R}(v - \frac{w_{,\theta}}{\sin\phi})$$

$$\varepsilon_{\phi\theta m} = \frac{1}{R\sin\phi}(u_{,\theta} - \cos\phi\, v + \sin\phi\, v_{,\phi}) + \beta_\phi\beta_\theta \tag{4.1-2}$$

$$k_\phi = \frac{1}{R^2}(u_{,\phi} - w_{,\phi\phi})$$

$$k_\theta = \frac{1}{R^2\sin\phi}(\cos\phi\, u + v_{,\theta} - \cos\phi\, w_{,\phi} - \frac{w_{,\theta\theta}}{\sin\phi})$$

$$k_{\phi\theta} = \frac{1}{2R^2 \sin\phi}(u_{,\theta} - \cos\phi \, v + \sin\phi \, v_{,\phi} + 2\cot\phi \, w_{,\theta} - 2w_{,\phi\theta})$$

where $()_{,\phi}$ and $()_{,\theta}$ indicate partial derivatives with respect to the variables ϕ and θ, and β_ϕ, β_θ are the prebuckling rotations along the ϕ and θ coordinate directions, respectively.

Donnell-Mushtari-Vlasov (D.M.V.) assumed that the terms u and v in the prebuckling rotations β_ϕ and β_θ are of negligible influence for shell segments (curved plates) that are almost flat and for shells whose displacement components in the deformed configuration are rapidly varying functions of the shell coordinates [27]. They also assumed that transverse shear forces have negligible effect in equilibrium and stability equations and can be discarded [28]. This theory is called quasi-shallow shell theory. Further, it has been expected that the boundary conditions cannot play a critical role in buckling of the shells and only a shallow portion of the shells should be controlled for stability. This means that if the length of the shell meridian is large enough for the buckled wavelength to develop, the boundary conditions cannot affect the buckling wave [1]. Most researchers investigated the buckling of shells based on shallow shell theory due to its simple form for solution.

A thin spherical shell can be regarded as shallow if the ratio of its rise (H) at center of the shell to the base supporting radius (a) is less than 0.25 [7]. In other words, a spherical shell is shallow if supporting latitude angle ϕ_L is not greater than $\pi/6$. According to the shallow shell theory, the nonlinear D.M.V kinematic relations for spherical shells can be written as following which are simplified form of Eqs. (4.1-2)[27].

$$\varepsilon_{\phi m} = \frac{1}{R}(u_{,\phi} + w) + \frac{1}{2}\beta_\phi^2 \quad , \qquad \beta_\phi = -\frac{w_{,\phi}}{R}$$

$$\varepsilon_{\theta m} = \frac{1}{R\sin\phi}(\cos\phi \, u + v_{,\theta} + \sin\phi \, w) + \frac{1}{2}\beta_\theta^2 \quad , \qquad \beta_\theta = -\frac{w_{,\theta}}{R\sin\phi}$$

$$\varepsilon_{\phi\theta m} = \frac{1}{R\sin\phi}(u_{,\theta} - \cos\phi \, v + \sin\phi \, v_{,\phi}) + \beta_\phi\beta_\theta$$

$$k_\phi = -\frac{w_{,\phi\phi}}{R^2} \tag{4.1-3}$$

$$k_\theta = -\frac{1}{R^2 \sin\phi}(\cos\phi \, w_{,\phi} + \frac{w_{,\theta\theta}}{\sin\phi})$$

$$k_{\phi\theta} = \frac{1}{R^2 \sin\phi}(\cot\phi \, w_{,\theta} - w_{,\phi\theta})$$

The stress-strain relations (Hooke's law) for thin spherical shells under thermal loading are

$$\sigma_\phi = \frac{E}{1-\nu^2}(\varepsilon_\phi + \nu\varepsilon_\theta) - \frac{E\alpha_T \Delta T}{1-\nu}$$

$$\sigma_\theta = \frac{E}{1-\nu^2}(\varepsilon_\theta + \nu\varepsilon_\phi) - \frac{E\alpha_T \Delta T}{1-\nu} \tag{4.1-4}$$

$$\sigma_{\phi\theta} = \frac{E}{2(1+\nu)}\varepsilon_{\phi\theta}$$

where E, ν, and α_T are the modulus of elasticity, Poisson's ratio, and linear thermal expansion coefficient, respectively. The material constants are assumed to be independent of the temperature and the temperature difference from reference temperature is considered as ΔT.

According to the first-order shell theory, the forces (N_{ij}) and the moments (M_{ij}) per unit length of the shell are related to the stresses through the following relations

$$N_\phi = \int_{-\frac{h}{2}}^{\frac{h}{2}} \sigma_\phi \, dz \quad , \quad N_\theta = \int_{-\frac{h}{2}}^{\frac{h}{2}} \sigma_\theta \, dz \quad , \quad N_{\phi\theta} = \int_{-\frac{h}{2}}^{\frac{h}{2}} \sigma_{\phi\theta} \, dz$$

$$M_\phi = \int_{-\frac{h}{2}}^{\frac{h}{2}} \sigma_\phi \, z \, dz \quad , \quad M_\theta = \int_{-\frac{h}{2}}^{\frac{h}{2}} \sigma_\theta \, z \, dz \quad , \quad M_{\phi\theta} = \int_{-\frac{h}{2}}^{\frac{h}{2}} \sigma_{\phi\theta} \, z \, dz \tag{4.1-5}$$

Introducing Eqs. (4.1-4) into Eq. (4.1-5) and carrying out the integration, gives the forces and moments in terms of the strains as

$$N_\phi = C(\varepsilon_{\phi m} + \nu\varepsilon_{\theta m}) - \frac{E\alpha_T T_1}{1-\nu}$$

$$N_\theta = C(\varepsilon_{\theta m} + \nu\varepsilon_{\phi m}) - \frac{E\alpha_T T_1}{1-\nu}$$

$$N_{\phi\theta} = C(\frac{1-\nu}{2})\,\varepsilon_{\phi\theta m} \tag{4.1-6}$$

$$M_\phi = D(k_\phi + \nu k_\theta) - \frac{E\alpha_T T_2}{1-\nu}$$

$$M_\theta = D(k_\theta + \nu k_\phi) - \frac{E\alpha_T T_2}{1-\nu}$$

$$M_{\phi\theta} = D(1-\nu)\,k_{\phi\theta}$$

where

$$T_1 = \int_{-\frac{h}{2}}^{\frac{h}{2}} \Delta T \, dz \quad , \quad T_2 = \int_{-\frac{h}{2}}^{\frac{h}{2}} \Delta T \, z \, dz \tag{4.1-7}$$

Here, the constants C and D are the membrane and flexural stiffness of the shell wall, respectively, and are defined as

$$C = \frac{Eh}{1-\nu^2} \quad , \quad D = \frac{Eh^3}{12(1-\nu^2)} \tag{4.1-8}$$

In accordance with the deep shell theory, transverse shear strains $\varepsilon_{\phi z}$ and $\varepsilon_{\theta z}$ shall also be taken into consideration in strain-stress relations. Their average value can be written as [6]

$$\varepsilon_{\phi z} = \frac{2}{1-\nu}\frac{Q_{\phi z}}{C} \qquad , \qquad \varepsilon_{\theta z} = \frac{2}{1-\nu}\frac{Q_{\theta z}}{C} \qquad (4.1\text{-}9)$$

where $Q_{\phi z}$ and $Q_{\theta z}$ are the transverse shear forces per unit length given as

$$
\begin{aligned}
Q_{\phi z} &= \frac{1}{R\sin\phi}[(\sin_\phi \; M_\phi)_{,\phi} + M_{\phi\theta,\theta} - \cos\phi \; M_\theta] \\
Q_{\theta z} &= \frac{1}{R}[M_{\theta,\theta} + (\sin_\phi \; M_{\phi\theta})_{,\phi} + \cos\phi \; M_{\phi\theta}] \qquad (4.1\text{-}10)
\end{aligned}
$$

Then, the average transverse shear stresses are determined as

$$\sigma_{\phi z} = \frac{Q_{\phi z}}{h} \qquad , \qquad \sigma_{\theta z} = \frac{Q_{\theta z}}{h} \qquad (4.1\text{-}11)$$

The equilibrium equations of spherical shells for both deep and shallow shell theories may be derived on the basis of stationary potential energy criterion. The total potential energy (π) of a shell subjected to thermal loading is the sum of membrane strain energy (π_m), bending strain energy (π_b), and thermal strain energy (π_T). The strain energies of a thin spherical shell made from the isotropic elastic material may be written as

$$
\begin{aligned}
\pi &= \pi_m + \pi_b + \pi_T \\
\pi_m &= \frac{C}{2}\int_\phi\int_\theta[\varepsilon_{\phi m}^2 + \varepsilon_{\theta m}^2 + 2\nu\varepsilon_{\phi m}\varepsilon_{\theta m} + (\frac{1-\nu}{2})\varepsilon_{\phi\theta m}^2]\; R^2\sin\phi\; d\theta d\phi \\
\pi_b &= \frac{D}{2}\int_\phi\int_\theta[k_\phi^2 + k_\theta^2 + 2\nu k_\phi k_\theta + 2(1-\nu)k_{\phi\theta}^2]\; R^2\sin\phi\; d\theta d\phi \quad (4.1\text{-}12) \\
\pi_T &= -\frac{\alpha_T E}{1-\nu}\int_\phi\int_\theta[(\varepsilon_{\phi m} + \varepsilon_{\theta m})T_1 + (k_\phi + k_\theta)T_2 + \alpha_T T_3]\; R^2\sin\phi\; d\theta d\phi
\end{aligned}
$$

where

$$T_3 = \int_{-\frac{h}{2}}^{\frac{h}{2}}(\Delta T)^2\; dz \qquad (4.1\text{-}13)$$

Collection of the various components of the total strain energy function gives

$$\pi = \int_\phi\int_\theta F(u,v,w,u_{,\phi},u_{,\theta},v_{,\phi},v_{,\theta},w_{,\phi},w_{,\theta},w_{,\phi\phi},w_{,\theta\theta},w_{\phi\theta})\; d\theta d\phi \qquad (4.1\text{-}14)$$

where

$$
\begin{aligned}
F =\ & \{[\varepsilon_{\phi m}^2 + \varepsilon_{\theta m}^2 + 2\nu\varepsilon_{\phi m}\varepsilon_{\theta m} + (\tfrac{1-\nu}{2})\varepsilon_{\phi\theta m}^2]\} \\
& + [k_\phi^2 + k_\theta^2 + 2\nu k_\phi k_\theta + 2(1-\nu)k_{\phi\theta}^2] \\
& + [(\varepsilon_{\phi m} + \varepsilon_{\theta m})T_1 + (k_\phi + k_\theta)T_2 + \alpha_T T_3]\}\, R^2 \sin\phi \quad (4.1\text{-}15)
\end{aligned}
$$

For equilibrium condition, the strain energy must be stationary i.e., its first variation (δF) must be set equal to zero. Accordingly, the integrand π must satisfy the following Euler equations [27] in the calculus of variations.

$$
\begin{aligned}
&\frac{\partial F}{\partial u} - \frac{\partial}{\partial\phi}\frac{\partial F}{\partial u_{,\phi}} - \frac{\partial}{\partial\theta}\frac{\partial F}{\partial u_{,\theta}} = 0 \\
&\frac{\partial F}{\partial v} - \frac{\partial}{\partial\phi}\frac{\partial F}{\partial v_{,\phi}} - \frac{\partial}{\partial\theta}\frac{\partial F}{\partial v_{,\theta}} = 0 \qquad\qquad (4.1\text{-}16) \\
&\frac{\partial F}{\partial w} - \frac{\partial}{\partial\phi}\frac{\partial F}{\partial w_{,\phi}} - \frac{\partial}{\partial\theta}\frac{\partial F}{\partial w_{,\theta}} + \frac{\partial^2}{\partial\phi^2}\frac{\partial F}{\partial w_{,\phi\phi}} + \frac{\partial^2}{\partial\theta^2}\frac{\partial F}{\partial w_{,\theta\theta}} + \frac{\partial^2}{\partial\phi\partial\theta}\frac{\partial F}{\partial w_{,\phi\theta}} = 0
\end{aligned}
$$

Introducing functional F given by Eq. (4.1-15) into Eqs. (4.1-16), the nonlinear equilibrium equations of thin spherical shells under thermal loading are obtained. Depending on the kinematical relations, Eq. (4.1-2) or Eq. (4.1-3), which may be used for the functional F, the equilibrium equations according to the deep or shallow shell theories can be derived. Using the Sanders kinematic relations, Eq. (4.1-2), the equilibrium equations for spherical shells on the basis of the deep shell theory are obtained as

$$
\begin{aligned}
&(\sin\phi M_\phi + R\sin\phi N_\phi)_{,\phi} + (RN_{\phi\theta} + M_{\phi\theta})_{,\theta} - R\sin\phi\,\beta_\phi N_\phi - R\cos\phi N_\theta \\
&- R\sin\phi\,\beta_\theta N_{\phi\theta} - \cos\phi M_\theta = 0 \\
&(RN_\theta + M_\theta)_{,\theta} + (R\sin\phi N_{\phi\theta} + \sin\phi M_{\phi\theta})_{,\phi} + R(\cos\phi - \sin\phi\,\beta_\phi)N_{\phi\theta} \\
&- R\sin\phi\,\beta_\theta N_\theta + \cos\phi M_{\phi\theta} = 0 \qquad\qquad (4.1\text{-}17)\\
&(\sin\phi M_\phi)_{,\phi\phi} + \frac{1}{\sin\phi}M_{\theta,\theta\theta} - [R\sin\phi\,(\beta_\phi N_\phi + \beta_\theta N_{\phi\theta}) + \cos\phi M_\theta]_{,\phi} \\
&+ 2(M_{\phi\theta,\phi\theta} + \cot\phi M_{\phi\theta,\theta}) - R\sin\phi\,(N_\phi + N_\theta) - R(\beta_\theta N_\theta + \beta_\phi N_{\phi\theta})_{,\theta} = 0
\end{aligned}
$$

The equilibrium equations for spherical shells on the basis of shallow shell theory are also obtained in the same manner with the replacement of D.M.V. kinematical relations (Eqs. (4.1-3)) instead of Sanders kinematical relations (Eqs. (4.1-2)). Comparing the outcome equilibrium equations shows that omitting the underlined terms from Eqs. (4.1-17) gives equilibrium equations for spherical shells based on shallow shell theory. Disappearance of the underlined terms from the equilibrium equations is due to shallow shell theory assumptions.

To derive the stability equations, the Trefftz criterion is used. According to this criterion, a structure is in a configuration of stable equilibrium if and only if the change in total potential energy corresponding to any sufficiently small, kinematically admissible displacement is positive [27]. A displacement is kinematically admissible if the displacement function satisfies continuity and boundary conditions. The critical load is defined as the smallest load at which the equilibrium of the structure fails to be stable as the load is slightly increased from its value. Therefore, the critical load for a continuous structural system is the lowest load for which the definite integral $\delta^2 F$ is non-positive for at least one possible variation. At this load, the equilibrium changes from stable to unstable. To obtain the corresponding expression for the second variation of the total potential energy it is assumed that

$$
\begin{aligned}
u &= u_0 + u_1 \\
v &= v_0 + v_1 \\
w &= w_0 + w_1
\end{aligned}
\qquad (4.1\text{-}18)
$$

where (u_0, v_0, w_0) represents a configuration on the primary equilibrium path. The virtual increment of displacements (u_1, v_1, w_1) is infinitesimally small. The subscripts 0 and 1 are two adjacent equilibrium configurations. Introducing Eqs. (4.1-18) into the kinematical relations, Eqs. (4.1-2) or (4.1-3), gives terms that are linear, quadratic, and cubic in u_0, v_0, w_0 and u_1, v_1, w_1 displacement components . In the new equations, the terms in u_0, v_0, w_0 alone add to zero because they are related to the stationary equilibrium configuration and have no variations. The terms that are quadratic and cubic in u_1, v_1, w_1 can be omitted because of smallness of the incremental displacement. Thus, the resulting equations are homogeneous and linear in u_1, v_1, w_1 with variable coefficients in u_0, v_0, w_0 which are governed by the original nonlinear equilibrium equations (Eq. (4.1-17)). In stability applications, the displacement u_0, v_0, w_0 are commonly called prebuckling deformations and u_1, v_1, w_1 are called buckling modes [27]. The increments u_1, v_1, w_1 cause corresponding change in the forces and moments as

$$
\begin{aligned}
N_\phi &= N_{\phi 0} + N_{\phi 1} & , && M_\phi &= M_{\phi 0} + M_{\phi 1} \\
N_\theta &= N_{\theta 0} + N_{\theta 1} & , && M_\theta &= M_{\theta 0} + M_{\theta 1} \\
N_{\phi\theta} &= N_{\phi\theta 0} + N_{\phi\theta 1} & , && M_{\phi\theta} &= M_{\phi\theta 0} + M_{\phi\theta 1}
\end{aligned}
\qquad (4.1\text{-}19)
$$

With these provisions, the stability equations of thin spherical shells based on the deep and shallow shell theories are obtained (Eqs. (4.1-20)). Similar to the equilibrium equations, omitting the underlined terms gives the stability equations according to the shallow shell theory.

$$
(\sin\phi M_{\phi1} + R\sin\phi N_{\phi1})_{,\phi} + (RN_{\phi\theta1} + \underline{M_{\phi\theta1}})_{,\theta} - R\cos\phi N_{\theta1} - \underline{\cos\phi M_{\theta1}}
$$
$$
-R\sin\phi\,[(\underline{\beta_{\phi1}N_{\phi0}} + \underline{\beta_{\theta1}N_{\phi\theta0}}) + (\underline{\beta_{\phi0}N_{\phi1}} + \underline{\beta_{\theta0}N_{\phi\theta1}})] = 0
$$
$$
(RN_{\theta1} + \underline{M_{\theta1}})_{,\theta} + (R\sin\phi N_{\phi\theta1} + \underline{\sin\phi M_{\phi\theta1}})_{,\phi} + R\cos\phi N_{\phi\theta1} + \underline{\cos\phi M_{\phi\theta1}}
$$
$$
-R\sin\phi\,[(\underline{\beta_{\phi1}N_{\phi\theta0}} + \underline{\beta_{\theta1}N_{\theta0}}) + (\underline{\beta_{\phi0}N_{\phi\theta1}} + \underline{\beta_{\theta0}N_{\theta1}})] = 0 \qquad (4.1\text{-}20)
$$
$$
(\sin\phi M_{\phi1})_{,\phi\phi} + \frac{1}{\sin\phi}M_{\theta1,\theta\theta} + 2(M_{\phi\theta1,\phi\theta} + \cot\phi M_{\phi\theta1,\theta}) - R\sin\phi\,(N_{\phi1} + N_{\theta1})
$$
$$
-\{R\sin\phi\,[(\beta_{\phi0}N_{\phi1} + \beta_{\theta0}N_{\phi\theta1}) + (\beta_{\phi1}N_{\phi0} + \beta_{\theta1}N_{\phi\theta0})] + \cos\phi M_{\theta1}\}_{,\phi}
$$
$$
-R[(\beta_{\theta0}N_{\theta1} + \beta_{\phi0}N_{\phi\theta1}) + (\beta_{\theta1}N_{\theta0} + \beta_{\phi1}N_{\phi\theta0})]_{,\theta} = 0
$$

where

$$
\begin{aligned}
N_{\phi1} &= C[(e_{\phi\phi1} + \beta_{\phi0}\beta_{\phi1}) + \nu(e_{\theta\theta1} + \beta_{\theta0}\beta_{\theta1})]\\
N_{\theta1} &= C[(e_{\theta\theta1} + \beta_{\theta0}\beta_{\theta1}) + \nu(e_{\phi\phi1} + \beta_{\phi0}\beta_{\phi1})]\\
N_{\phi\theta1} &= C(\frac{1-\nu}{2})(e_{\phi\theta1} + \beta_{\phi0}\beta_{\theta1} + \beta_{\theta0}\beta_{\phi1})\\
M_{\phi1} &= D(k_{\phi1} + \nu k_{\theta1})\\
M_{\theta1} &= D(k_{\theta1} + \nu k_{\phi1})\\
M_{\phi\theta1} &= D(1-\nu)\,k_{\phi\theta1}
\end{aligned} \qquad (4.1\text{-}21)
$$

where the strain-displacement relations for the state of stability conditions are linearized as

$$
\begin{aligned}
e_{\phi\phi1} &= \frac{1}{R}(u_{1,\phi} + w_1)\\
e_{\theta\theta1} &= \frac{1}{R\sin\phi}(\cos\phi\,u_1 + v_{1,\theta} + \sin\phi\,w_1) \qquad (4.1\text{-}22)\\
e_{\phi\theta1} &= \frac{1}{R\sin\phi}(u_{1,\theta} - \cos\phi\,v_1 + \sin\phi\,v_{1,\phi})
\end{aligned}
$$

The expressions for the $\beta_{\phi1}, \beta_{\theta1}, k_{\phi1}, k_{\theta1}$, and $k_{\phi\theta1}$ are the same as given in Eqs. (4.1-2) and Eqs. (4.1-3) for the deep and shallow shell theories, respectively. Substituting the linearized strain-displacement relations into Eqs. (4.1-21) for the forces and moments, and then into the Eqs. (4.1-20), the stability equations in terms of the displacement components are obtained as

1.3 Deep shell theory

The stability equations of a deep spherical shell in terms of the displacement components are

$$(\frac{D}{R^2}+C)[\sin\phi\ u_{1,\phi\phi}+(\frac{1-\nu}{2})\frac{1}{\sin\phi}\ u_{1,\theta\theta}+\cos\phi\ u_{1,\phi}-(\nu\sin\phi+\cos\phi\cot\phi)u_1$$

$$+(\frac{1+\nu}{2})v_{1,\phi\theta}-(\frac{3-\nu}{2})\cot\phi\ v_{1,\theta}]+(\frac{D}{R^2})[-\sin\phi\ w_{1,\phi\phi\phi}-\cos\phi\ w_{1,\phi\phi}-\frac{1}{\sin\phi}w_{1,\phi\theta\theta}$$

$$+2\frac{\cot\phi}{\sin\phi}w_{1,\theta\theta}+(\nu\sin\phi+\cos\phi\cot\phi)w_{1,\phi}]+C(1+\nu)\sin\phi\ w_{1,\phi}-\sin\phi\ (u_1-w_{1,\phi})N_{\phi0}$$

$$-(\sin\phi\ v_1-w_{1,\theta})N_{\phi\theta0}=0$$

$$(\frac{D}{R^2}+C)[\frac{1}{\sin\phi}v_{1,\theta\theta}+(\frac{1-\nu}{2})\sin\phi\ v_{1,\phi\phi}+(\frac{1-\nu}{2})\cos\phi\ v_{1,\phi}$$

$$+(\frac{1-\nu}{2})(\sin\phi-\cos\phi\cot\phi)v_1+(\frac{1+\nu}{2})u_{1,\phi\theta}+(\frac{3-\nu}{2})\cot\phi\ u_{1,\theta}]$$

$$+(\frac{D}{R^2})[-\frac{1}{\sin^2\phi}w_{1,\theta\theta\theta}-w_{1,\phi\phi\theta}-\cot\phi\ w_{1,\phi\theta}-(1-\nu)w_{1,\theta}]$$

$$+C(1+\nu)w_{1,\theta}+(-\sin\phi\ v_1+w_{1,\theta})N_{\theta0}+\sin\phi\ (w_{1,\phi}-u_1)N_{\phi\theta0}=0$$

$$(4.1\text{-}23)$$

$$(\frac{D}{R^2})\{-\sin\phi\ w_{1,\phi\phi\phi\phi}-\frac{1}{\sin^3\phi}w_{1,\theta\theta\theta\theta}-\frac{2}{\sin\phi}\ w_{1,\phi\phi\theta\theta}-2\cos\phi\ w_{1,\phi\phi\phi}+2\frac{\cot\phi}{\sin\phi}w_{1,\phi\theta\theta}$$

$$+[(1+\nu)\sin\phi+\cos\phi\cot\phi]w_{1,\phi\phi}+(\frac{-3+\nu-4\cot^2\phi}{\sin\phi})\ w_{1,\theta\theta}+\cos\phi\ (-2+\nu-\cot^2\phi)w_{1,\phi}$$

$$+v_{1,\phi\phi\theta}+\frac{1}{\sin^2\phi}v_{1,\theta\theta\theta}-\cot\phi\ v_{1,\phi\theta}+(2-\nu+\cot^2\phi)v_{1,\theta}+\frac{1}{\sin\phi}u_{1,\phi\theta\theta}+\sin\phi\ u_{1,\phi\phi\phi}$$

$$+2\cos\phi\ u_{1,\phi\phi}+\frac{\cot\phi}{\sin\phi}u_{1,\theta\theta}-[(1+\nu)\sin\phi+\cos\phi\cot\phi]u_{1,\phi}+\cos\phi\ (2-\nu+\cot^2\phi)u_1\}$$

$$-C(1+\nu)(2\sin\phi\ w_1+v_{1,\theta}+\sin\phi\ u_{1,\phi}+\cos\phi\ u_1)+(\frac{1}{\sin\phi}w_{1,\theta\theta}-v_{1,\theta})N_{\theta0}$$

$$+(\sin\phi\ w_{1,\phi\phi}+\cos\phi\ w_{1,\phi}-\cos\phi\ u_1-\sin\phi\ u_{1,\phi})N_{\phi0}$$

$$+(2w_{1,\phi\theta}-\sin\phi\ v_{1,\phi}-\cos\phi\ v_1-u_{1,\theta})N_{\phi\theta0}=0$$

1.4 Shallow shell theory

The stability equations of a shallow spherical shell in terms of the displacement components are

$$\sin\phi\ u_{1,\phi\phi}+(\frac{1-\nu}{2})\frac{1}{\sin\phi}\ u_{1,\theta\theta}+\cos\phi\ u_{1,\phi}-(\nu\sin\phi+\cos\phi\cot\phi)u_1+(\frac{1+\nu}{2})v_{1,\phi\theta}$$

$$-(\frac{3-\nu}{2})\cot\phi\ v_{1,\theta}+(1+\nu)\sin\phi\ w_{1,\phi}=0$$

$$\frac{1}{\sin\phi}v_{1,\theta\theta} + (\frac{1-\nu}{2})\sin\phi \; v_{1,\phi\phi} + (\frac{1-\nu}{2})\cos\phi \; v_{1,\phi} + (\frac{1-\nu}{2})(\sin\phi - \cos\phi\cot\phi)v_1$$

$$+(\frac{1+\nu}{2})u_{1,\phi\theta} + (\frac{3-\nu}{2})\cot\phi \; u_{1,\theta} + (1+\nu)w_{1,\theta} = 0$$

$$(\frac{D}{R^2})\{-\sin\phi \; w_{1,\phi\phi\phi\phi} - \frac{1}{\sin^3\phi}w_{1,\theta\theta\theta\theta} - \frac{2}{\sin\phi} \; w_{1,\phi\phi\theta\theta} - 2\cos\phi \; w_{1,\phi\phi\phi} + 2\frac{\cot\phi}{\sin\phi}w_{1,\phi\theta\theta}$$

$$+[(1+\nu)\sin\phi + \cos\phi\cot\phi]w_{1,\phi\phi} + (\frac{-3+\nu-4\cot^2\phi}{\sin\phi}) \; w_{1,\theta\theta}$$

$$+ \cos\phi \; (-2+\nu-\cot^2\phi)w_{1,\phi}\} - C(1+\nu)(2\sin\phi \; w_1 + v_{1,\theta} + \sin\phi \; u_{1,\phi} + \cos\phi \; u_1)$$

$$+\frac{1}{\sin\phi}w_{1,\theta\theta}N_{\theta0} + (\sin\phi \; w_{1,\phi\phi} + \cos\phi \; w_{1,\phi})N_{\phi0} + 2w_{1,\phi\theta}N_{\phi\theta0} = 0 \qquad (4.1\text{-}24)$$

Since the thermal loading is assumed to be symmetric ($N_{\phi\theta0} = 0$), terms containing $\beta_{\phi0}$ and $\beta_{\theta0}$ are ignored in the stability equations and the resulting deformations are also symmetric. Therefore, the stability equations depend to the equilibrium configuration of shell via the prebuckling forces $N_{\phi0}$ and $N_{\theta0}$, only.

Considering the assumed simply supported boundary conditions, the edge conditions may be written as [3]

$$w_1 = w_{1,\phi\phi} = v_1 = u_{1,\phi} = 0 \qquad (4.1\text{-}25)$$

An approximate series expansion form solution satisfying the simply supported boundary conditions, is [29]

$$\begin{aligned} u_1 &= A_1\cos n\theta \; \cos\lambda\phi \\ v_1 &= A_2\sin n\theta \; \sin\lambda\phi \\ w_1 &= A_3\cos n\theta \; \sin\lambda\phi \end{aligned} \qquad (4.1\text{-}26)$$

where $\lambda = m\pi/\phi_L$ and ϕ_L being the latitude angle measured from the top of the sphere to the boundary (Fig. (4.1-1)). The coefficients A_1, A_2, and A_3 are constants, m is the buckling wave numbers in ϕ-direction, and n is the buckling wave numbers in θ-direction. Substituting Eqs. (4.1-26) into the stability equations, Eqs. (4.1-23) or Eqs. (4.1-24), the following general form is obtained.

$$\begin{aligned} R_{11}A_1 + R_{12}A_2 + R_{13}A_3 &= E_1 \\ R_{21}A_1 + R_{22}A_2 + R_{23}A_3 &= E_2 \\ R_{31}A_1 + R_{32}A_2 + R_{33}A_3 &= E_3 \end{aligned} \qquad (4.1\text{-}27)$$

where R_{ij} is function of $m, n, \phi_L, \Delta T$, and the material constants. The variables ϕ and θ also appear in R_{ij}. The value of E_1, E_2, and E_3 are the expected errors due to the proposed approximate solution (Eqs. (4.1-26)). These errors may be minimized using the Bubnov-Galerkin method as

$$\int_0^{2\pi}\int_0^{\phi_L} E_1 \cos n\theta \cos \lambda\phi \, dA = 0$$

$$\int_0^{2\pi}\int_0^{\phi_L} E_2 \sin n\theta \sin \lambda\phi \, dA = 0 \qquad (4.1\text{-}28)$$

$$\int_0^{2\pi}\int_0^{\phi_L} E_3 \cos n\theta \sin \lambda\phi \, dA = 0$$

where $dA = R\sin\phi \, d\theta d\phi$. Substituting from Eqs. (4.1-27) into Eqs. (4.1-28), a set of homogeneous equations for the constant coefficients A_1, A_2, and A_3 are obtained as

$$
\begin{aligned}
a_{11}A_1 + a_{12}A_2 + a_{13}A_3 &= 0 \\
a_{21}A_1 + a_{22}A_2 + a_{23}A_3 &= 0 \qquad (4.1\text{-}29)\\
a_{31}A_1 + a_{32}A_2 + a_{33}A_3 &= 0
\end{aligned}
$$

Evidently, Eqs. (4.1-28) are satisfied for all values of θ and its sine and cosine factors can be divided out. For a nontrivial solution, the determinant of the coefficients a_{ij} shall be set equal to zero to find the smallest eigenvalue corresponding to m and n. The values of prebuckling forces (N_{ij}) and moments (M_{ij}) depend to the applied thermal loading. The uniform temperature change and the radial temperature difference loading are two cases considered in this paper.

A. Uniform temperature change (UTCL)

Consider a spherical shell of thickness h, radius R, and the supporting latitude angle ϕ_L. The sphere is initially at constant temperature T_i. Under simply supported boundary condition, the temperature is uniformly increased (or decreased) to a final value T_f such that shell buckles. This temperature change $\Delta T = T_f - T_i$ is called the critical temperature change. The prebuckling forces are assumed to be the membrane solution of the equilibrium equations. For both deep and shallow shell theories the prebuckling solution is

$$
\begin{aligned}
N_{\phi 0} &= N_{\theta 0} = -\frac{E\alpha_T T_1}{1-\nu} = -\frac{E\alpha_T}{1-\nu}\int_{-\frac{h}{2}}^{\frac{h}{2}} \Delta T \, dz = -\frac{Eh\varepsilon_T}{1-\nu} \\
N_{\phi\theta 0} &= 0 \qquad (4.1\text{-}30)\\
M_{ij0} &= 0
\end{aligned}
$$

where $\varepsilon_T = \alpha_T \Delta T$.

B. Radial temperature difference (RTDL)

Consider a spherical shell of thickness h, radius R, and the supporting latitude angle ϕ_L. The temperature of the inside and outside surfaces of the sphere are at constant temperatures T_a and T_b, respectively. Therefore, a linear temperature variation across the shell wall may be assumed as

$$\Delta T = T_b - T(z) = (T_b - T_a)(\frac{1}{2} - \frac{z}{h}) \qquad (4.1\text{-}31)$$

Similar to uniform temperature change loading, the prebuckling forces are assumed to be the membrane solution of the equilibrium equations. For both deep and shallow shell theories the prebuckling solution is

$$\begin{aligned} N_{\phi 0} &= N_{\theta 0} = -\frac{E\alpha_T T_1}{1-\nu} = -\frac{E\alpha_T}{1-\nu}\int_{-\frac{h}{2}}^{\frac{h}{2}} \Delta T \, dz = -\frac{Eh\varepsilon_T}{2(1-\nu)} \\ N_{\phi\theta} &= 0 \\ M_{ij0} &= 0 \end{aligned}$$
$$(4.1\text{-}32)$$

where $\varepsilon_T = \alpha_T \Delta T$. Comparing Eq. (4.1-30) and Eq. (4.1-32) reveals that the prebuckling forces $N_{\phi 0}$ and $N_{\theta 0}$ for UTCL is twice for that of RTDL. That is, for numerically identical ΔT, the critical buckling strain associated with UTCL is twice the critical buckling strain of RTDL. Consequently, here the critical buckling strain related to the UTCL is discussed, and the results can be extended to the RTDL.

1.5 Results and discussion

Setting the determinant of the coefficients a_{ij} equal to zero, results in an equation relating ε_T to h/R. For shallow shell theory, this equation is quadratic and is

$$\varepsilon_T = f_1(\phi_L, \nu)(\frac{h}{R})^2 + f_2(\phi_L, \nu) \qquad (4.1\text{-}33)$$

For deep shell theory, the subject equation is more complicated and has the form of

$$\varepsilon_T^3 + f_1(\phi_L, \nu, \frac{h}{R})\,\varepsilon_T^2 + f_2(\phi_L, \nu, \frac{h}{R})\,\varepsilon_T + f_3(\phi_L, \nu, \frac{h}{R}) = 0 \qquad (4.1\text{-}34)$$

It is noted that for both shell theories when $h = 0$ then $\varepsilon_T \neq 0$. This error is produced due to the shell theories' approximations. The error for

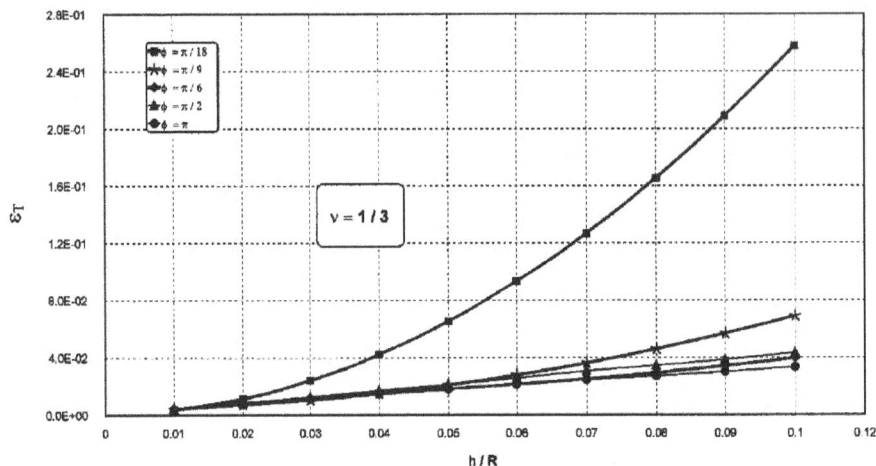

Figure 4.1-2: Critical uniform temperature rise for shallow shell.

shallow shell theory is equal to $f_2(\phi_L, \nu)$ and for deep shell theory is the least real root of Eq. (4.1-34) when $h/R = 0$. The error incurred in two cases, however, is negligible. It is also to be noted that the referred errors are not just for the spherical shells or the method of solution discussed in this section. One may refer to Eq. (5.48) of reference [27] which is the error incurred in the buckling analysis of cylindrical shells.

For both shell theories, the determinant of constant coefficients a_{ij} results in complicated mathematical expression and its minimum, providing critical buckling load, for different value of m and n, cannot easily be evaluated. Thus, a computer program is developed to obtain the buckling load. The critical values of ε_T in case of UTCL and on the basis of deep and shallow shell theories are shown in figures (4.1-2) and (4.1-3). In these figures $\nu = \frac{1}{3}$ and shell support angles of $\phi_L = \pi, \pi/2, \pi/6, \pi/9$, and $\pi/18$ are considered. It is further noticed, from Eq. (4.1-23) or (4.1-24), that the modulus of elasticity does not influence thermal buckling of the spherical shells.

Figure (4.1-2) shows that based on the shallow shell theory the critical buckling strain, for the range of spherical shells with supporting angles $\phi_L \geq \pi/9$, is nearly independent of ϕ_L and all fall inside a narrow band and vary almost linearly with the variation of h/R. In case of smaller supporting angle ϕ_L, the structure becomes more stable under applied thermal loading. Also, as h/R is increased, approaching to the range of thick shells, the critical buckling strain increases, as expected.

The results of the deep shell theory are shown in Figure (4.1-3). Although the stability equations associated with this theory involve more terms, the critical buckling strains predicted by this theory follow the

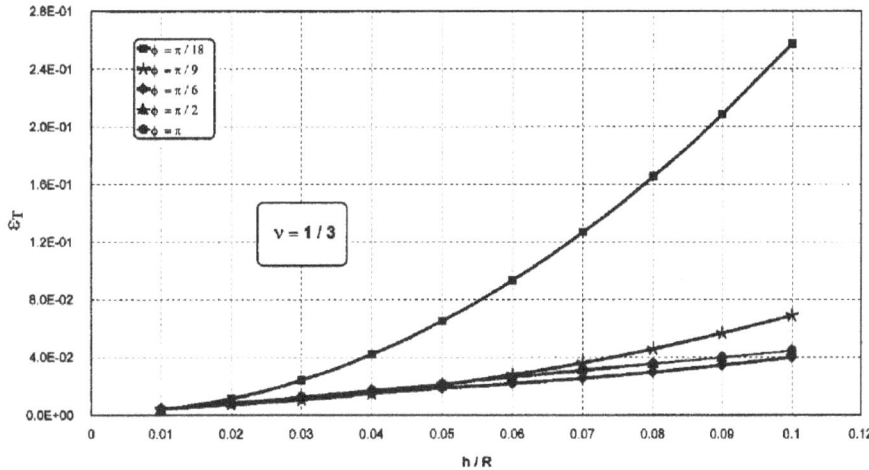

Figure 4.1-3: Critical uniform temperature rise for deep shell.

same values of shallow shell theory. In general, the difference between the results of two shell theories for spherical shells with $\phi_L \leq \frac{\pi}{2}$ is under 4 percent. The deep shell theory predicts higher values for buckling load compared to the shallow shell theory. The difference between the two theories, however, is significant for larger supporting angles. As ϕ_L increases, the difference increases and becomes quite noticeable. For example, for full sphere, $\phi_L = \pi$, the difference ranges from 6.4 percent for $h/R = 0.01$ to about 33.3 percent for $h/R = 0.10$. As a consequence;

1-The shallow shell theory predicts more conservative buckling loads than deep shell theory, as the results obtained by this theory suggest lower buckling loads.

2-As the ratio of h/R is increased the critical buckling strain ε_T is increased.

3-It is found that the buckling wave modes along the principal shell directions ϕ and θ, which are m and n respectively, are identical for both shallow and deep shell theories for $\phi_L \leq \frac{\pi}{2}$. However, as ϕ_L increases, and specially for a full spherical shell, the deep shell theory predicts buckling load along the latitude angle ϕ of higher m-values compared to the shallow shell theory which always predict the buckling at $m = 1$. This relationship between the supporting angle and buckling mode is vice versa for the tangential direction θ. That is, the deep shell theory predicts buckling load at $n = 1$ where the shallow shell theory predicts buckling at higher n-values. In addition, when h/R is increased the buckling occurs at lower modes and as the supporting angle ϕ_L is increased, the buckling occurs at higher modes.

To compare the results of this section with other known references,

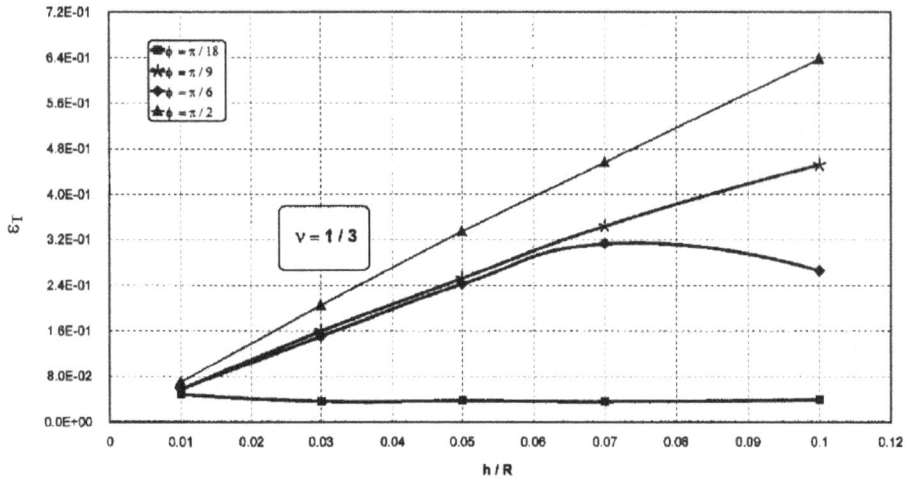

Figure 4.1-4: Buckling under UTCL for different supporting angles ϕ_L and $\nu = 0.3$.

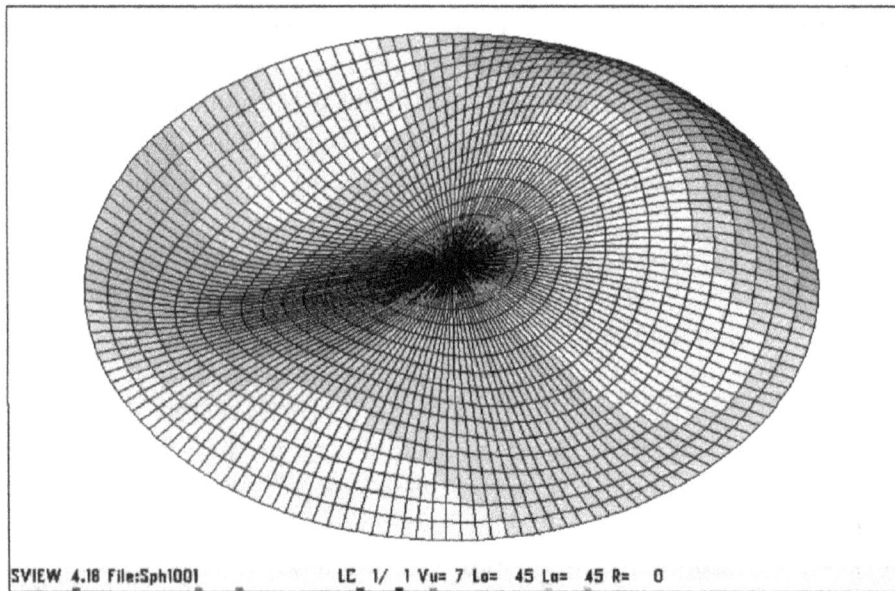

Figure 4.1-5: Buckling configuration of normalized buckling shape for the spherical shells of $\phi_L = \pi/18$

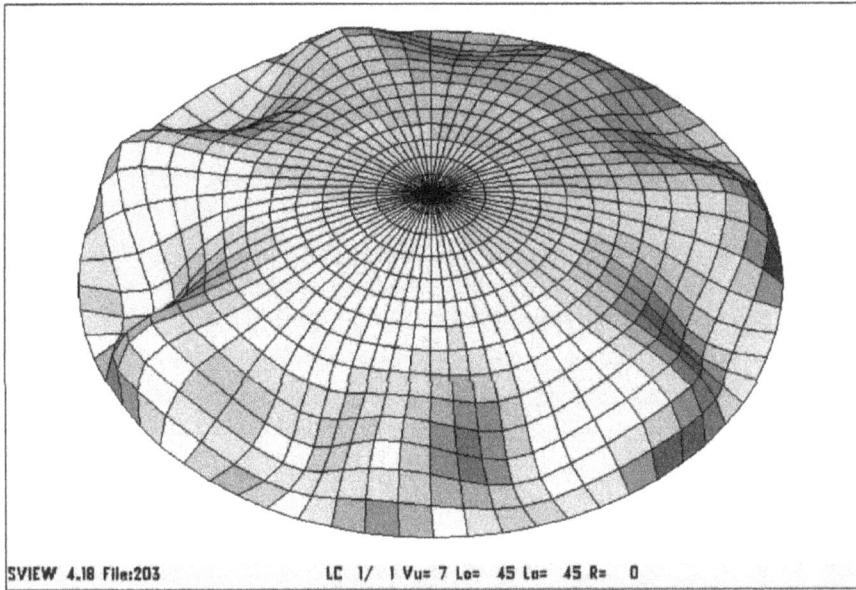

Figure 4.1-6: Buckling configuration of normalized buckling shape for the spherical shells of $\phi_L = \pi/9$

due to the lack of references in the literature, the Algor computer program is used. The buckling results of spherical shells under UTCL for different supporting angles ϕ_L and $\nu = 0.3$ are shown in figure (4.1-4). The Algor analysis uses the stress in the shell to form the geometrical stiffness matrix. This means that displacements prior to buckling are ignored. The shell theory used by Algor is defined in Ref. [30]. As noticed from figure (4.1-4), it may not be concluded that ε_T increase as ϕ_L increase. This is unlike the conclusion reached by figures (4.1-2) and (4.1-3). In addition, as h/R is increased the buckling load for $\phi_L = \frac{\pi}{18}, \frac{\pi}{9}$, and $\frac{\pi}{2}$ increases but for $\phi_L = \frac{\pi}{6}$ it reaches to a maximum and then decreases. The comparison of the results of Algor computer program with the figures (4.1-3) and (4.1-4) reveals that

1-For $\phi_L = \frac{\pi}{18}$ the value of critical ε_T in the range of $h/R \leq 0.035$ predicted by Algor is smaller than the shallow and deep shell theories. As h/R is increased beyond 0.035, this relationship becomes vice versa, i.e. the critical ε_T predicted by the theories is smaller than what is predicted by Algor.

2-At $\phi_L = \frac{\pi}{9}$ and $\frac{\pi}{6}$ (the limiting value between the shallow and deep spherical shells), value of ε_T given by Algor is much higher than the theoretical values and unlike the trend of theoretical curves, reaches to a maximum and then drops.

3-For hemispherical shells ($\phi_L = \frac{\pi}{2}$), the relation between ε_T and h/R

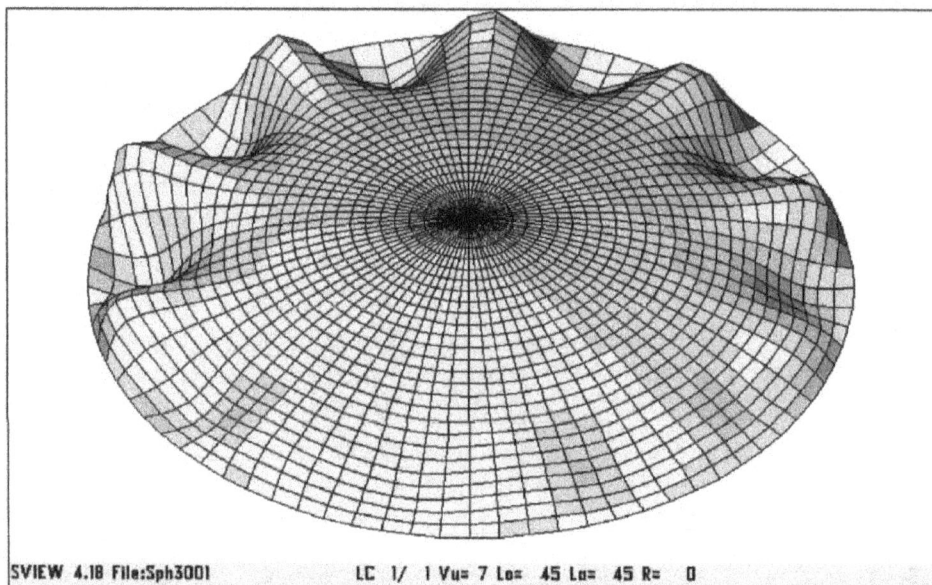

SVIEW 4.18 File:Sph3001   LC 1/ 1 Vu= 7 Lo= 45 La= 45 R=   0

Figure 4.1-7: Buckling configuration of normalized buckling shape for the spherical shells of $\phi_L = \pi/6$

is almost linear, and the value of ε_T is much higher than the theoretical values.

4-For $\phi_L \geq \frac{\pi}{2}$, the results of Algor approach to large values.

5- The buckling modes m and n are not given in the Algor output, explicitly. However, the buckling configurations of some normalized buckling shapes are geometrically shown in figures (4.1-5) to (4.1-8) for the spherical shells of $\phi_L = \pi/18, \pi/9, \pi/6$, and $\pi/2$.

For the comparison purpose, the critical buckling strain of hemispherical shell, circular flat plate, and long and short cylindrical shells under UTCL versus h/R are shown in figure (4.1-5). In all cases simply supported edge and $\nu = \frac{1}{3}$ are considered.

The critical buckling strain of a circular flat plate of simply supported edge under UTCL is [23]

$$\varepsilon_T = \frac{2}{3} \frac{1}{3 + \nu} (\frac{h}{R})^2 \tag{4.1-35}$$

where R is the radius of circular plate (identical with a in Figure (4.1-1)), and h is the plate thickness. The critical buckling strain of simply supported short and long cylindrical shells under UTCL are given by equation (4.1-36) and (4.1-37), respectively [25].

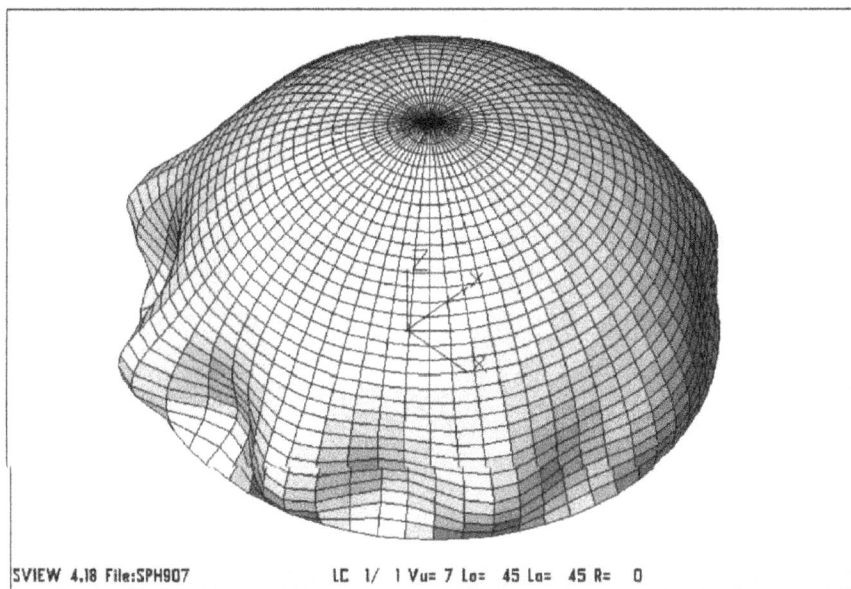

Figure 4.1-8: Buckling configuration of normalized buckling shape for the spherical shells of $\phi_L = \pi/2$.

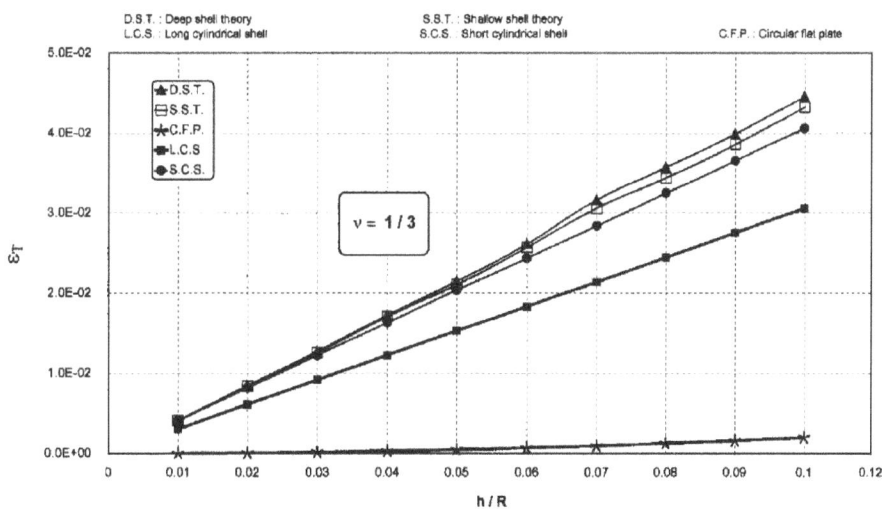

Figure 4.1-9: Critical buckling strain of hemispherical shell, circular flat plate, and long and short cylindrical shells under UTCL versus h/R.

$$\varepsilon_T = [\frac{(1-\nu)}{3(1+\nu)}]^{\frac{1}{2}}(\frac{h}{R})^2 \tag{4.1-36}$$

$$\varepsilon_T = \frac{3}{4}[\frac{(1-\nu)}{3(1+\nu)}]^{\frac{1}{2}}(\frac{h}{R})^2 \tag{4.1-37}$$

where R is again identical with a in Figure (4.1-1) and h is the thickness of cylindrical shell. It is noted that spherical shells, even with consideration of shallow shell theory's result which predicts the least buckling values, are the most stable structures compared to the others. The circular flat plate has the lowest state of stability of all the aforementioned structures. The stability of all the structures shown in figure (4.1-9) increase as the ratio of h/R is increased.

The conclusion of this section may be drawn as:

1-The analysis done in this section indicates that the thermal buckling of spherical shells is independent of the modulus of elasticity and depends on the shell geometric properties, thickness, Poisson's ratio, and the coefficient of thermal expansion.

2-In spherical shells of supporting angles $\phi_L \geq \pi/9$, the thermal buckling load is almost independent of ϕ_L. That is, for a constant ratio of $\frac{h}{R}$, ε_T does not change significantly with the variation of ϕ_L, when $\phi_L \geq \frac{\pi}{9}$.

3-The thermal buckling load of spherical shells increases as the ratio of h/R increases.

4-The deep shell theory predicts higher values for buckling load compared to the shallow shell theory. The maximum difference between the thermal buckling load predicted by deep and shallow shell theories for spherical shells with $\phi_L \leq \frac{\pi}{2}$ is about 4 percent. The difference between the two theories, however, is significant for larger supporting angles and reaches a maximum value of 33.3 percent for full spherical shell with $h/R = 0.10$.

5-In general, the buckling wave numbers along the principal shell directions ϕ and θ, decreases as the ratio of h/R is increased, and increases as ϕ_L is increased.

6-Circular flat plates under thermal loads are much less stable compared to the spherical or cylindrical shells with the same R, h, ν, and α_T values.

2 Imperfect Shallow Spherical Shell

2.1 Introduction

An early buckling study of spherical shells under external pressure was done by Zoelly [31]. He considered the membrane shell theory and assumed the linear strain-displacement relations. The buckling load thus obtained is called the linear classical critical buckling load. In 1939 Von Karman and Tsien [4] introduced the nonlinear load-displacement relations and, using nonlinear analysis, obtained the critical buckling load of the spherical shells. The buckling load obtained by this analysis was called the nonlinear classical critical buckling load. Many studies are carried out on the buckling analysis of spherical shells since then, of which we may refer to the work of Kaplan [2] in 1947. He considered the symmetrical geometry and, using the energy method, obtained the static and dynamic buckling loads. In 1962 Budiansky and Roth [32] studied the dynamic and static buckling of shallow spherical shells and, assuming geometrical symmetry and using the Galerkin method, obtained the critical buckling load of the shell under external pressure. This paper has become a basic reference for many later research works. Weinitschke [6] studied the buckling of shallow spherical shells under mechanical loads, using power series and expansion of the displacement components. In 1964 Huang [7] studied the nonlinear buckling of shallow spherical shells using the finite difference method. His work reduced the gap between the experimental results and the results which were reported by Budiansky [5]. Famili and Archer [33] confirmed the results which were obtained by Huang [7]. Many studies on the mechanical buckling of shallow spherical shells including the effects of the geometrical imperfection are reported during the 90's such as Eggwerta and Samuelson [34], Goncalves [35], and Kiyuan and Weiping [36]. These studies are based on the linear and nonlinear buckling theories for isotropic materials.

Studies on buckling of composite shells were reported during the 1980's. In 1982 Ganapathi and Varadan [37] studied the dynamic buckling of orthotropic (shallow and deep) spherical shells under external pressure. In 1990 Chao and Lin [38] considered the initial geometric imperfection and using the finite difference method, obtained the buckling of shallow single layer orthotropic spherical shells under external pressure. In the latter study, the Love-Kirchhoff hypothesis is considered. In 1995 Ganapathi and Varadan [39] studied the static and dynamic buckling of composite shallow and deep spherical shells using the finite element method. They considered the effects of the shear deformations and rotary inertia in their formulations and used the nonlinear von Karman equations for multilayer composite shells. Aleksander Muc in 1989 [29] and 1992 [40] studied the buckling of composite orthotropic multilayer spherical shells under external pressure using the first order Love-Kirchhoff theory with the inclusion of the transverse shear deformation. He investigated the effect of ply-angles on the critical buckling load of shallow spherical shells using the Donnell stability equation. Eslami et al. [25] studied the thermal buckling of perfect spherical shells made of isotropic and homogeneous material. The analysis is based on first order shell theory with Donnell-Mushtari-Velasov (DMV) hypothesis and the Sanders nonlinear strain-displacement relations. The imperfection analysis of a cylindrical shell is studied by Eslami et al. and may be found in references [41-44].

In this section, the thermal and mechanical buckling loads of a cap of shallow spherical shell of isotropic material and geometrically imperfect shell are considered. The Sanders nonlinear strain-displacement relations are used. The shell is under external pressure for mechanical loading and uniform temperature rise and radial temperature difference for thermal loading. Simply supported boundary condition is assumed. The expressions for the thermal and mechanical buckling loads are obtained analytically and are given in closed form solutions.

2.2   Basic equations

Consider a shell of revolution made of isotropic and homogeneous material. The principal coordinates are designated by x and y which are located in the shell middle surface. The shell thickness is assumed to be h. The coordinate z is measured across the shell thickness from the reference middle plane and is considered to be positive outward. According to the Love-Kirchhoff hypothesis, the strains at a point z away from the middle plane is related to the strains on the middle plane as [27]

$$\epsilon_x = \epsilon_{xm} + z k_x$$
$$\epsilon_y = \epsilon_{ym} + z k_y \qquad (4.2\text{-}1)$$
$$\epsilon_{xy} = \epsilon_{xym} + 2z k_{xy}$$

where ϵ_{ij} are the strains at point z away from the middle plane, ϵ_{ijm} are strains on the middle plane, and k_{ij} are the curvatures. The strains on the middle plane are related to the displacement components through the Sanders nonlinear assumption [27]

$$\epsilon_{xm} = e_{xx} + \frac{1}{2}\beta_x^2$$
$$\epsilon_{ym} = e_{yy} + \frac{1}{2}\beta_y^2$$
$$\gamma_{xym} = e_{xy} + \beta_x \beta_y$$
$$k_x = \chi_{xx} \qquad (4.2\text{-}2)$$
$$k_y = \chi_{yy}$$
$$k_{xy} = \chi_{xy}$$

In these equations, e_{ij} are the linear strains, β_x and β_y are the rotations of normal vector to the middle plane, and χ_{ij} are linear functions of displacement components, and are all defined in terms of displacements as

$$e_{xx} = \frac{u_{,x}}{A} + \frac{A_{,y}v}{AB} + \frac{w}{R_x}$$

$$e_{yy} = \frac{v_{,y}}{B} + \frac{B_{,x}u}{AB} + \frac{w}{R_y}$$

$$e_{xy} = \frac{v_{,x}}{A} + \frac{u_{,y}}{B} - \frac{B_{,x}v + A_{,y}u}{AB}$$

$$\beta_x = \frac{-w_{,x}}{A} + \frac{u}{R_x} \qquad\qquad (4.2\text{-}3)$$

$$\beta_y = \frac{-w_{,y}}{B} + \frac{v}{R_y}$$

$$\chi_{xx} = \frac{1}{A}\beta_{x,x} + \frac{1}{AB}A_{,y}\beta_y$$

$$\chi_{yy} = \frac{1}{B}\beta_{y,y} + \frac{1}{AB}B_{,x}\beta_x$$

$$2\chi_{xy} = \frac{1}{A}\beta_{y,x} + \frac{1}{B}\beta_{x,y} - \frac{1}{AB}(A_{,y}\beta_x + B_{,x}\beta_y)$$

where (,) defines partial derivative, $u, v,$ and w are the displacement components along the x, y and z-directions, respectively. The terms A and B are the Lame parameters and are given as [27]

$$A = \{(\frac{\partial X}{\partial x})^2 + (\frac{\partial Y}{\partial x})^2 + (\frac{\partial Z}{\partial x})^2\}^{\frac{1}{2}}$$

$$\qquad\qquad (4.2\text{-}4)$$

$$B = \{(\frac{\partial X}{\partial y})^2 + (\frac{\partial Y}{\partial y})^2 + (\frac{\partial Z}{\partial y})^2\}^{\frac{1}{2}}$$

Here, $X, Y,$ and Z are the global coordinates related to the curvilinear coordinates $x, y,$ and z [27]. For shallow shells or quasi shallow shells, we ignore the terms u and v in the expression of β_x and β_y, and thus the Donnell-Mushtari-Velasov relations for the shallow shell becomes [27]

$$\beta_x = -\frac{w_{,x}}{A}$$

$$\beta_y = -\frac{w_{,y}}{B}$$

$$\chi_{xx} = \frac{-w_{,xx}}{A^2} + \frac{A_{,x}w_{,x}}{A^3} - \frac{A_{,y}w_{,y}}{AB^2} \qquad (4.2\text{-}5)$$

$$\chi_{yy} = \frac{-w_{,yy}}{B^2} + \frac{B_{,y}w_{,y}}{B^3} - \frac{B_{,x}w_{,x}}{A^2B}$$

$$\chi_{xy} = \frac{-w_{,xy}}{AB} + \frac{A_{,y}w_{,x}}{A^2B} + \frac{B_{,x}w_{,y}}{AB^2}$$

For the spherical shells, considering the variables ϕ and θ in the longitudinal and circumferential directions, the Lame parameters are

$$A = R_\phi = R_\theta = R \qquad (4.2\text{-}6)$$
$$B = Rsin\phi$$

and the relations (4.2-1) through (4.2-5) for shallow spherical shells reduce to

$$\epsilon_\phi = \epsilon_{\phi m} + zk_\phi$$
$$\epsilon_\theta = \epsilon_{\theta m} + zk_\theta$$
$$\gamma_{\phi\theta} = \gamma_{\phi\theta m} + 2zk_{\phi\theta}$$
$$\epsilon_{\phi m} = e_{\phi\phi} + \frac{1}{2}\beta_\phi^2$$
$$\epsilon_{\theta m} = e_{\theta\theta} + \frac{1}{2}\beta_\theta^2$$
$$\gamma_{\phi\theta m} = e_{\phi\theta} + \beta_\phi\beta_\theta \qquad (4.2\text{-}7)$$
$$k_\phi = \chi_{\phi\phi}$$
$$k_\theta = \chi_{\theta\theta}$$
$$k_{\phi\theta} = \chi_{\phi\theta}$$
$$e_{\phi\phi} = \frac{u_{,\phi}}{R} + \frac{w}{R}$$
$$e_{\theta\theta} = \frac{v_{,\theta}}{Rsin\phi} + cot\phi\frac{u}{R} + \frac{w}{R}$$
$$e_{\phi\theta} = \frac{v_{,\phi}}{R} + \frac{u_{,\theta}}{Rsin\phi} - cot\phi\frac{v}{R}$$
$$\beta_\phi = \frac{-w_{,\phi}}{R}$$
$$\beta_\theta = \frac{-w_{,\theta}}{Rsin\phi}$$
$$\chi_{\phi\phi} = \frac{1}{R}\beta_{\phi,\phi}$$
$$\chi_{\theta\theta} = \frac{1}{R\sin\phi}\beta_{\theta,\theta} + \frac{\cot\phi}{R}\beta_\phi$$
$$2\chi_{\phi\theta} = \frac{1}{R}\beta_{\theta,\phi} + \frac{1}{Rsin\phi}\beta_{\phi,\theta} - \frac{cot\phi}{R}\beta_\theta$$

From Eqs. (4.2-7) the strains and curvatures in terms of displacement components and their derivatives become

$$k_\phi = \frac{-w_{,\phi\phi}}{R^2}$$
$$k_\theta = \frac{-1}{R^2\sin\phi}\left(\frac{w_{,\theta\theta}}{\sin\phi} + w_{,\phi}\cos\phi\right)$$

$$k_{\phi\theta} = \frac{1}{R^2 \sin\phi}(w,_\theta \cot\phi - w,_{\phi\theta})$$

$$\epsilon_{\phi m} = \frac{1}{R}(u,_\phi + w) + \frac{w,_\phi^2}{2R^2}$$

$$\epsilon_{\theta m} = \frac{1}{R \sin\phi}(v,_\theta + u\cos\phi + w\sin\phi) + \frac{w,_\theta^2}{2R^2 \sin^2\phi}$$

$$\gamma_{\phi\theta m} = \frac{1}{R \sin\phi}(v,_\phi \sin\phi + u,_\theta - v\cos\phi) + \frac{w,_\phi w,_\theta}{R^2 \sin\phi} \tag{4.2-8}$$

$$\epsilon_\theta = \frac{1}{R \sin\phi}(v,_\theta + u\cos\phi + w\sin\phi) + \frac{w,_\theta^2}{2R^2 \sin^2\phi} - \frac{z}{R^2 \sin\phi}\left(\frac{w,_{\theta\theta}}{\sin\phi} + w,_\phi \cos\phi\right)$$

$$\epsilon_\phi = \frac{1}{R}(u,_\phi + w) + \frac{w,_\phi^2}{2R^2} - z\frac{w,_{\phi\phi}}{R^2}$$

$$\gamma_{\phi\theta} = \frac{1}{R \sin\phi}(v,_\phi \sin\phi + u,_\theta - v\cos\phi) + \frac{w,_\phi w,_\theta}{R^2 \sin\phi} + \frac{2z}{R^2 \sin\phi}(w,_\theta \cot\phi - w,_{\phi\theta})$$

The strain-stress relations for thermal loading, following Hooke's law for plane-stress condition, are

$$\epsilon_\phi = \frac{1}{E}(\sigma_\phi - \nu\sigma_\theta) + \alpha\Delta T$$

$$\epsilon_\theta = \frac{1}{E}(\sigma_\theta - \nu\sigma_\phi) + \alpha\Delta T \tag{4.2-9}$$

$$\gamma_{\phi\theta} = \frac{2(1+\nu)}{E}\tau_{\phi\theta}$$

The forces and moments per unit length based on the first order shell theory are

$$N_\phi = \int\int \sigma_\phi dz d\theta \quad , \quad M_\phi = \int\int \sigma_\phi z dz d\theta$$

$$N_\theta = \int\int \sigma_\theta dz d\theta \quad , \quad M_\theta = \int\int \sigma_\theta z dz d\theta \tag{4.2-10}$$

$$N_{\phi\theta} = \int\int \tau_{\phi\theta} dz d\theta \quad , \quad M_{\phi\theta} = \int\int \tau_{\phi\theta} z dz d\theta$$

Using Eqs. (4.2-8) and (4.2-9) and (4.2-10), the forces and moments, based on the first order shell theory, become

$$N_\phi = C\left\{\frac{1}{R}(u,_\phi + w) + \frac{w,_\phi^2}{2R^2} + \frac{\nu}{R \sin\phi}(v,_\theta + u\cos\phi + w\sin\phi) + \frac{\nu w,_\theta^2}{2R^2 \sin^2\phi}\right\} - \frac{E\alpha T_1}{1-\nu}$$

$$N_\theta = C\left\{\frac{1}{R \sin\phi}(v,_\theta + u\cos\phi + w\sin\phi) + \frac{w,_\theta^2}{2R^2 \sin^2\phi} + \frac{\nu}{R}(u,_\phi + w) + \frac{\nu w,_\phi^2}{2R^2}\right\} - \frac{E\alpha T_1}{1-\nu}$$

$$N_{\phi\theta} = \frac{C(1-\nu)}{2}\left\{\frac{v,_\phi \sin\phi + u,_\theta - v\cos\phi}{R\sin\phi} + \frac{w,_\phi w,_\theta}{R^2 \sin\phi}\right\} \tag{4.2-11}$$

$$M_\phi = D\{\frac{-w,_{\phi\phi}}{R^2} - \frac{\nu}{R^2 \sin\phi}(\frac{w,_{\theta\theta}}{\sin\phi} + \cos\phi w,_\phi)\} - \frac{E\alpha T_2}{1-\nu}$$

$$M_\theta = D\{\frac{-1}{R^2 \sin\phi}(\frac{w_{\theta\theta}}{\sin\phi} + \cos\phi w,_\phi) - \frac{\nu w,_{\phi\phi}}{R^2}\} - \frac{E\alpha T_2}{1-\nu}$$

$$M_{\phi\theta} = \frac{D(1-\nu)}{R^2 \sin\phi}\{w,_\theta \cot\phi - w,_{\phi\theta}\}$$

where

$$C = \frac{Eh}{1-\nu^2} \qquad D = \frac{Eh^3}{12(1-\nu^2)}$$

$$T_1 = \int_{-h/2}^{h/2} T dz \qquad T_2 = \int_{-h/2}^{h/2} Tz dz \qquad (4.2\text{-}12)$$

To obtain the equilibrium equations, the functional of total potential energy is considered. The functional for a thin shallow spherical shell is

$$F = \frac{C}{2}\int_\phi\int_\theta(\epsilon_{\phi m}^2 + \epsilon_{\theta m}^2 + 2\nu\epsilon_{\phi m}\epsilon_{\theta m} + \frac{(1-\nu)}{2}\gamma_{\phi\theta m}^2)R^2 \sin\phi d\phi d\theta$$

$$+\frac{D}{2}\int_\phi\int_\theta(k_\phi^2 + k_\theta^2 + 2\nu k_\phi k_\theta + 2(1-\nu)k_{\phi\theta}^2)R^2 \sin\phi d\phi d\theta -$$

$$\frac{E\alpha}{1-\nu}\int_\phi\int_\theta\{(\epsilon_{\phi m} + \epsilon_{\theta m})T_1 + (k_\phi + k_\theta)T_2 + \alpha T'''\}R^2 \sin\phi d\phi d\theta$$

$$(4.2\text{-}13)$$

where $T''' = \int\int T^2 dz d\theta$. Substituting for strains and curvatures from Eqs. (4.2-8) into Eq. (4.2-13) and using the Euler equations, the functional of Eq. (4.2-13) is minimized resulting in the equilibrium equations of shallow spherical shell as

$$\cos\phi N_\theta - (\sin\phi N_\phi),_\phi - N_{\phi\theta,\theta} = 0$$

$$N_{\theta,\theta} + (\sin\phi N_{\phi\theta}),_\phi + \cos\phi N_{\phi\theta} = 0 \qquad (4.2\text{-}14)$$

$$(\sin\phi M_\phi),_{\phi\phi} + \frac{1}{\sin\phi}M_{\theta,\theta\theta} - [R\sin\phi(N_\phi\beta_\phi + N_{\phi\theta}\beta_\theta) + \cos\phi M_\theta],_\phi +$$

$$2[M_{\phi\theta,\phi\theta} + \cot\phi M_{\phi\theta,\theta}] - R\sin\phi(N_\phi + N_\theta) - R(N_\theta\beta_\theta + N_{\phi\theta}\beta_\phi),_\theta = 0$$

The stability equations are obtained from the second variation of the total potential energy function [42]. Ignoring the prebuckling rotations β_{θ_0} and β_{ϕ_0} and using the linearized strain-displacement relations, the Donnell-Mushtari-Velasov stability equations for the shallow spherical shell reduce to

$$(\sin\phi N_{\phi_1})_{,\phi} + N_{\phi\theta_1,\theta} - N_{\theta_1}\cos\phi = 0$$

$$(\sin\phi N_{\phi\theta_1})_{,\phi} + N_{\theta_1,\theta} + N_{\phi\theta_1}\cos\phi = 0$$

$$(\sin\phi M_{\phi_1})_{,\phi\phi} + 2(M_{\phi\theta_1,\phi\theta} + \cot\phi M_{\phi\theta_1,\theta}) + \frac{1}{\sin\phi}M_{\theta_1,\theta\theta}$$

$$-R\sin\phi(N_{\phi_1}+N_{\theta_1}) - (R\sin\phi N_{\phi_0}\beta_{\phi_1} + R\sin\phi N_{\phi\theta_0}\beta_{\theta_1} + M_{\theta_1}\cos\phi)_{,\phi}$$

$$-(RN_{\theta_0}\beta_{\theta_1} + RN_{\phi\theta_0}\beta_{\phi_1})_{,\theta} = 0$$

$$(4.2\text{-}15)$$

where the terms with the index (0) refer to the state of equilibrium and the terms with index (1) refer to the state of stability. In the stability equations, forces and moments related to the state of stability condition are

$$N_{\phi_1} = C\left[\frac{1}{R}(u_{1,\phi}+w_1) + \frac{\nu}{R\sin\phi}(v_{1,\theta}+u_1\cos\phi+w_1\sin\phi)\right]$$

$$N_{\theta_1} = C\left[\frac{1}{R\sin\phi}(v_{1,\theta}+u_1\cos\phi+w_1\sin\phi) + \frac{\nu}{R}(u_{1,\phi}+w_1)\right]$$

$$N_{\phi\theta_1} = \frac{C(1-\nu)}{2}\left[\frac{1}{R\sin\phi}(v_{1,\phi}\sin\phi+u_{1,\theta}-v_1\cos\phi)\right] \quad (4.2\text{-}16)$$

$$M_{\theta_1} = D\left[\frac{-1}{R^2\sin\phi}(\frac{w_{1,\theta\theta}}{\sin\phi}+\cos\phi w_{1,\phi}) - \frac{\nu w_{1,\phi\phi}}{R^2}\right]$$

$$M_{\phi_1} = D\left[\frac{-\nu w_{1,\phi\phi}}{R^2} - \frac{1}{R^2\sin\phi}\left(\frac{w_{1,\theta\theta}}{\sin\phi}+\cos\phi w_{1,\phi}\right)\right]$$

$$M_{\phi\theta_1} = \frac{D(1-\nu)}{R^2\sin\phi}(w_{1,\theta}\cot\phi - w_{1,\phi\theta})$$

For mechanical loading, the temperature terms in Eqs. (4.2-9), (4.2-11), and (4.2-13) are ignored.

For the shallow spherical cap, the rise H of the shell is assumed to be much smaller than the base radius a, see Fig. (4.2-1).

Form Fig. (4.2-1), $\sin\phi = r/R$ and $r_\phi d\phi = dr$. We further assume $\cos\phi \simeq 1$. Introduction of these values into Eqs. (4.2-15) gives

$$(rN_{r_1})_{,r} + N_{r\theta_1,\theta} - N_{\theta_1} = 0$$

$$(rN_{r\theta_1})_{,r} + N_{\theta_1,\theta} + N_{r\theta_1} = 0 \quad (4.2\text{-}17)$$

$$(rM_{r_1})_{,rr} + 2(M_{r\theta_1,r\theta} + \frac{1}{r}M_{r\theta_1,\theta}) + (\frac{1}{r}M_{\theta_1,\theta\theta} - M_{\theta_1,r}) - \frac{r}{R}(N_{r_1}+N_{\theta_1})$$

$$-[(rN_{r_0}\beta_{r_1} + rN_{r\theta_0}\beta_{\theta_1})_{,r} + (N_{r\theta_0}\beta_{r_1} + N_{\theta_0}\beta_{\theta_1})_{,\theta}] = 0$$

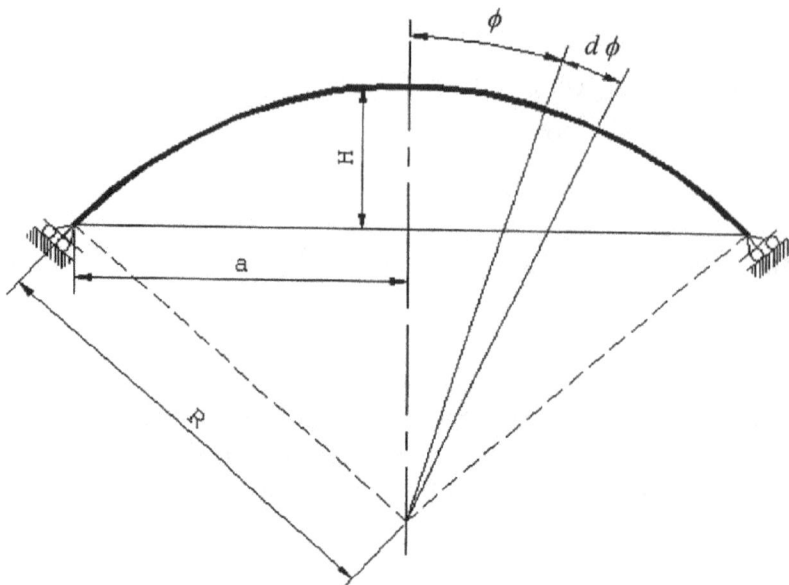

Figure 4.2-1: Shallow spherical cap.

where, from Eqs. (4.2-16)

$$N_{r_1} = C \left\{ u_{1,r} + \frac{w_1}{R} + \nu \left(\frac{v_{1,\theta} + u_1}{r} + \frac{w_1}{R} \right) \right\}$$

$$N_{\theta_1} = C \left\{ \frac{v_{1,\theta} + u_1}{r} + \frac{w_1}{R} + \nu \left(u_{1,r} + \frac{w_1}{R} \right) \right\} \qquad (4.2\text{-}18)$$

$$N_{r\theta_1} = C \frac{(1-\nu)}{2} \left\{ r \left(\frac{v_1}{r} \right)_{,r} + \frac{u_{1,\theta}}{r} \right\}$$

$$M_{r_1} = D \left\{ \beta_{r_1,r} + \frac{\nu}{r} (\beta_{\theta_1,\theta} + \beta_{r_1}) \right\}$$

$$M_{\theta_1} = D \left\{ \frac{1}{r} (\beta_{\theta_1,\theta} + \beta_{r_1}) + \nu \beta_{r_1,r} \right\}$$

$$M_{r\theta_1} = D \frac{(1-\nu)}{2} \left\{ r \left(\frac{\beta_{\theta_1}}{r} \right)_{,r} + \frac{\beta_{r_1,\theta}}{r} \right\}$$

$$\beta_{r_1} = -w_{1,r} \quad , \quad \beta_{\theta_1} = -\frac{w_{1,\theta}}{r}$$

Introducing the constitutive and kinematic relations for $M_{r_1}, M_{r\theta_1}$, and M_{θ_1} from Eqs. (4.2-18) in Eqs. (4.2-17) and decoupling the resulting equations in terms of w gives

$$D\nabla^4 w_1 + \frac{N_{r_1} + N_{\theta_1}}{R} - \frac{1}{r} \left\{ (rN_{r_0}w_{1,r} + N_{r\theta_0}w_{1,\theta})_{,r} + (N_{r\theta_0}w_{1,r} + N_{\theta_0}\frac{w_{1,\theta}}{r})_{,\theta} \right\} = 0$$

$$(4.2\text{-}19)$$

where

$$\nabla^2(\ ) \equiv \left\{(\ )_{,rr} + \frac{1}{r}(\ )_{,r} + \frac{1}{r^2}(\ )_{,\theta\theta}\right\}$$

$$\nabla^4(\ ) \equiv \nabla^2\nabla^2(\ ) \tag{4.2-20}$$

Equation (4.2-19) is identical for the mechanical and thermal loading.

2.3 Thermal and mechanical buckling, Wan-Donnell model

The Wan-Donnell model for the radial imperfection is [27]

$$w^* = \frac{K-1}{2}w \tag{4.2-21}$$

where the coefficient K is a constant value $0 \leq K \leq 1$. The value of $K = 1$ represents a perfect shell. The imperfection w^* is thus defined as a function of w, the lateral deflection of the shallow spherical cap. Considering an axisymmetric imperfection, due to the dependency of w to w^*, the lateral deflection w must be assumed to be axisymmetric. This assumption results in an axisymmetric buckling mode of the shallow spherical cap. The total strains at the middle surface of the shell in terms of the displacement components are

$$\epsilon^t_{rm_1} = u_{1,r} + \frac{w_1}{R}$$

$$\epsilon^t_{\theta m_1} = \frac{u_1 + v_{1,\theta}}{r} + \frac{w_1}{R} \tag{4.2-22}$$

$$\gamma^t_{r\theta m_1} = r(\frac{v_1}{r})_{,r} + \frac{u_{1,\theta}}{r}$$

The initial strains due to the initial axisymmetric imperfection w^* are

$$\epsilon^*_{rm_1} = \frac{w^*}{R}$$

$$\epsilon^*_{\theta m_1} = \frac{w^*}{R} \tag{4.2-23}$$

$$\gamma^*_{r\theta m_1} = 0$$

Thus, the net strains are obtained by replacing w_1 by $w_1 + w^*_1$ in Eqs. (4.2-22) and subtracting Eqs. (4.2-23) to yield

$$\epsilon_{rm_1} = u_{1,r} + \frac{w_1}{R}$$

$$\epsilon_{\theta m_1} = \frac{u_1 + v_{1,\theta}}{r} + \frac{w_1}{R} \qquad (4.2\text{-}24)$$

$$\gamma_{r\theta m_1} = r(\frac{v_1}{r})_{,r} + \frac{u_{1,\theta}}{r}$$

Considering a shallow spherical cap with simply supported edge condition and under axisymmetric mechanical and thermal loading, $N_{r\theta_0} = 0$. Since axisymmetric imperfection is considered, therefore $w_{1,\theta} = 0$, and Eq. (4.2-19) reduces to

$$D\nabla^4 w_1 + \frac{N_{r_1} + N_{\theta_1}}{R} - \frac{1}{r}(N_{r_0} w_{1,r} + r N_{r_0} w_{1,rr}) = 0 \qquad (4.2\text{-}25)$$

Substituting

$$w_1 = w_1 + w^*$$

$$N_{r_1} = \frac{1}{r}\Phi_{,r} + \frac{1}{r^2}\Phi_{,\theta\theta} \qquad (4.2\text{-}26)$$

$$N_{\theta_1} = \Phi_{,rr}$$

into Eq. (4.2-25), yields

$$D\nabla^4 w_1 + \frac{1}{R}\nabla^2\Phi - \frac{N_{r_0}}{r}(w_{1,r} + r w_{1,rr} + w^*_{,r}) = 0 \qquad (4.2\text{-}27)$$

In Eqs. (4.2-26), Φ is the Airy stress function. To obtain a second equation relating the dependent function w_1 and Φ, the compatibility equation may be used, as follows

$$\frac{1}{r^2}\epsilon_{rm_1,\theta\theta} - \frac{1}{r}\epsilon_{rm_1,r} + \frac{1}{r^2}(r^2\epsilon_{\theta m_1,r})_{,r} - \frac{1}{r^2}(r\gamma_{r\theta m_1})_{,r\theta} = 0 \qquad (4.2\text{-}28)$$

Substituting for strains from Eqs. (4.2-24) and using the Hooke's law, the compatibility equation in terms of the Airy stress function reduces to

$$\nabla^4\Phi = \frac{Eh}{R}\nabla^2 w_1 \qquad (4.2\text{-}29)$$

Equations (4.2-27) and (4.2-29) are the basic equations used to obtain the critical buckling loads of imperfect spherical cap. Based on these equations, thermal and mechanical buckling loads of imperfect spherical cap of isotropic and homogeneous material are determined.

I - Mechanical buckling load

A thin shallow spherical cap of thickness h and base radius a with rise H is considered. The edge is assumed to be simply supported, where the boundary conditions at $r = a$ are

$$w = w_{,rr} = 0 \tag{4.2-30}$$

We consider a spherical cap subjected to uniform external pressure P_e. The prebuckling loads, using the shell membrane equations, are

$$N_{r_0} = N_{\theta_0} = \frac{-P_e R}{2} \tag{4.2-31}$$

Substituting Eq. (4.2-31) into Eqs. (4.2-27) and (4.2-29) gives

$$\begin{cases} S\nabla^4 \overline{w}_1 + \dfrac{\nabla^2 \psi}{hR} + \dfrac{P_e R}{2Erh^3}(\overline{w}_{1,r} + \overline{w}^*_{,r} + r\overline{w}_{1,rr}) = 0 \\[4mm] hR\nabla^4 \psi = \nabla^2 \overline{w}_1 \end{cases} \tag{4.2-32}$$

where

$$S = \frac{1}{12(1-\nu^2)} \quad \overline{w}_1 = \frac{w_1}{h} \quad \overline{w}^* = \frac{w^*}{h}$$

$$\psi = \frac{\Phi}{Eh^3} \qquad \overline{N}_{r_0} = \frac{N_{r_0}}{Eh} \tag{4.2-33}$$

To solve the system of Eqs. (4.2-32), with the consideration of the boundary conditions (4.2-30), the approximate solutions are assumed as

$$\begin{aligned} \overline{w}_1 &= A\sin\frac{m\pi r}{a} \\ \psi &= B\sin\frac{m\pi r}{a} \end{aligned} \qquad 0 \le r \le a \tag{4.2-34}$$

where A and B are constant coefficients. The system of Eqs. (4.2-32) are made orthogonal with respect to the approximate solution (4.2-34) according to the Galerkin method

$$\int_r \left(S\nabla^4 \overline{w}_1 + \frac{\nabla^2 \psi}{hR} + \frac{P_e R}{2Erh^3}[\overline{w}_{1,r} + \overline{w}^*_{,r} + r\overline{w}_{1,rr}] \right) \times \sin\frac{m\pi r}{a} dr = 0 \tag{4.2-35}$$

$$\int_r \left(hR\nabla^4 \psi - \nabla^2 \overline{w}_1 \right) \times \sin\frac{m\pi r}{a} dr = 0$$

The determinant of the system of Eqs. (4.2-35) for the coefficients A and B is set equal to zero and the resulting equation is solved and minimized with respect to m for $(P_e)_{cr}$ to give

$$\boxed{(P_e)_{cr} = 0.25E(\frac{h}{R})^2(\frac{a}{R})^2H_1^{-1}} \qquad (4.2\text{-}36)$$

where $H_1 = \dfrac{K+1}{9}\pi^2 - \dfrac{K}{2}$ and $\nu = 0.3$

II - Thermal buckling load

A. Uniform temperature rise

The spherical cap is initially at uniform temperature T_i and is raised to a uniform temperature T_f, such that the temperature rise is $\Delta T = T_f - T_i$. The prebuckling forces are obtained from the membrane shell equations and are

$$N_{r_0} = N_{\theta_0} = \beta\Delta T \qquad (4.2\text{-}37)$$

where $\beta = \dfrac{-E\alpha h}{1 - \nu}$. Introducing Eq. (37) into Eqs. (27) and (29) gives

$$S\nabla^4\overline{w}_1 + \frac{\nabla^2\psi}{hR} - \frac{\beta'\Delta T}{rh^2}(\overline{w}_{1,r} + \overline{w}^*_{,r} + r\overline{w}_{1,rr}) = 0 \qquad (4.2\text{-}38)$$

$$hR\nabla^4\psi = \nabla^2\overline{w}_1$$

where

$$\beta' = \frac{\beta}{Eh} \qquad (4.2\text{-}39)$$

The same approximate solutions (4.2-34) are considered and the system of Eqs. (4.2-38) is made orthogonal with respect to the approximate solutions (4.2-34) according to the Galerkin method

$$\int_r (S\nabla^4\overline{w}_1 + \frac{\nabla^2\psi}{hR} - \frac{\beta'\Delta T}{rh^2}[\overline{w}_{1,r} + \overline{w}^*_{,r} + r\overline{w}_{1,rr}]) \times \sin\frac{m\pi r}{a}dr = 0 \qquad (4.2\text{-}40)$$

$$\int_r (hR\nabla^4\psi - \nabla^2\overline{w}_1) \times \sin\frac{m\pi r}{a}dr = 0$$

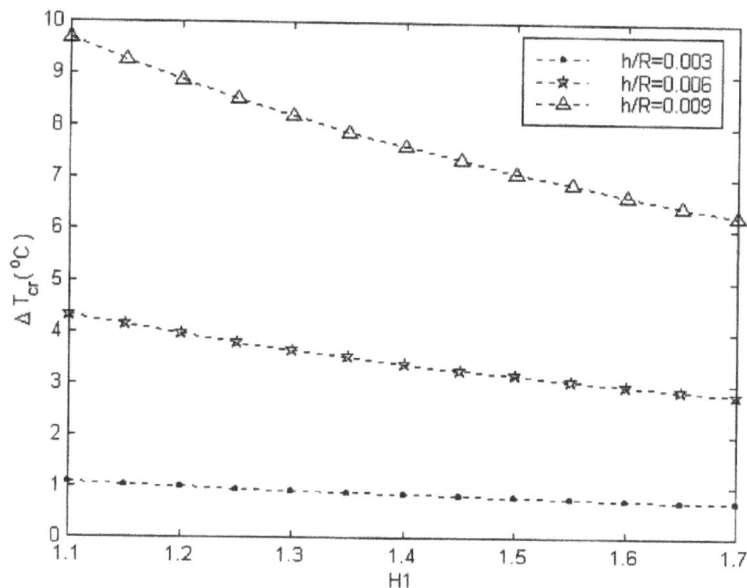

Figure 4.2-2: Variations of the critical uniform temperature rise based on Wan-Donnell model versus H_1 for different values of h/R.

The determinant of the system of Eqs. (4.2-40) for the coefficients A and B is set equal to zero and the resulting equation is solved and minimized with respect to m for $(\Delta T)_{cr}$ to give

$$\boxed{(\Delta T)_{cr} = \frac{1.54}{H_1 \alpha} (\frac{h}{R})^2}$$

(4.2-41)

B. Radial temperature difference

Consider a linear radial temperature distribution across the shallow spherical cap thickness as

$$T(z) = \frac{\Delta T}{h}(z + \frac{h}{2}) + T_a \qquad -\frac{h}{2} \le z \le \frac{h}{2}$$

(4.2-42)

where $\Delta T = T_b - T_a$, T_a is temperature of inner surface and T_b is temperature of outer surface of the spherical cap. The prebuckling forces developed in the shallow spherical cap due to this temperature distribution are

$$N_{r_0} = N_{\theta_0} = \frac{-E\alpha h}{1-\nu}\frac{\Delta T}{2} = \beta\frac{\Delta T}{2}$$

(4.2-43)

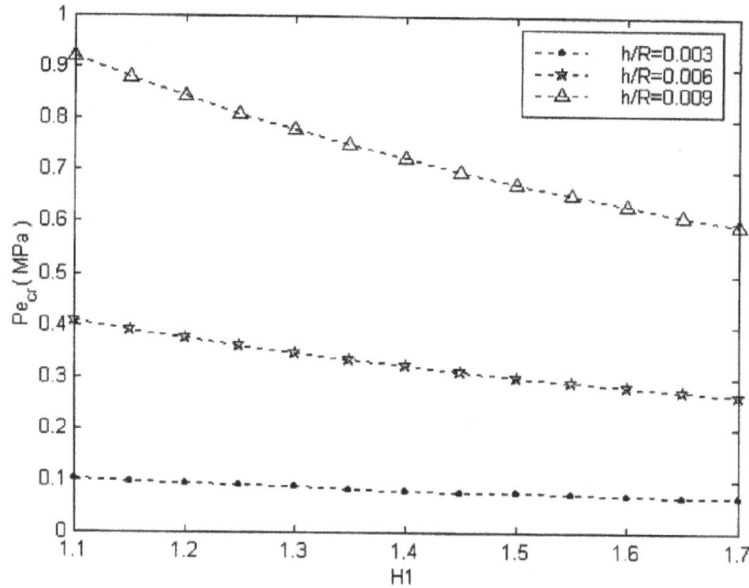

Figure 4.2-3: Variations of the critical uniform external pressure based on Wan-Donnell model versus H_1 for different values of h/R.

Comparing Eqs. (4.2-37) and (4.2-43), the critical buckling thermal load associated with the radial temperature rise for $\nu = 0.3$ is

$$\Delta T_{cr} = \frac{3.08}{H_1 \alpha} (\frac{h}{R})^2 \qquad (4.2\text{-}44)$$

It is noted that the buckling thermal load of radial temperature difference is twice that of uniform temperature rise.

2.4   Results and discussion

The results of this section may be summarized. Thermal and mechanical buckling analysis of isotropic imperfect shallow spherical cap based on Wan-Donnell model with simply supported edge conditions for one type of mechanical loading and two types of thermal loading result in closed form solutions. The thermal and mechanical buckling loads are directly proportional to E and $(h/R)^2$ and $(a/R)^2$ and inversely proportional to the parameters α and H_1.   Figure (4.2-2) shows the critical buckling temperature of an imperfect shallow spherical cap under uniform temperature rise based on the Wan-Donnell imperfection model in terms of the parameters H_1, h/R. The buckling load related to the uniform temperature rise versus the imperfection parameter H_1 for different values of

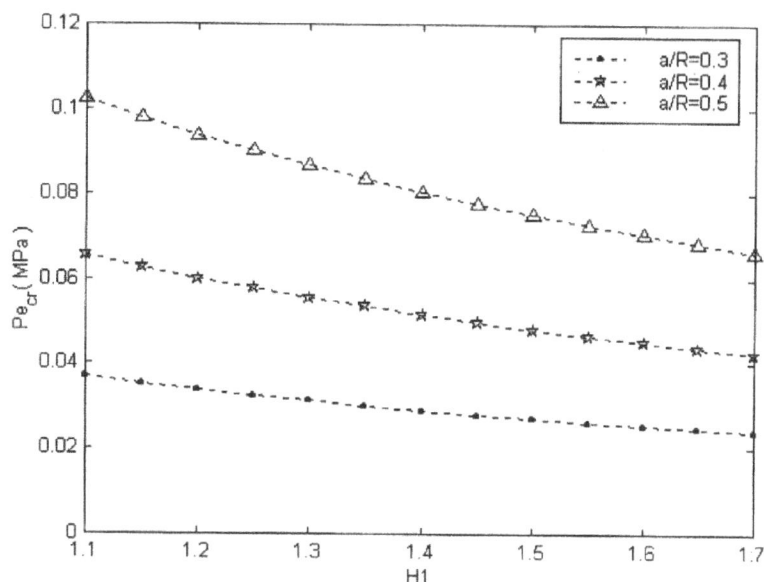

Figure 4.2-4: Variations of the critical uniform external pressure based on Wan-Donnell model versus H_1 for different values of a/R.

h/R is shown in this figure. It is seen that the buckling temperature rise decreases by the increase of H_1 and increases by the increase of h/R. In this example $\alpha = 11.7 \times 10^{-6} 1/C$ is considered. The buckling temperature associated with the gradient though the thickness is obtained using Fig. (4.2-2) with the values of $\Delta T_{radial} = 2\Delta T_{uniform}$.

Figures (4.2-3) and (4.2-4) show the critical buckling loads of an imperfect shallow spherical cap under uniform external pressure based on the Wan-Donnell imperfection model in terms of the parameters $H_1, h/R$, $a/R, E$. Figure (4.2-3) shows the critical buckling load versus the imperfection parameter H_1 for different values of h/R. In this example, $E = 200$ GPa and $\frac{a}{R} = 0.5$ are considered. Figure (4.2-4) shows the critical buckling load versus the imperfection parameter H_1 for different values of a/R. In this example, $E = 200$ GPa and $\frac{h}{R} = 0.003$ are considered. It is seen that the buckling load decreases by the increase of H_1 and increases by the increases of $(h/R)^2$ and $(a/R)^2$.

3   References

1 - Morris, N.F., Shell Stability; The Long Road from Theory to practice, J. Engineering Structures, Vol 18, pp. 801-806, 1996.

2 - Kaplan, A., *Buckling of Spherical Shells, Thin Shell Structures; Theory, Experiment and Design*, Fung Y.C. and Sechler E.E. (Eds), Prentice Hall, Englewood Cliffs, pp. 247-288, 1974.

3 - Timoshenko, S.P., and Gere, J.M., *Theory of Elastic Stability*, McGraw-Hill, 1961.

4 - Karman, Th. von, and Tsien, H.S., Buckling of Spherical Shells by External Pressure, J. of Aerospace Science, Vol. 7, pp. 43-50, 1939.

5 - Budiansky, B., Buckling of Clamped Shallow Spherical Shells, Proc. IUTAM Symp. on the Theory of Thin Elastic Shells, Delt, Netherlands, Koiter, W.T., Ed., pp. 64-94, 1959.

6 - Weinitschke, H.J., Asymmetric Buckling of Clamped Shallow Spherical Shells, NASA TN D-1510, pp. 481-490, 1962.

7 - Huang, N.C., Unsymmetrical Buckling of Thin Shallow Spherical Shells, J. Applied Mechanics, Trans. ASME, Vol. 1.31, pp. 447-457, 1964.

8 - Krenzke, M.A., and Kiernan, I.J., Elastic Stability of Near Perfect Shallow Spherical Shells, AIAA J., Vol. 1, pp. 2855-2857, 1963.

9 - Penning, F.A., and Thurston, C.A., The Stability of Shallow Spherical Shells under Concentrated Load, NASA Contract Rept. 265, 1965.

10 - Bushnell, D., Bifurcation Phenomena in Spherical Shells under Concentrated Loads, AIAA J., Vol. 5, pp. 2034-2040, 1967.

11 - Fitch, J.R., The Buckling and Postbuckling Behavior of Spherical Caps under Concentrated Load, Int. J. Solids and Structures, Vol. 4, pp. 421-466, 1968.

12 - Fitch, J.R., and Budiansky, B., Buckling and Postbuckling Behavior of Spherical Caps under Axisymmetric Load, AIAA J., Vol. 8, pp. 686-693, 1970.

13 - Akkas, N., Asymmetric Buckling Behavior of Spherical Caps under Uniform Step Pressures, J. Applied Mechanics, Trans. ASME, Vol. 39, pp. 293-294, 1972.

14 - Hoff, N., Buckling at High Temperature, J. Royal Aeronautical Sci-

ences, Vol. 61, pp. 756-774, 1957.

15 - Hoff, N., Buckling of Thin Cylindrical Shells under Hoop Stress Varying in the Axial Direction, J. Applied Mechanics, Trans. ASME, Vol. 24, pp. 405-412, 1957.

16 - Abir, D., and Nardo, S.Y., Thermal Buckling of Circular Cylindrical Shells under Circumferential Temperature Gradients, J. Aeronautical Sciences, Vol. 26, pp. 803-808, 1959.

17 - Hill, D.W., Buckling of a Thin Circular Cylindrical Shell Heated along an Axial Strip, SUDAER 88, Stanford University, 1959.

18 - Anderson, M.S., Thermal Buckling of Cylinders, Collected Papers on Instability of Shell Structures, NASA TN-D-1510, pp. 255-265, 1962.

19 - Johns, D.J., Local Circumferential Buckling of Thin Circular Cylindrical Shells, Collected Papers on Instability of Shell Structures, NASA TN-D-1510, pp. 267-276, 1962.

20 - Zuk, W., Thermal Buckling of Clamped Cylindrical Shells, J. Aeronautical Sciences, Vol. 24, pp. 247-257, 1962.

21 - Ross, B., Mayers, J., and Jaworski, A., Buckling of a Thin Cylindrical Shell Heated along an Axial Strip, Experimental Mechanics, pp. 255-265, 1965.

22 - Lu, S.Y., and Chang, L.K., Thermal Buckling of Conical Shells, J. Applied Mechanics, Trans. ASME, Vol. 5, 1967.

23 - Kovalenko, A.D., and Alblas, J.B., *Thermoelasticity, Basic Theory and Applications*, Appendix on Thermoelastic Stability, Wolters-Noordhoff Publications, 1969.

24 - Chen, L.W., and Chen, L.Y., Thermal Buckling Analysis of Cylindrical Plates by Finite Element Analysis, Composite Structures, Elsevier Science Publishers Ltd, Vol. 34, **5,** pp. 71-78, 1990.

25 - Eslami, M.R., Ziaii, A.R., and Ghorbanpour, A., Thermo-Elastic Buckling of Thin Cylindrical Shells Based on Improved Stability Equations, J. Thermal Stresses, Vol. 19, No. 4, PP. 299-316, 1996.

26 - Eslami, M.R. and Rafeeyan. M., Thermal and Mechanical Buckling

of Conical Shells, Montreal Proceeding of 8th Int. Conf. on Pressure Vessel Tech., ICPVT-8, Part 2, 1996.

27 - Brush, D.O. and Almroth, B.O., *Buckling of Bars, Plates and Shells*, McGraw Hill, New York, 1975.

28 - Donnell, L.H., *Beams, Plates and Shells*, McGraw-Hill, New York, 1977.

29 - Muc, A., Transverse Shear Effects in Stability Problems of Laminated Shallow Shells, Composite Structures, Elsevier Science Publishers Ltd., Vol. 12, pp. 171-180, 1989.

30 - Balourchi, S., On Finite Element Nonlinear Analysis of General Shell Structure, Ph.D. diss., M.I.T., 1979.

31 - Zoelly, R., Ueber ein Krichungra Problem und der Kugclschale, Thesis, Switzerland, 1915.

32 - Budiansky, B., and Roth R.S., Axisymmetric Dynamic Buckling of Clamped Shallow Spherical Caps, NASA TN, Washington, 1962.

33 - Famili, J., and Archer, R.R., Finite Asymmetric Deformation of Shallow Spherical Shells, J. AIAA, Vol. 3, pp. 100-120, 1965.

34 - Eggwerta, S.F., and Samuelson, L.A. Buckling Strength of Spherical Shells, J. Cons. Steel Research, Vol. 17, 1990.

35 - Goncalves, B., and Croll, G.A., Axisymmetric Buckling of Pressure Loaded Spherical Caps, J. Struc. Eng, Vol. 118, 1992.

36 - Yeh Kiyuan, S., and Weiping, S., Axisymmetric Buckling of Thin Shallow Spherical Shells under Uniform Pressure for Large Parameter λ, Int. J. Non-linear Mech., Vol. 29, 1994.

37 - Ganapathi, M., and Varadan, T.K., Dynamic Buckling of Orthotropic Shallow Spherical Shells, Computer and Structures, Vol. 15, 1982.

38 - Chao, C.C., and Lin, I.S., Static and Dynamic Snap-through of Orthotropic Spherical Caps, Composite Structures, Vol. 14, 1990.

39 - Ganapathi, M., and Varadan, T.K., Dynamic Buckling of Laminated Anisotropic Spherical Caps, J. Appl. Mech., Vol. 62, 1995.

40 - Aleksander, M., On the Buckling of Composite Shells of Revolution under External Pressure, Composite Structures, Vol. 12, 1992.

41 - Eslami, M.R., and Shariyat, M., Dynamic Buckling and Postbuckling of Imperfect Orthotropic Cylindrical Shells under Mechanical and Thermal Loads Based on the Three-dimensional Theory of Elasticity, J. Appl. Mech, Trans. ASME, Vol. 66, pp. 476-484. June 1999.

42 - Shariyat, M., Eslami, M.R., and Shakeri, M., Elastic-Plastic Dynamic Buckling Analysis of Imperfect Cylindrical Shells, Amirkabir Journal of Science and Tech., Vol. 11, No. 42, pp. 35-46, 1999.

43 - Eslami, M.R., and Shariyat, M., On Thermal Dynamic Buckling Analysis of Imperfect Laminated Cylindrical Shells, ZAMM, Vol. 80, No. 3, pp. 171-182, 2000.

44 - Eslami, M.R., and Shahsiah, R., Thermal Buckling of Imperfect Cylindrical Shells, J. Thermal Stresses, Vol. 24, No. 1, pp. 71-90, 2001.

Chapter 5

Dynamic Buckling of Cylindrical Shells

1   Introduction

Thin cylindrical shell elements employed in construction of the air frames of the high-speed aerospace vehicles, aircraft fuselages in supersonic flight and in many engineering applications, are subjected to significant transient heating. These elements experience higher temperature than the restraining frames and are subjected to thermal dynamic buckling.

Investigation of buckling phenomenon is often concentrated on studing the behavior of the flat structures such as beams and plates. Besides, most of the investigations made in buckling analysis of circular cylindrical shells are based on static buckling of perfect shells. Static buckling analysis of multilayer composite shells began from the 1970 decade (for example, Booton and Tennyson [1], Jones and Hennemann [2] and Sheinman, Shaw and Simitses [3,4]), Simitses and Anastasiadis [5], Shen and Williams [6], Song et al. [7], Kardomateas et al. [8-12], and Eslami and Shariyat [13]. However, the study of dynamic buckling of composite cylindrical shells were restricted on buckling under mechanical loads. Some of these studies are done for example by Saigal and Yang [14], Liaw and Yang [15], Argento and Scott [16] and Kasagi and Sridharan [17].

Laminated cylindrical shells are often modeled as equivalent single-layer shell using Love-Kirchhoff hypothesis in which transverse normal stress and shear deformations are neglected. In contrast to homogeneous isotropic materials, due to low transverse shear moduli of modern composite materials to their in-plane tensile moduli, transverse shear deformation effects may be significant, even in thin composite shells.

The classical shell theory suggested by Donnell [18] have formed the basis for dynamic stability analysis of many works in the literature. The transverse shear and normal stress and strains are disregarded in this approach. Results of Kardomateas and Philobos [12] for mechanical buckling of single-layer cylindrical shells, reveals that disregarding such effects can lead to approximately as much as 120% or more errors in prediction of the buckling loads and the discrepancy increases as the wall thickness of the shell increases.

Almost all of the above studies employed approximate strain-displacement expressions that are based on the classical shell theories assumptions (such as Love-Timoshenko or using some moderate displacement, Von Karman or Sanders assumptions).

Much less research has occurred for buckling of composite cylindrical shell in a thermal environment. Investigations regarding thermal buckling phenomenon are often restricted to studying the static behavior of the flat structures [19]. Generally, the existing studies on thermal buckling analysis of circular cylindrical shells are based on static buckling consideration of a perfect shell and belong to the classic equivalent single-layer theories (for example, Radhamohan and Venkataramana [20], Thangaratnam et al. [21], Ma and Wilcox [22], Huang and Tauchert [23], Kossira and Haupt [24], Weller and Patlashenko [25]). Thornton [26] summarizes researches done in thermal buckling of plates and shells.

In many references effect of the transverse normal and shear stresses and strains are accounted for by means of higher order strain-displacement expressions [3,5,27-31]. The formulations proposed so far, are displacement based. Therefore, only displacement type boundary conditions are accurately satisfied in their solutions. These formulations are suitable for semi-analytical solution. Recently, a high order shear-deformation theory is proposed by Eslami and Shariyat [32] which is most suitable to satisfy the stress and displacement boundary conditions.

The mentioned works belong to a class of shell theories called *equivalent single-layer theories*. These theories are characterized by the continuity of the value and first derivatives of the displacement components. Although these theories provide a sufficiently accurate description of the global laminate response, they are not capable of accurately determining interlaminar stresses near discontinuities such as holes, traction-free edges and layer interfaces. For determining the three dimensional stress field at the ply level, another class of theories, called *layer-wise theories* are provided. Improvement of this type of theories is mainly due to Reddy [33,34]. Although the proposed formulations can produce much more accurate results, they are displacement-based and suffer from the point that moment, force or stress boundary conditions (for example, simply-supported edge conditions) can not be incorporated accurately in these formulations. Recently, the author have proposed a new general layer-wise theory which can overcome this problem [35].

More recently, efforts for introducing a three dimensional elasticity solution for the static buckling of perfect, single-layered thick composite cylindrical shells under mechanical loads alone, are done by Kardomateas et al. [8-12]. In the formulations of References [8-12] the transverse shear deformability and normal stress are automatically incorporated and the dependency on a particular shell theory is eliminated. The proposed formulations are rather simple, linear, static and disregard the thermal effects and are suitable for single-layer thick composite shells. Besides, one of the Fourier triple series terms is adopted in the solution (which can be justified for thick shells only) and the boundary conditions are only satisfied geometrically. Thus, the problem is reduced to a standard eigenvalue problem that is solved by discretizing the shell thickness and using the finite difference method in this direction. A technique proposed by Eslami and Shariyat [42] which overcome the above shortcomings and give more accurate results.

In the present chapter, the three-dimensional theory of elasticity in curvilinear coordinates is employed to investigate the dynamic buckling of an imperfect orthotropic circular cylindrical shell under mechanical and thermal loading. Mechanical loads may be composed of axial compression, external pressure, torsion or a combination of them. For this purpose, instead of employing the conventional infinitesimal displacements or using the classic assumptions, Green's strain tensor for large deflections in curvilinear coordinates is used and the strain expressions of an imperfect circular cylindrical shell are derived, first. Then, an efficient and relatively accurate solution algorithm for treatment of the highly nonlinear equations resulting from this approach are introduced. It is noted that the solution of a cylindrical shell under mechanical loads is boundary conditions driven whereas it is constitutive equation driven in the case of thermal loading. Based on the energy form of the accurate three dimensional theory of elasticity, a high-order shear deformation, a high-order deformation and a generalized layer-wise theory are developed. Since the governing equations are proposed in a hybrid form consists of both stress (or its resultants) and displacement components, the displacement as well as the stress (or force or moment) boundary conditions can be satisfied exactly.

In each section of this chapter, a few examples of some well-known references done on the basis of the other theories in the mechanical and thermal buckling field are reexamined and a sensitivity analysis against the initial geometric deviation, dimensional changes, mechanical and thermal properties of the materials, is accomplished.

2   Three-Dimensional Theory of Elasticity

In this section, equations of the three-dimensional theory of elasticity in curvilinear coordinates that are also the basis for preceedind sections are introduced and employed to investigate the dynamic buckling of an imperfect orthotropic circular cylindrical shell under mechanical and thermal loading.

2.1   Formulation

Strain-displacement and stress-strain expressions

As it will be explained later, the response of the cylindrical shell under a time varying load is investigated to find the buckling point. Then, this approach requires that behavior of the shell slightly beyond the buckling point (postbuckling) to be studied, too. For this reason, in this chapter, a general formulation based on considering the possibility of large deflection occurrence is adopted.

The position vector of a point before deformation is given by

$$\vec{R}^{(0)} = \vec{R}^{(0)}(\alpha^1, \alpha^2, \alpha^3) \tag{5.2-1}$$

where ($\alpha_1, \alpha_2, \alpha_3$) are the curvilinear coordinates of the point. After deformation, the point moves to a new position, the position vector to which is denoted by

$$\vec{R} = \vec{R}(\alpha^1, \alpha^2, \alpha^3) \tag{5.2-2}$$

If the covariant base vectors associated with the initial configuration and after deformation are denoted by $\vec{g}_\lambda$ and $\vec{G}_\lambda$, respectively, then we have

$$\vec{g}_\lambda = \vec{R}^{(0)}_{,\lambda} \qquad\qquad \vec{G}_\lambda = \vec{R}_{,\lambda} \tag{5.2-3}$$

Defining the displacement vector, $\vec{U}(\alpha^1, \alpha^2, \alpha^3)$ by

$$\vec{R} = \vec{R}^{(0)} + \vec{U} \tag{5.2-4}$$

we have

$$\vec{G}_\lambda = \frac{\partial x^k}{\partial \bar{x}^\lambda}\vec{g}_k = (\delta^k_\lambda + u^k_{;\lambda})\vec{g}_k \tag{5.2-5}$$

where x and $\bar{x}$ are the coordinate systems corresponding to initial and final configurations, respectively, δ is the Kronecker symbol and the symbol ";" stands for the covariant derivative. Since the initial dimensions are considered in calculations, Green's strain tensor for large deformations is used in its general form which can be written in curvilinear coordinates as (Washizu [37])

$$f_{ij} = \frac{1}{2}(G_{ij} - g_{ij}) \tag{5.2-6}$$

where g_{ij} and G_{ij} are the Euclidian metric tensors defined as

$$g_{ij} = \vec{g}_i.\vec{g}_j \qquad\qquad G_{ij} = \vec{G}_i.\vec{G}_j \tag{5.2-7}$$

Substituting Eq. (5.2-5) into Eq. (5.2-6), the following expressions of the strain in terms of the displacement components are obtained. After rearrangement

$$f_{ij} = \frac{1}{2}(U_{i;j} + U_{j;i} + U_{\alpha;i}U^{\alpha}_{;j}) \tag{5.2-8}$$

The tensorial and physical components of the displacements and strains can be related through the following expressions (Flugge [38])

$$u_i = \sqrt{g^{ii}}U_i = \sqrt{g_{ii}}U^i$$
$$\epsilon_{ij} = \frac{f_{ij}}{\sqrt{g_{ii}}\sqrt{g_{jj}}} \tag{5.2-9}$$

where g^{ij} are the associated metric tensors. Expansion of Eq. (5.2-8) leads to the strain-displacement relations of a perfect circular cylindrical shell. These equations, after rearrangement and employing Eq. (5.2-9), are

$$\epsilon_{rr} = w_{,r} + \frac{1}{2}[(w_{,r})^2 + (v_{,r})^2 + (u_{,r})^2]$$

$$\epsilon_{\theta\theta} = \frac{1}{r}(v_{,\theta} + w) + \frac{1}{2r^2}[(v_{,\theta} + w)^2 + (w_{,\theta} - v)^2 + (u_{,\theta})^2]$$

$$\epsilon_{xx} = u_{,x} + \frac{1}{2}[(w_{,x})^2 + (v_{,x})^2 + (u_{,x})^2]$$

$$\epsilon_{r\theta} = \frac{1}{2r}[rv_{,r} + (w_{,\theta} - v)(w_{,r} + 1) + (v_{,\theta} + w)v_{,r} + u_{,r}u_{,\theta}]$$

$$\epsilon_{rx} = \frac{1}{2}[w_{,x}(1 + w_{,r}) + u_{,r} + v_{,r}v_{,x} + u_{,r}u_{,x}]$$

$$\epsilon_{x\theta} = \frac{1}{2}(v_{,x} + \frac{1}{r}u_{,\theta}) + \frac{1}{2r}[v_{,x}(v_{,\theta} + w) + u_{,\theta}u_{,x} + w_{,x}(w_{,\theta} - v)]$$

$$\tag{5.2-10}$$

where u, v, and w are the displacement components in the axial, circumferential and radial directions, respectively. For an imperfect shell, some initial strains exist due to the initial deformations of the shell. These imperfections are practically in radial direction (considering presence of initial axial and circumferential imperfections does not alter the generality of present discussion). Thus, the initial strains $\epsilon_{ij0} \equiv \epsilon_{ij0}(w_0)$, where w_0 is the initial radial imperfection. Expressions of these strains can be obtained from Eq. (5.2-10) by substituting u, $v = 0$, and $w = w_0$. Thus, strains due to loading are derived by subtracting the initial strains from the final strains

$$\epsilon_{ij} = \tilde{\epsilon}_{ij} - \epsilon_{ij0} \tag{5.2-11}$$

$\tilde{\epsilon}_{ij} \equiv \tilde{\epsilon}_{ij}(u, v, w + w_0)$ are the final strains. Therefore, referring to Eqs. (5.2-10) and (5.2-11), the strain-displacement relations for imperfect shells become

$$\epsilon_{rr} = w_{,r} + \frac{1}{2}[(w_{,r})^2 + 2w_{,r}w_{0,r} + (v_{,r})^2 + (u_{,r})^2]$$

$$\epsilon_{\theta\theta} = \frac{1}{r}(v_{,\theta} + w) + \frac{1}{2r^2}\{(v_{,\theta} + w + w_0)^2 + [(w + w_0)_{,\theta} - v]^2 + (u_{,\theta})^2 - w_0^2 - (w_{0,\theta})^2\}$$

$$\epsilon_{xx} = u_{,x} + \frac{1}{2}[(w_{,x})^2 + 2w_{,x}w_{0,x} + (v_{,x})^2 + (u_{,x})^2]$$

$$\epsilon_{r\theta} = \frac{1}{2}v_{,r} + \frac{1}{2r}[(w_{,\theta} - v)(w_{,r} + 1) - vw_{0,r} + w_{,r}w_{0,\theta} + w_{,\theta}w_{0,r} + (v_{,\theta} + w + w_0)v_{,r} + u_{,r}u_{,\theta}]$$

$$\epsilon_{rx} = \frac{1}{2}(w_{,x}(1 + w_{,r}) + w_{0,x}w_{,r} + w_{0,r}w_{,x} + u_{,r} + v_{,r}v_{,x} + u_{,r}u_{,x})$$

$$\epsilon_{x\theta} = \frac{1}{2}(v_{,x} + \frac{1}{r}u_{,\theta}) + \frac{1}{2r}[v_{,x}(v_{,\theta} + w + w_0) + u_{,\theta}u_{,x} + (w + w_0)_{,x}(w_{,\theta} - v) + w_{,x}w_{0,\theta}] \tag{5.2-12}$$

The stress-strain relation of an orthotropic body in the orthotropy axes can be expressed as

$$\{\sigma^*\} = [Q]\{\epsilon^*{}_M\} \tag{5.2-13}$$

where $\{\sigma^*\}$ and $\{\epsilon^*{}_M\}$ are the stress components and mechanical strain components and $[Q]$ is the stiffness matrix, in the orthotropy axes

$$[Q] = [S]^{-1} \tag{5.2-14}$$

and

$$[S] = \begin{pmatrix} \frac{1}{E_1} & \frac{-\nu_{21}}{E_2} & \frac{-\nu_{31}}{E_3} & 0 & 0 & 0 \\ \frac{-\nu_{12}}{E_1} & \frac{1}{E_2} & \frac{-\nu_{32}}{E_3} & 0 & 0 & 0 \\ \frac{-\nu_{13}}{E_1} & \frac{-\nu_{23}}{E_2} & \frac{1}{E_3} & 0 & 0 & 0 \\ 0 & 0 & 0 & \frac{1}{2G_{12}} & 0 & 0 \\ 0 & 0 & 0 & 0 & \frac{1}{2G_{13}} & 0 \\ 0 & 0 & 0 & 0 & 0 & \frac{1}{2G_{23}} \end{pmatrix} \quad (5.2\text{-}15)$$

Equations (5.2-13) to (5.2-15) are valid for all orthogonal coordinate systems, including the cylindrical coordinate system (Vinson [39]).

In the general case

$$\{\epsilon^*_M\} = \{\epsilon^* - \alpha^* \Delta T\} \quad (5.2\text{-}16)$$

and

$$\{\sigma^*\} = [Q]\{\epsilon^* - \alpha^* \Delta T\} \quad (5.2\text{-}17)$$

In these equations, α^*_{ij} are the coefficients of the thermal expansion tensor in the orthotropy axes. Transformation of the stress and strain tensors from the orthotropy coordinate of the orthotropic shell to the geometrical coordinates, yields the relation between the stress and strain components in these coordinates as

$$\{\epsilon\} = [\bar{T}]\{\epsilon^*\} \qquad \{\sigma\} = [\bar{T}]\{\sigma^*\} \quad (5.2\text{-}18)$$

or regarding Eqs. (5.2-13) to (5.2-18)

$$\{\sigma\} = [\bar{Q}]\{\epsilon - \alpha \Delta T\} \quad (5.2\text{-}19)$$

where the transformed stiffness and coefficients of thermal expansion matrices in geometrical coordinates, $[\bar{Q}]$ and $\{\alpha_{ij}\}$, are

$$[\bar{Q}] = [\bar{T}][Q][T]^{-1}$$
$$\{\alpha\} = [\bar{T}]\{\alpha^*\} \quad (5.2\text{-}20)$$

$[\bar{T}]$ is the transformation matrix which is defined as

$$[\bar{T}] = \begin{pmatrix} 0 & 0 & 1 & 0 & 0 & 0 \\ \sin^2\theta & \cos^2\theta & 0 & \sin 2\theta & 0 & 0 \\ \cos^2\theta & \sin^2\theta & 0 & -\sin 2\theta & 0 & 0 \\ 0 & 0 & 0 & 0 & \sin\theta & \cos\theta \\ 0 & 0 & 0 & 0 & \cos\theta & -\sin\theta \\ 0.5\sin 2\theta & -0.5\sin 2\theta & 0 & \cos 2\theta & 0 & 0 \end{pmatrix}$$

$$(5.2\text{-}21)$$

where θ is the ply angle.

Since fiber orientation varies during the buckling, the stress and strain components correspond to the deformed shape of the element, must be incorporated in the constitutive Eq. (5.2-19). For this purpose, strain components of the cylindrical coordinates must be transformed using the following equation

$$e_{ij}(\bar{x}) = f_{kl}\frac{\partial x^k}{\partial \bar{x}_i} \cdot \frac{\partial x^l}{\partial \bar{x}_j} = f_{kl}.(\delta_i^k + u_{;i}^k).(\delta_j^l + u_{;j}^l) \qquad (5.2\text{-}22)$$

2.2   Equations of motion

Following the manner described in Ref. [37], the equations of motion for large deflection are

$$(\sqrt{g}\vec{\tau}^k)_{,k} + \vec{P}\sqrt{g} = \rho\frac{d^2\vec{R}}{dt^2} \qquad (5.2\text{-}23)$$

where $\vec{R}$ is the position vector, $\vec{P}$ is the body force vector defined per unit volume of the body before deformation and $g = |g_{ij}|$. The traction vector $\vec{\tau}$ can be resolved in the direction of the base and lattice vectors $\vec{g}_\mu$ and $(\vec{G}_\mu)$ to obtain the first and second Piola-Kirchhoff stress tensors $(\hat{\tau}^{k\mu})$ and $(\tau^{k\mu})$, respectively, as

$$\vec{\tau}^k = \tau^{k\mu}\vec{G}_\mu = \hat{\tau}^{k\mu}\vec{g}_\mu \qquad (5.2\text{-}24)$$

Then, from Eqs. (5.2-5) and (5.2-24) it is immediately concluded that

$$\hat{\tau}^{k\mu} = \tau^{ki}(\delta_i^\mu + u_{;i}^\mu) \qquad (5.2\text{-}25)$$

The following expression holds in the curvilinear coordinates

$$\vec{g}^k.\vec{g}_\mu = \{{}^{\lambda}_{\mu\nu}\}\vec{g}_\lambda \qquad (5.2\text{-}26)$$

where the notation $\{{}^{\lambda}_{\mu\nu}\}$ is the Christoffel three-index symbol of the second kind. Therefore, Eq. (5.2-23) can be expressed in a scalar form by resolving it in the direction of the base vectors $\vec{g}_\lambda$, such as

$$[\sqrt{g}\hat{\tau}^{kj}]_{,j} + \sqrt{g}\hat{\tau}^{\rho j}\{{}^{k}_{j\rho}\} + \sqrt{g}P_g^k = \rho\frac{d^2R^k}{dt^2} \qquad (k = 1, 2, 3) \qquad (5.2\text{-}27)$$

or

$$[\sqrt{g}(\delta_i^k + u_{;i}^k)\tau^{ij}]_{,j} + \sqrt{g}(\delta_i^\rho + u_{;i}^\rho)\tau^{ij}\{{k \atop j\rho}\} + \sqrt{g}P_g^k = \rho\frac{d^2u^k}{dt^2} \quad (k=1,2,3)$$
$$(5.2\text{-}28)$$

where the suffix g in P_g^k denotes that the components are taken in the direction of the base vectors. In the absence of body forces, Eq. (5.2-28) becomes

$$[\sqrt{g}(\delta_i^k + u_{;i}^k)\tau^{ij}]_{,j} + \sqrt{g}(\delta_i^\rho + u_{;i}^\rho)\tau^{ij}\{{k \atop j\rho}\} = \rho\frac{d^2u^k}{dt^2} \quad (k=1,2,3) \quad (5.2\text{-}29)$$

and its counterpart in infinitesimal deformations is

$$\tau_{;j}^{ij} = \rho\frac{d^2u^i}{dt^2} \qquad (i=1,2,3) \quad (5.2\text{-}30)$$

The above equations can be written in terms of the physical components of the stress through using the following relation

$$\sigma^{ij} = \tau^{ij}\cdot\sqrt{\frac{g_{jj}}{g^{ii}}} \qquad (5.2\text{-}31)$$

The boundary conditions are satisfied by means of the following equation

$$\vec{F} = (\vec{g}_k.\vec{\nu})\vec{\tau}^k \qquad (5.2\text{-}32)$$

where $\vec{F}$ is boundary traction vector and $\vec{\nu}$ is the unit normal vector on the boundary surface before deformation

$$\vec{\nu} = \nu_k\vec{g}^k \qquad (5.2\text{-}33)$$

It can be noted that resolution of $\vec{F}$ in the directions of the base vectors reads

$$\vec{F} = \tau^{k\mu}\nu_k(\delta_\mu^\lambda + u_{;\mu}^\lambda)\vec{g}_\lambda \qquad (5.2\text{-}34)$$

2.3 Numerical solution and buckling criteria

Since in the resulting equations of motion the stress terms and their derivatives are coupled and are nonlinear functions of the displacement components, they are highly nonlinear equations of displacements. Nonlinear problems of mechanics are often solved by adopting an incremental formulation. By the incremental solution procedure, the real time-variant system is approximated in a step by step way assuming time-invariance within each time step.

Equation (5.2-28) in terms of the physical components can be rewritten in the following form

$$f_1(\sigma_{ij}) = \ddot{W}$$
$$f_2(\sigma_{ij}) = \ddot{V}$$
$$f_3(\sigma_{ij}) = \ddot{U} \qquad (5.2\text{-}35)$$

so that the following perturbed equations hold for the time interval Δt (Eslami and Shariyat [32,35,40,41])

$$f_1(\tilde{\sigma}^{ij}) = \Delta\ddot{W}$$
$$f_2(\tilde{\sigma}^{ij}) = \Delta\ddot{V}$$
$$f_3(\tilde{\sigma}^{ij}) = \Delta\ddot{U} \qquad (5.2\text{-}36)$$

in which the incremented terms are demonstrated by a *tilde* symbol. The displacements have to satisfy the initial conditions, for $(x = 0, L$ and $\theta = 0, 2\pi)$, during each time step

$$\Delta u = 0 \qquad \Delta v = 0 \qquad \Delta w = 0$$
$$\Delta \dot{u} = \dot{u} \qquad \Delta \dot{v} = \dot{v} \qquad \Delta \dot{w} = \dot{w} \qquad (5.2\text{-}37)$$

Furthermore, the continuity of ($\sigma_r, \sigma_{r\theta}$ and σ_{rx}) and ($\epsilon_x, \epsilon_\theta$ and $\epsilon_{\theta x}$) must be considered in the numerical solution.

The numerical solution procedure is accomplished through the following steps

1- The numerical solution begins with discretizing the cylindrical shell into m×n×k grid points in the axial, circumferential and radial directions, respectively.

2- The initial deviations of the shell at the grid points are defined and the initial values of the displacement components $(u,\ v,\ w)$ are set to zero.

3- Time is incremented.

4- Corresponding increments of the mechanical and thermal loads are found.

5- According to the temperature distribution described in the previous step, the mechanical properties in geometrical coordinates of the shell are related through Eqs. (5.2-20) and (45.2-21) to those in the orthotropy directions of the materials. In this way, the dependency of E_1, E_2, ν_{12} and α_{ij}^* coefficients on temperature can be expressed as prescribed functions.

6- Derivative terms of u,v and w with respect to the spatial coordinates that appeared in Eq. (5.2-12), are approximated by a second-order finite difference method (the central difference method) and the strain components along the geometrical coordinates, are computed in each grid point.

7- The tensorial strain components are computed from their physical components calculated at the previous step by means of Eq. (5.2-9) and are related to those of the lattice coordinates through Eq. (5.2-22). The physical components of the latter strains are computed from Eq. (5.2-9) by substituting g_{ij} by G_{ij}.

8- Based on the constitutive equation (5.2-19) the physical components of the second Piola-Kirchhoff stress tensor and therefore, the corresponding tensorial components at each grid point are calculated.

9- Derivatives of the stress terms of Eq. (5.2-28) are approximated by a second-order finite difference method and derivatives involving multiplications of the radial distance of the points in the stresses are substituted by a fourth-order finite difference approximation.

While kinematic and stress boundary conditions may be directly incorporated in the formulations, force and moment boundary conditions are satisfied through relating axial or circumferential stresses of the grid points of the longitudinal edges by means of a numerical integrating method, such as Simpson's method. Thus, in contrast to the governing equations between the force/moment and displacement of the grid points, stresses at the longitudinal edges are related through linear equations to the applied loads. Therefore, even the force/moment boundary conditions are incorporated in a numerically exact procedure.

10- In the process of solution, time-invariance for terms appeared in the first sides of Eq. (5.2-36) is assumed during each time step. Thus, a set of second order differential equations are derived that can be solved by employing the fourth-order Runge-Kutta method subjected to the initial, Eq. (5.2-37), and boundary conditions. To improve the convergence of the proposed procedure, it is advisable to complete steps (6) to (10) for each individual point, before proceeding to the remaining points instead of considering all the points simultaneously.

11- When all equations in each iterative step of the current time interval are solved, the maximum value of the lateral displacements (w_{max}) is determined.

12- In each grid point, the displacement increments ($\Delta u, \Delta v$, and Δw) are added to displacement components obtained at the end of the previous time interval. To improve the results, solution is continued by using more iterations starting from step 6, until difference of the successive values of w_{max} of the same time interval becomes negligible.

13- The corrected values of u, v, w, $\dot{u}, \dot{v}$ and $\dot{w}$ obtained in this manner are considered as initial values for the next time interval.

14- Beginning from step 4, results corresponding to the next time increments are obtained.

15- Possibility of dynamic buckling occurrence is checked. For this purpose, variations of w_{max} versus time or versus applied load (external pressure, axial load, temperature gradient, etc.) are plotted.

16- In the case of no buckling point occurrence, amplitude of the applied loads are increased and calculations are continued starting from step 2.

Buckling load can be determined using one of the two equivalent stability criteria stated below:

1- The generalized concept of dynamic buckling proposed by Budiansky [42]. This concept is associated with dynamic buckling of a structure where small changes in the magnitude of loading lead to large changes in the structure response. According to this criterion, abrupt reduction in slope of the maximum lateral displacement versus load curve (minimum slope) indicates a dynamic buckling state.

2- Large increase of the displacements amplitude with time. Qualitatively, if the time histories of the maximum lateral deformation or other modes of deformation are given for several amplitudes of the applied load, the critical buckling load can simply be defined as the load at which large increase in the amplitude of the deformations is seen to occur (Saigal et al. [14]).

2.4   Calculations and results

Results are presented for pure mechanical or pure thermal loads and the interaction between these loads is not considered here. Dynamic behavior of the shells, except where mentioned,is investigated under step loads. To ensure incorporation of the higher buckling modes effect, the mesh is chosen sufficiently fine and time interval is adopted small enough. Therefore, the improvement in the results due to increasing the mesh density or decreasing time step is negligibly small. For this purpose, in the following results, a mesh composed of approximately 150000 three dimensional grid points is chosen. To avoid round off errors, the calculations are carried out in the double-precision mode. For example, the improvement noticed in the displacement component values by increasing this number of grid points to 200000 in the first example, is of the order of 10^{-12}. To reduce influence of the solution procedure on the accuracy of the final results, fourth order approximations with respect to both spatial variables and time is employed, and integrating time interval of the order of 10^{-6} is adopted.

Buckling under mechanical loads

As a first example, static buckling results of the present accurate three dimensional elasticity formulation are compared with those of the latest 3-D analysis proposed by Ref. (1) and the modified Donnell's equations. Formulations of the above reference are suitable for static buckling of thick perfect single layer circular cylindrical shells under mechanical loads and are based on infinitesimal strains assumption (linear strain-displacement expressions) . Furthermore, in deriving the final results, one of the triple Fourier series is chosen (which means separation of variables in all coordinate directions and neglecting the coupling of the buckling modes which usually occurs), boundary conditions are satisfied in average and the results are obtained using a finite difference method. Therefore, the results are less accurate than the layer-wise theories, especially that proposed previously by the author [35] which assumes independency of the variables in the radial direction only. None of the above simplifications are made in the present analysis.

For comparison purposes, Boron/Epoxy cylindrical shells are considered for two fiber orientations: Circumferential reinforcement ($E_2 > E_1, E_3$) and axial reinforcement ($E_3 > E_1, E_2$). The geometric and mechanical properties of the shells (for the axial reinforcement case) are

$$E_1 = E_2 = 18.6159[GPa] \qquad E_3 = 206.844[GPa]$$
$$G_{12} = 2.55107[GPa] \qquad G_{13} = G_{23} = 4.48162[GPa]$$
$$\nu_{21} = 0.45 \qquad \nu_{31} = \nu_{32} = 0.3$$
$$b = 1[m] \qquad l/b = 5$$

where b is the external radius of the shells.

The critical loads are defined by normalizing the external pressure and the axial load ($\bar{p}$ and $\bar{P}$) as

$$\bar{p} = \frac{pb^3}{E_2 h^3} \qquad \bar{P} = \frac{P.b}{\pi E_3 h(b^2 - a^2)} \qquad (5.2\text{-}38)$$

where a is the internal radius of the shell and the load interaction parameter, S, is defined as:

$$\frac{P}{2\pi} = S.p.b^2 \qquad (5.2\text{-}39)$$

Table (5.2-1) shows the static critical loads as predicted by the modified Donnell's equations, preliminary 3-D theory approach of reference (1) as well as the results of the present accurate approach for static and

Table 5.2-1: Comparison of various theories results for cylindrical shells under combined external pressure and axial load. For each case, the first row gives $(\bar{p})$ and the second row gives $(\bar{P})$.

b/a	Donnell	P3D	Present Static	Present Dynamic
Load interaction parameter, S=1				
1.03	0.4134	0.3899	0.3927	0.3142
	0.1880	0.1773	0.1788	0.1432
1.05	0.3090	0.2834	0.2905	0.2334
	0.2319	0.2127	0.2180	0.1752
1.1	0.2793	0.2352	0.2569	0.2046
	0.4092	0.3446	0.3765	0.2998
1.15	0.2880	.2140	0.2346	0.1846
	0.6183	0.4593	0.4869	0.3872
Load interaction parameter, S=5				
1.03	0.2845	0.2511	0.2617	0.2083
	0.6467	0.5708	0.5949	0.4739
1.05	0.2137	0.1826	0.1945	0.1546
	0.8019	0.6852	0.7297	0.5817
1.1	0.1519	0.1125	0.1321	0.1046
	1.1125	0.8245	0.9367	0.7478
1.15	0.1092	0.0754	0.0906	0.0715
	1.1719	0.8089	0.9611	0.7628

Table 5.2-2: Comparison of various theories results for cylindrical shells under external pressure

Geometry	FOSD	HOSD	P3D	GLW	Present Static	Present Dynamic
Circumferential reinforcement						
h=6.35[mm], h/R=0.03,L/R=100	0.9668	0.9637	0.9694	0.9608	0.9613	.7698
h=12.7[mm], h/R=0.07,L/R=100	0.9050	0.8933	0.9148	0.8892	0.894	.7352
Axial reinforcement						
h=6.35[mm], h/R=0.03,L/R=100	0.9822	0.9822	0.9817	0.9816	0.9812	.7792
h=12.7[mm], h/R=0.07,L/R=100	0.9588	0.9556	0.9605	0.9528	0.9534	.7589

dynamic buckling (step loading). Results of the preliminary 3-D analysis is denoted by (P3D).

Table (5.2-2) gives a comparison of the predicted critical loads for a very long shell, under external pressure based on static buckling results of a first order shear-deformation shell theory [43] (FOSD), a Higher order shear-deformation equivalent-layer theory [43] (HOSD), the preliminary 3-D theory approach [12], the general layer-wise theory of the author [35] (GLW), and the static and dynamic buckling analysis of the present theory. The critical pressures are normalized with respect to the classical theory (Donnell's shell theory) results.

In the next example, buckling of a 4-ply laminated imperfect cylindrical shell under uniform axial compression of references [3] and [15] is reexamined. The cylinder is considered to be simply supported with radius $R = 19.0$ [cm] and $L/R = 2$. Thickness of each layer is .0135 [cm] and the laminate construction is $[0°/30°/60°/90°]$. The material of each lamina is assumed as Boron/Epoxy, AVCO 550 with the following properties

$$E_x = 207[GPa], \qquad E_\theta = E_x = 18.6[GPa]$$
$$G_{\theta x} = 4.48[GPa] \qquad \nu_{x\theta} = \nu_{xz} = \nu_{z\theta} = .21$$

The initial geometric imperfections of the shell are defined as

$$w_0(x,\theta) = W_0 h.sin(\frac{\pi_x}{L}).cos(8\theta) \qquad (5.2\text{-}40)$$

where W_0 is the amplitude of imperfection and h is the total thickness of the shell.

Results of static and dynamic buckling analysis performed by references [3] and [15] as well as the present results are illustrated in Fig. (5.2-1). In the earlier results, an equivalent layer approach is adopted and von Karman type expressions of strains are chosen. Results of reference [3] are based on Hoff-Simitses criterion (the total potential energy approach) that is explained in details in reference [44], so that the static and dynamic loads for $W_0 = 0$ are identical. In reference [15], Love-Kirchhoff assumptions are employed and the nonlinear equations of motion are linearized and solved using an incremental method previously described in reference [14]. As seen from this figure, the reduction in buckling loads is diminished as imperfection amplitude increases. In other words, imperfection sensitivity is more noticeable in small imperfection amplitudes. In comparison with the other results, the present formulation gives smaller critical loads. This is due to incorporation of the exact strain-displacement relations of the imperfect shell in the exact nonlinear three dimensional elasticity analysis presented here.

In Fig. (5.2-2), comparison among the predictions of the more accurate approaches, recently proposed by the author, namely, a higher-order shear deformation equivalent-layer theory (HOSDEL) [32], a general layer-wise theory (GLW) [35] and the present approach is presented. As it may be noticed from this figure, the present results are more conservative.

Figure (5.2-3) reveals the effect of various impulsive loads (rectangular, triangular and parabolic impulsive loads) and their time duration on the predicted critical loads. For this purpose, the following nondimensional time is used

$$t^* = \frac{\bar{\tau}}{\bar{\tau}_0} \qquad (5.2-41)$$

where $\bar{\tau}$ is the pulse duration and $\bar{\tau}_0$ is the free vibration period of the shell. In Fig. (5.2-3), P_{cr} is the static buckling load. These results indicates that shell stiffness increases for short duration of loading (especially if the pulse duration is comparable to the hoop breathing mode). For short time duration, the dynamic buckling loads are larger than the static buckling load. It is believed that this phenomenon is due to the inertia wave vibration between the impacted and fixed ends of the shell [45-48]. For a given amplitude, the step load has a maximum curve area. Thus, as it may be expected, this type of loading has the worst influence on the strength of the shell. The critical loads corresponding to the triangular impulsive loads is the largest.

In order to investigate the large deflection behavior of imperfect multilayered cylindrical shells, a potbuckling analysis is carried out for three-layered cross ply ($0°90°0°$) shells with the following properties

$$E_1 = 209.5[GPa] \quad E_2 = E_3 = 7[GPa] \qquad G_{12} = G_{13} = 3.5[GPa]$$
$$G_{23} = 1.4[GPa] \quad \nu_{12} = \nu_{13} = \nu_{23} = 0.3$$
$$h_i = h/3 \qquad R = 914[mm] \qquad L = 2540[mm]$$

Two sets are considered; thin shells with total thickness of $h = 2.54[mm]$ and moderately thick shells with $h = 25.4[mm]$. The initial imperfection is assumed to have the form

$$w_0 = \sum_{m=1}^{M} \sum_{n=0}^{N} W_0^{mn} sin(\frac{m\pi x}{L}) cos(n\theta) \qquad (5.2\text{-}42)$$

where m and n are the numbers of axial and circumferential half waves, respectively.

Static axial load versus axial deflection curves of the thin and thick shells are shown simultaneously in Fig. (5.2-4) for two different amplitude imperfections. As it stated before, points where abrupt change in displacement modes (such as axial deflection) due to small increase in the applied loads is noticed, indicate buckling occurrence. In Fig. (5.2-4), the axial load is normalized with respect to the classical static buckling loads. The deformed shape and buckling modes of the thin and thick shells are illustrated in Figs. (5.2-5) and (5.2-6), respectively. It may be easily seen that the imperfection sensitivity is higher for thinner shells. Comparison of Figs. (5.2-5) and (5.2-6) reveals that in thick cylindrical shells, in contrast to thin shells, lower buckling modes are dominant. So that very thick shells buckle in an axisymmetric manner. In Fig. (5.2-7) the dynamic postbuckling response of the mentioned cylindrical shells is compared with the static response.

Thermal buckling

Numerical results for materially orthotropic clamped cylindrical shells of reference [21] are reexamined. Though the material properties are functions of temperature, as a first approximation, the assumed properties are representative in the temperature range under consideration. These properties are

$$E_1 = 13.17[GN/m^2] \qquad\qquad E_2 = E_3 = 34.61[GN/m^2]$$
$$G_{12} = G_{13} = G_{23} = 5.17[GN/m^2] \qquad \nu_{12} = \nu_{13} = \nu_{23} = 0.25$$

The shell is subjected to uniform temperature rise throughout. Results of static thermal buckling of reference [20] which are based on the parametric differentiation technique (with temperature as a parameter) is depicted along with the static and dynamic thermal buckling of the present formulation in Fig. (5.2-8). In this figure, the thermal buckling parameter $\alpha_2 T_{cr}$ chosen in reference [20] is shown for various curvature parameter $Z = l^2/\{R.h.\sqrt{1-\nu_1\nu_2}\}$. As it may be expected, the static results of the present study are lower.

As a second example, thermal buckling analysis of shells constructed from materials with the following mechanical and thermal properties are considered

$$E_{11} = 1.5 \times 10^5 [N/cm^2] \qquad E_{22} = E_{33} = 0.1E_{11}$$
$$G_{12} = G_{13} = G_{23} = 0.5E_{22} \qquad \nu_{12} = \nu_{13} = \nu_{23} = 0.25$$
$$\alpha_1 = \alpha_2 = \alpha_3 = 10^{-6}[C^{\circ-1}] \qquad L^2/Rh = 400$$

In reference [21], static thermal buckling analysis of composite cylindrical shells with the above properties, is accomplished through extending the linear semiloof shell element formulated by Irons which is based on the classical assumptions and infinitesimal strains. The results obtained using one term of the double Fourier series. It is proved that a large number of modes are amplified during the buckling, so that this form of solution can cause a notable errors. The shells are considered to undergo uniform temperature rise. The predicted static critical temperature (T_{cr}) proposed by reference [21] for angle-ply cylindrical shells undergoing axisymmetric buckling modes, along with the results of the static and dynamic buckling analysis developed in the current chapter, are illustrated in Fig. (5.2-9). For this purpose, two staking sequence ($[\phi/-\phi/\phi]$ and $[\phi/-\phi/\phi/-\phi/\phi]$) are used. As it can be seen, the critical temperature is considerably dependent on the inclination angle of the fibers (ϕ), which is measured from the cylinder cross section. Its maximum is attained in $(\phi = 90°)$, where the least axial strength of the shell is obtained. Indeed, in the cylindrical shells, the predominant stress is the axial one and circumferential stress is developed mainly in the axial edges. It is shown (for example, in reference [15]) that in the case of mechanical buckling, the critical loads vary symmetrically about $(\phi = 45°)$. As is shown in Fig. (5.2-10), effect of the nonlinearity of the strain-displacement relations in critical thermal loads of the antisymmetric shells is larger. Thus, the differences between curves corresponding to these two approaches is larger in this case. In Fig. (5.2-10), staking sequences are taken to be $([\phi/-\phi], [\phi/-\phi/\phi/-\phi]$ and $[\phi/-\phi/\phi/-\phi/\phi/-\phi])$. Figures (5.2-9) and (5.2-10) indicate that increasing number of the layers while fixing the

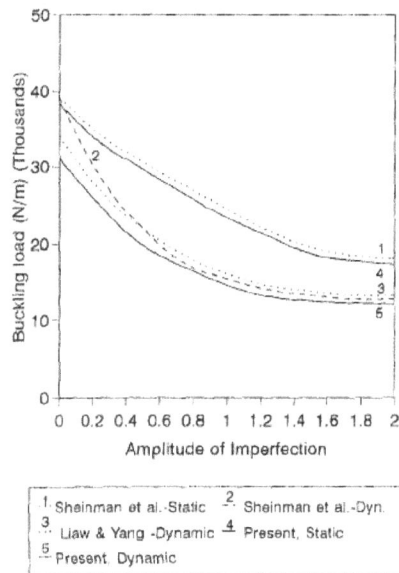

Figure 5.2-1: Effect of imperfection amplitude on the dynamic buckling load.

overall thickness of the shell, leads to a higher strength for the shell. This variation is more pronounced for the two-layer shell. This conclusion is in agreement with the report of reference [47] for mechanical buckling. Linear analysis done by Eslami [49] et al. reveals that the critical buckling temperatures for isotropic materials are independent of modulus of elasticity and are inversely proportional to the coefficient of thermal expansion. Whereas, in the nonlinear analysis introduced in this chapter, all of the mechanical properties are coupled together. Hence, the critical load is a function of all of mechanical and thermal properties of the materials.

3 High-Order Shear Deformation Theory

The three-dimensional theory of elasticity introduced in Section 2 is employed to develop a high-order shear deformation theory with a hybrid formulations. The main assumptions done for reducing the general equations obtained in the Section 2 to those of the present section, are:
- Transverse normal stress and strain are often negligible.
- Radial component of the displacements is constant trough thickness of the shell.
- The global behaviour of the shell can be modeled by an equivalent

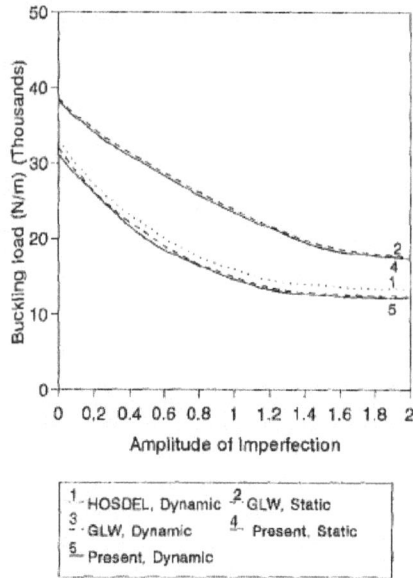

Figure 5.2-2: Comparison of the results of the high-order shear-deformation equivalent-layer theory, general layer-wise theory, and the present 3-D theory.

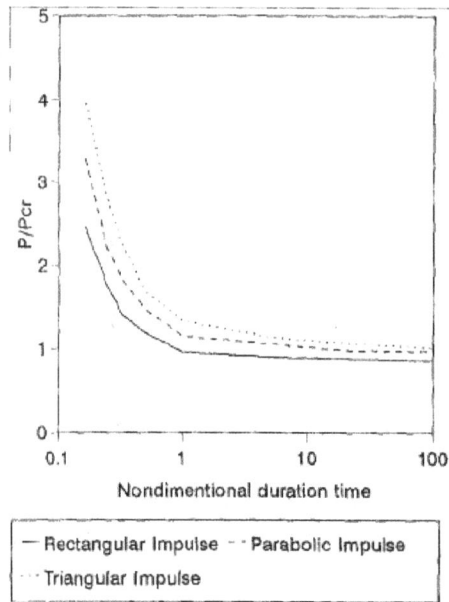

Figure 5.2-3: Effect of different types of impulsive loads and various loading duration on the buckling load of the cylindrical shell.

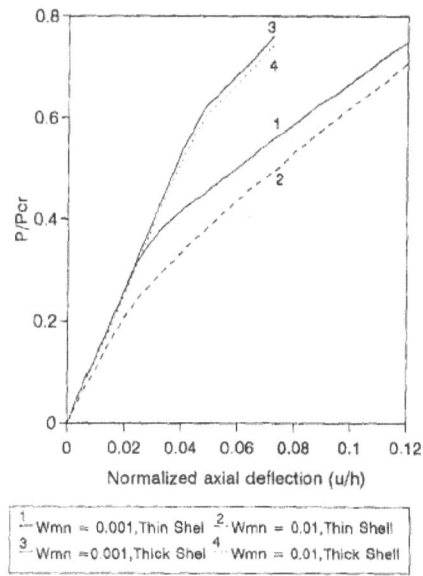

Figure 5.2-4: Comparison of the static post-buckling analysis results of the thin shells. The illustrated results correspond to the internal surface at $(x = 1)$.

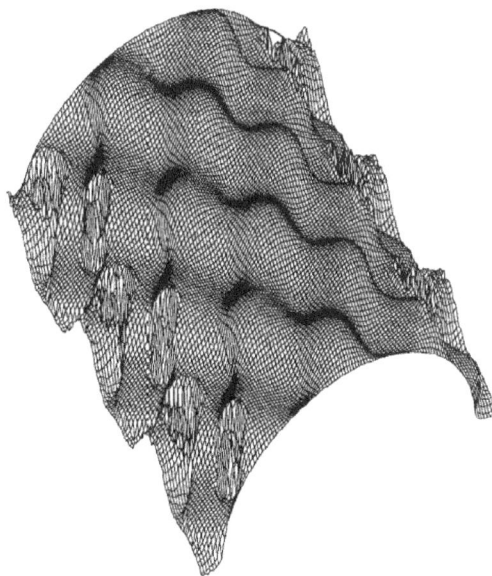

Figure 5.2-5: Buckling modes of a thin laminated cylindrical shell.

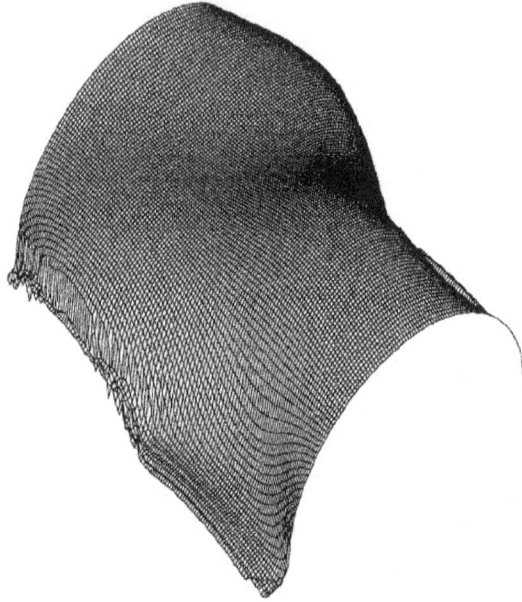

Figure 5.2-6: Buckling modes of a thick laminated cylindrical shell.

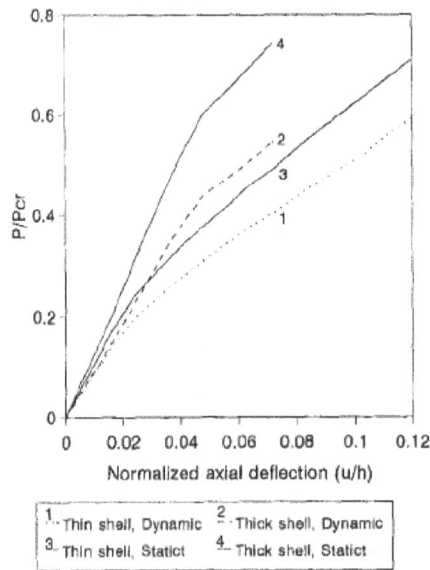

Figure 5.2-7: Variation of critical temperature with curvature as predicted by the present approach and the conventional approaches.

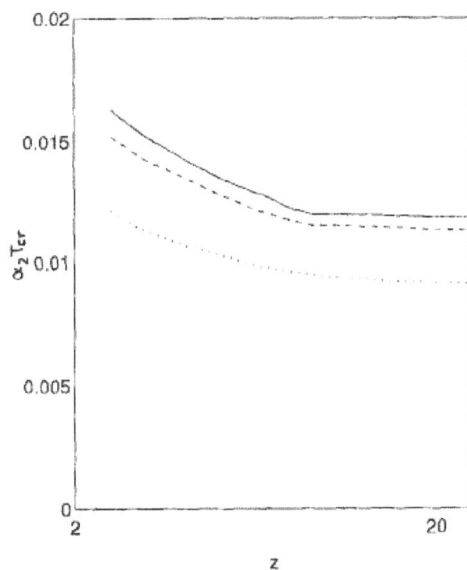

Figure 5.2-8: A comparison between the dynamic and static buckling result of the thin and the thick shells. The illustrated result correspond to the internal surface at $(x = 1)$ and $(w_{mm} = 0.01)$

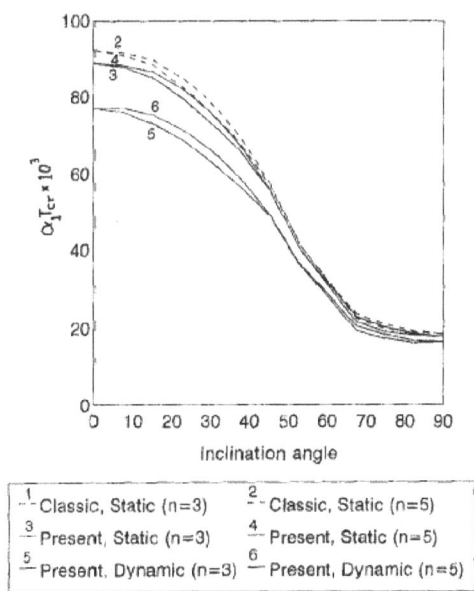

Figure 5.2-9: Comparison of classical and present critical temperatures of symmetric, angle-ply cylindrical shells.

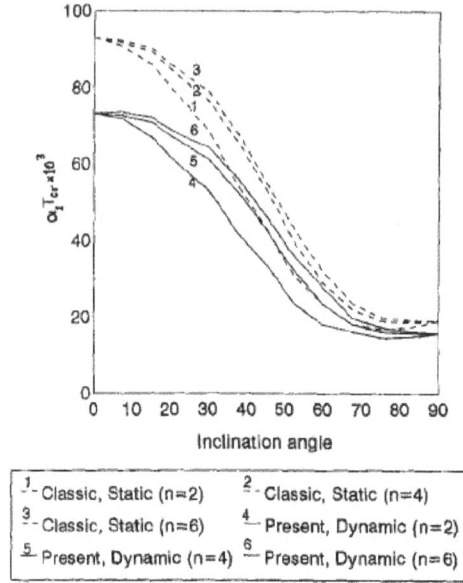

Figure 5.2-10: Comparison of classical and present critical temperatures of antisymmetric, angle-ply cylindrical shells.

single-layer shell.

3.1 Basic equations

In the first order Reissner-Mindlin shear-deformation theory, the displacement components u, v, and w at any point of the cross section of the shell, Fig. (5.3-1), are expressed as follows [50]

$$u = \bar{u} + z\psi_x$$
$$v = \bar{v} + z\psi_\theta$$
$$w = \bar{w} \qquad (5.3\text{-}1)$$

where $\bar{u}$, $\bar{v}$, and $\bar{w}$ are the midplane displacements, z is the distance from the mid-surface and ψ_x and ψ_θ are rotations of the normal to the mid-surface in the $x - z$ and $\theta - z$ planes, respectively. To generalize this formulation to a higher order one, the displacement expressions are written as

$$u = \bar{u} + \sum_{k=1}^{m} z^k \psi_k$$
$$v = \bar{v} + \sum_{l=1}^{n} z^l \phi_l$$

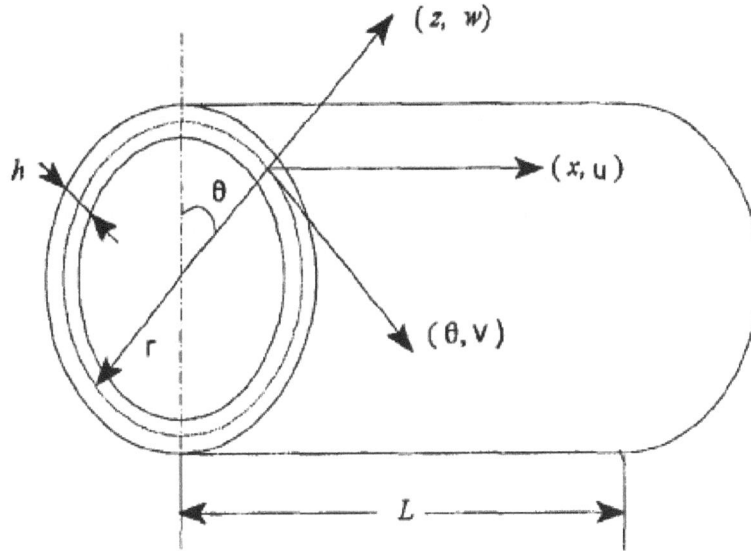

Figure 5.3-1: Coordinate axes and displacement components used in the present formulations.

$$w = \bar{w} \tag{5.3-2}$$

or in a compact form

$$u = \Sigma_{k=0}^{m} z^k \psi_k \qquad\qquad v = \Sigma_{l=0}^{n} z^l \phi_l \qquad\qquad w = \bar{w} \tag{5.3-3}$$

where

$$\psi_0 \equiv \bar{u} \qquad\qquad\qquad \phi_0 \equiv \bar{v}$$

and ψ_k and ϕ_l are continuous functions of x and θ. Thus, in contrast to the well-known shear-deformation theories [33,34], shear deformations through each layer is not constant, but their variations differ from a layer to another.

Remembering the assumption that w and w_0 are independent of z coordinate, the strain-displacement expressions of Eq. (5.2-12) become

$$\epsilon_{xx} = u_{,x} + \frac{1}{2}[(w_{,x})^2 + 2w_{,x}w_{0,x} + (v_{,x})^2 + (u_{,x})^2]$$

$$\epsilon_{\theta\theta} = \frac{1}{r}(v_{,\theta} + w) + \frac{1}{2r^2}\{(v_{,\theta} + w + w_0)^2 + [(w + w_0)_{,\theta} - v]^2 + (u_{,\theta})^2$$

$$-w_0^2 - (w_{0,\theta})^2\}$$

$$\epsilon_{x\theta} = \frac{1}{2}(v_{,x} + \frac{1}{r}u_{,\theta}) + \frac{1}{2r}[v_{,x}(v_{,\theta} + w + w_0) + u_{,\theta}u_{,x} + (w + w_0)_{,x}$$

$$(w_{,\theta} - v) + w_{,x}w_{0,\theta}]$$

$$\epsilon_{zx} = \frac{1}{2}(w_{,x} + u_{,z} + v_{,z}v_{,x} + u_{,z}u_{,x})$$

$$\epsilon_{z\theta} = \frac{1}{2}v_{,z} + \frac{1}{2r}[(w_{,\theta} - v) + (v_{,\theta} + w + w_0)v_{,z} + u_{,z}u_{,\theta}] \qquad (5.3\text{-}4)$$

3.2 Equations of motion

Derivation of equations of motion of the imperfect cylindrical shell is accomplished by employing Hamilton's principle. According to this principle, the motion equations are derived when the following equation holds

$$\delta I = \int_{t_1}^{t_2} \delta L.dt = 0 \qquad (5.3\text{-}5)$$

where L is the Lagrangian function and t_1 and t_2 are arbitrary time limits. For elastic shells, neglecting the damping effects and body forces, Eq. (5.2-12) reduces to [20]

$$\int_{t_1}^{t_2}(\int_V \sigma_{ij}\delta\epsilon_{ij}dV + \int_V \rho\ddot{u}_i\delta u_i dV - \delta\Omega)dt = 0 \qquad (5.3\text{-}6)$$

where ρ is the material density and Ω is the potential energy due to the traction forces. In the case of external fluid pressure, Ω can be determined from

$$\Omega = p\int_S[\bar{w} + \frac{1}{2r}(\bar{v}^2 - \bar{v}\bar{w}_{,\theta} + \bar{v}_{,\theta}\bar{w} + \bar{w}^2)]dS \qquad (5.3\text{-}7)$$

where S is the area of the surface of the body. In the case of external pressure, only the first term in the bracket of Eq. (5.3-7) is nonzero. For cylindrical shells, Eq. (54.3-6) can be written as

$$\int_{t_1}^{t_2}\{\int_S\int_{-h/2}^{h/2}(\sigma_{ij}\delta\epsilon_{ij} + \rho\ddot{u}_i\delta u_i)dz.dS - \delta\Omega\}dt = 0 \qquad (5.3\text{-}8)$$

According to Eq. (5.3-4), the first variation of the strains are

$$\delta\epsilon_x = (1 + u_{,x})\delta u_{,x} + v_{,x}\delta v_{,x} + (w + w_0)_{,x}\delta w_{,x}$$

$$\delta\epsilon_\theta = \frac{1}{r^2}\{u_\theta\delta u_\theta - [(w + w_0)_{,\theta} - v]\delta v + r[1 + \frac{1}{r}(v_{,\theta} + w + w_0)]$$

$$\delta v_{,\theta} + r[1 + \frac{1}{r}(v_{,\theta} + w + w_0)]\delta w + [(w + w_0)_{,\theta} - v]\delta w_{,\theta}\}$$

$$\delta\epsilon_{x,\theta} = \frac{1}{2r}\{u_{,\theta}\delta u_{,x} + (1 + u_{,x})\delta u_{,\theta} - (w + w_0)_{,x}\delta v + r[1 + \frac{1}{r}(v_{,\theta} + w + w_0)]\delta v_{,x}$$

$$+v_{,x}\delta v_{,\theta}+v_{,x}\delta w+[(w+w_0)_{,\theta}-v]\delta w_{,x}+(w+w_0)_{,x}\delta w_{,\theta}\}$$

$$\delta\epsilon_{zx}=\frac{1}{2}[u_{,z}\delta u_{,x}+(1+u_{,x})\delta u_{,z}+v_{,z}\delta v_{,x}+v_{,x}\delta v_{,z}+\delta w_{,x}]$$

$$\delta\epsilon_{z\theta}=\frac{1}{2r}\{u_{,z}\delta u_{,\theta}+u_{,\theta}\delta u_{,z}-\delta v+v_{,z}\delta v_{,\theta}+(v_{,\theta}+w+w_0)\delta v_{,z}+v_{,z}\delta w+\delta w_{,\theta}\}$$

$$(5.3\text{-}9)$$

Recall that ρ and σ vary through the shell thickness. Defining the following symbols

$$M_{ij}^k=\int_{-h/2}^{h/2}\sigma_{ij}z^kdz\qquad N_{ij}=M_{ij}^0\qquad I^k=\int_{-h/2}^{h/2}\rho z^kdz$$
$$(5.3\text{-}10)$$

and substituting Eq. (5.3-3) in Eq. (5.3-9) and substituting the resulted equations into Eq. (5.3-8), we have

$$\int_{t_1}^{t_2}\int_S\{[\sum_{i=0}^m(M_x^i+\sum_{j=0}^mM_x^{i+j}\psi_{j,x})\delta\psi_{i,x}+\sum_{i=0}^n\sum_{j=0}^nM_x^{i+j}\phi_{j,x}\delta\phi_{i,x}+N_x(w+w_0)_{,x}\delta w_{,x}]$$

$$+\frac{1}{r^2}\sum_{i=0}^m\sum_{j=0}^mM_\theta^{i+j}\psi_{j,\theta}\delta\psi_{i,\theta}-\frac{1}{r^2}\sum_{i=0}^n[M_\theta^i(w+w_0)_{,\theta}-\sum_{j=0}^nM_\theta^{i+j}\phi_j]\delta\phi_i+$$

$$\frac{1}{r}\sum_{i=0}^n\{[1+\frac{1}{r}(w+w_0)]M_\theta^i+\frac{1}{r}\sum_{j=0}^nM_\theta^{i+j}\phi_{j,\theta}\}\delta\phi_{i,\theta}+\frac{1}{r}\{[1+\frac{1}{r}(w+w_0)]N_\theta$$

$$+\frac{1}{r}\sum_{i=0}^nM_\theta^i\phi_{i,\theta}\}\delta w+\frac{1}{r^2}[(w+w_0)_{,\theta}N_\theta-\sum_{i=0}^nM_\theta^i\phi_i]\delta w_{,\theta}+\frac{1}{r}\sum_{i=0}^m\sum_{j=0}^mM_{x\theta}^{i+j}$$

$$\psi_{j,\theta}\delta\psi_{i,x}+\frac{1}{r}\sum_{i=0}^m(M_{x\theta}^i+\sum_{j=0}^mM_{x\theta}^{i+j}\psi_{j,x})\delta\psi_{i,\theta}-\frac{1}{r}(w+w_0)_{,x}\sum_{i=0}^nM_{x\theta}^i\delta\phi_i$$

$$+\sum_{i=0}^n\{[1+\frac{1}{r}(w+w_0)]M_{x\theta}^i+\frac{1}{r}\sum_{j=0}^nM_{x\theta}^{i+j}\phi_{j,\theta}\}\delta\phi_{i,x}+\frac{1}{r}\sum_{i=0}^n\sum_{j=0}^nM_{x\theta}^{i+j}$$

$$\phi_{j,x}\delta\phi_{i,\theta}+\frac{1}{r}\sum_{i=0}^nM_{x\theta}^i\phi_{i,x}\delta w+\frac{1}{r}[N_{x\theta}(w+w_0)_{,\theta}-\sum_{i=0}^nM_{x\theta}^i\phi_i]\delta w_{,x}+\frac{1}{r}$$

$$(w+w_0)_{,x}N_{x\theta}\delta w_{,\theta}+\sum_{i=0}^m\sum_{j=0}^mjM_{zx}^{i+j-1}\psi_j\delta\psi_{i,x}+\sum_{i=0}^mi(M_{zx}^{i-1}+\sum_{j=0}^m$$

$$M_{zx}^{i+j-1}\psi_{j,x})\delta\psi_i+\sum_{i=0}^n\sum_{j=0}^njM_{zx}^{i+j-1}\phi_j\delta\phi_{i,x}+\sum_{i=0}^n\sum_{j=0}^niM_{zx}^{i+j-1}$$

$$\phi_{j,x}\delta\phi_i+N_{zx}\delta w_{,x}+\frac{1}{r}\sum_{i=0}^m\sum_{j=0}^mjM_{z\theta}^{i+j-1}\psi_j\delta\psi_{i,\theta}+\frac{1}{r}\sum_{i=0}^m\sum_{j=0}^miM_{z\theta}^{i+j-1}$$

$$\psi_{j,\theta}\delta\psi_i-\frac{1}{r}\sum_{i=0}^nM_{z\theta}^i\delta\phi_i+\frac{1}{r}\sum_{i=0}^n\sum_{j=0}^njM_{z\theta}^{i+j-1}\phi_j\delta\phi_{i,\theta}+\frac{1}{r}\sum_{i=0}^ni[M_{z\theta}^{i-1}$$

$$(r + w + w_0) + \sum_{j=0}^{n} M_{z\theta}^{i+j-1}\phi_{j,\theta}]\delta\phi_i + \frac{1}{r}\sum_{i=0}^{n} iM_{z\theta}^{i-1}\phi_i\delta w + \frac{1}{r}N_{z\theta}\delta w_{,\theta}$$

$$+\sum_{i=0}^{m}\sum_{j=0}^{m} I^{i+j}\ddot{\psi}_j\delta\psi_i + \sum_{i=0}^{n}\sum_{j=0}^{n} I^{i+j}\ddot{\phi}_j\delta\phi_i + I^0\ddot{w}\delta w - p[\delta w + \frac{1}{2r}$$

$$(2\phi_0\delta\phi_0 - w_{,\theta}\delta\phi_0 - \phi_0\delta w_{,\theta} + w\delta\phi_{0,\theta} + \phi_{0,\theta}\delta w + 2w\delta w)]\}dS.dt = 0 \quad (5.3\text{-}11)$$

Integrating terms containing derivatives of displacement variations and setting the summation of terms multiplied by $\delta\psi_i$, $\delta\phi_i$ and δw in the final form of Eq. (5.3-11) to zero, the equations of motion are obtained as

$$M_{x,x}^i + \frac{1}{r}M_{x\theta,\theta}^i - iM_{zx}^{i-1} + \sum_{j=0}^{m}\{(M_x^{i+j}\psi_{j,x})_{,x} + \frac{1}{r^2}(M_\theta^{i+j}\psi_{j,\theta})_{,\theta}$$

$$+\frac{1}{r}[(M_{x\theta}^{i+j}\psi_{j,\theta})_{,x} + (M_{x\theta}^{i+j}\psi_{j,x})_{,\theta}] + j(M_{zx}^{i+j-1}\psi_j)_{,x} - i(M_{zx}^{i+j-1}$$

$$\psi_{j,x}) + \frac{1}{r}[j(M_{z\theta}^{i+j-1}\psi_j)_{,\theta} - iM_{z\theta}^{i+j-1}\psi_{j,\theta}]\} = \sum_{j=0}^{m} I^{i+j}\ddot{\psi}_j$$

$$\frac{1}{r^2}M_\theta^i(w + w_0)_{,\theta} + \frac{1}{r}\{[1 + \frac{1}{r}(w + w_0)]M_\theta^i\}_{,\theta} + \frac{1}{r}M_{x\theta}^i(w + w_0)_{,x}$$

$$+\{[1 + \frac{1}{r}(w + w_0)]M_{x\theta}^i\}_{,x} + \frac{1}{r}[M_{z\theta}^i - iM_{z\theta}^{i-1}(r + w + w_0)] + \delta_{i0}.$$

$$p\frac{1}{r}(\phi_0 - w_{,\theta}) + \sum_{j=0}^{n}\{(M_x^{i+j}\phi_{j,x})_{,x} - \frac{1}{r^2}[M_\theta^{i+j}\phi_j - (M_\theta^{i+j}\phi_{j,\theta})_{,\theta}]$$

$$+\frac{1}{r}[(M_{x\theta}^{i+j}\phi_{j,\theta})_{,x} + (M_{x\theta}^{i+j}\phi_{j,x})_{,\theta}] + j(M_{zx}^{i+j-1}\phi_j)_{,x} - i(M_{zx}^{i+j-1}\phi_{j,x})$$

$$+\frac{1}{r}[j(M_{z\theta}^{i+j-1}\phi_j)_{,\theta} - iM_{z\theta}^{i+j-1}\phi_{j,\theta}]\} = \sum_{j=0}^{m} I^{i+j}\ddot{\phi}_j$$

$$[N_x(w + w_0)_{,x}]_{,x} - \frac{1}{r}[1 + \frac{1}{r}(w + w_0)]N_\theta + \frac{1}{r^2}[(w + w_0)_{,\theta}N_\theta]_{,\theta}$$

$$+\frac{1}{r}\{[N_{x\theta}(w + w_0)_{,\theta}]_{,x} + [N_{x,\theta}(w + w_0)_{,x}]_{,\theta}\} + N_{zx,x}$$

$$+\frac{1}{r}N_{z\theta,\theta} + p[1 + \frac{1}{r}(\phi_{0,\theta} + w)] + \sum_{j=0}^{n}\{\frac{-1}{r^2}[M_\theta^j\phi_{j,\theta} + (M_\theta^j\phi_j)_{,\theta}]$$

$$-\frac{1}{r}[M_{x\theta}^j\phi_{j,x} + (M_{x\theta}^j\phi_j)_{,x} + jM_{z\theta}^{j-1}\phi_j]\} = I^0\ddot{w} \qquad (5.3\text{-}12)$$

3.3 Constitutive equations

Substituting Eq. (5.3-3) in Eq. (5.3-4), the strain equations become

$$
\epsilon_{xx} = \sum_{i=0}^{m} z^i \psi_{i,x} + \frac{1}{2}[(w_{,x})^2 + 2w_{,x}w_{0,x} + \sum_{i=0}^{n}\sum_{j=0}^{n} z^{i+j}\phi_{i,x}\phi_{j,x}
$$

$$
+ \sum_{i=0}^{m}\sum_{j=0}^{m} z^{i+j}\psi_{i,x}\psi_{j,x}]
$$

$$
\epsilon_{\theta\theta} = \frac{1}{r}(\sum_{i=0}^{n} z^i \phi_{i,\theta} + w) + \frac{1}{r^2}\{\sum_{i=0}^{n}\sum_{j=0}^{n} z^{i+j}\phi_{i,\theta}\phi_{j,\theta} + 2(w+w_0)
$$

$$
\sum_{i=0}^{n} z^i \phi_{i,\theta} + (w+w_0)^2 + (w+w_0)_{,\theta}^2 - 2(w+w_0)_{,\theta} \sum_{i=0}^{n} z^i \phi_i
$$

$$
+ \sum_{i=0}^{n}\sum_{j=0}^{n} z^{i+j}\phi_i\phi_j + \sum_{i=0}^{m}\sum_{j=0}^{m} z^{i+j}\psi_{i,\theta}\psi_{j,\theta} - w_0^2 - (w_{0,\theta})^2\}
$$

$$
\epsilon_{x\theta} = \frac{1}{2}(\sum_{i=0}^{n} z^i \phi_{i,x} + \frac{1}{r}\sum_{i=0}^{m} z^i \psi_{i,\theta}) + \frac{1}{2r}[\sum_{i=0}^{n}\sum_{j=0}^{n} z^{i+j}\phi_{i,x}\phi_{j,\theta}
$$

$$
+(w+w_0)\sum_{i=0}^{n} z^i \phi_{i,x} + \sum_{i=0}^{m}\sum_{j=0}^{m} z^{i+j}\psi_{i,\theta}\psi_{j,x} + (w+w_0)_{,x}
$$

$$
(w_{,\theta} - \sum_{i=0}^{n} z^i \phi_i) + w_{,x}w_{0,\theta}]
$$

$$
\epsilon_{zx} = \frac{1}{2}[w_{,x} + \sum_{i=0}^{m}\sum_{i=0}^{m} i z^{i-1}\psi_i + \sum_{i=0}^{n}\sum_{j=0}^{n} i z^{i+j-1}\phi_i\phi_{j,x}
$$

$$
+ \sum_{i=0}^{n}\sum_{j=0}^{n} i z^{i+j-1}\psi_i\psi_{j,x}]
$$

$$
\epsilon_{z\theta} = \frac{1}{2}\sum_{i=0}^{n} i z^{i-1}\phi_i + \frac{1}{2r}[(w_{,\theta} - \sum_{i=0}^{n} i z^i \phi_i) + \sum_{i=0}^{n} i z^{i+j-1}\phi_i\phi_{j,\theta}
$$

$$
+(w+w_0)\sum_{i=0}^{n} z^{i-1}\phi_i + \sum_{i=0}^{n}\sum_{j=0}^{n} i z^{i+j-1}\psi_i\psi_{j,\theta}] \tag{5.3-13}
$$

According to Eqs. (5.3-10) and (5.2-19), one can write

$$
\{M^k\} = \int_{-h/2}^{h/2} z^k \{\sigma\} dz = \int_{-h/2}^{h/2} z^k [\bar{Q}]\{\epsilon - \bar{\alpha}\Delta T\} dz \tag{5.3-14}
$$

so that the k-th order stress resultant vector can be separated into two components

$$
\{M^k\} = \{M^k\}_{Mechanical} - \{M^k\}_{Thermal} \tag{5.3-15}
$$

in which

$$\{M^k\}_{Thermal} = \sum_{s=1}^{N}[\bar{Q}]_s \int_{h_{s-1}}^{h_s} z^k\{\bar{\alpha}\}_s \Delta T dz$$

$$\{M^k\}_{Mechanical} = \sum_{s=1}^{N}[\bar{Q}]_s\{e\}_s \qquad (5.3\text{-}16)$$

where N is the number of layers. In the last of the above equations, e_{ij} is obtained by replacement of z^k in Eq. (5.3-13) by a new parameter $\bar{z}^k$ defined as

$$\bar{z}^k = \int_{h_{s-1}}^{h_s} z^k dz = \frac{1}{k+1}(h_s^{k+1} - h_{s-1}^{k+1}) \qquad (5.3\text{-}17)$$

3.4   Numerical solution and buckling criteria

Due to the nonlinearity of the motion equations and coupling of the displacement terms and their derivatives in these equations, an incremental formulation is adopted. By this incremental solution procedure, the real time-variant system is approximated in a step by step way assuming time-invariance within each time step.

Further arrangement of Eq. (5.3-12) can led to the equations

$$f_1^r(M_{ij}^k, \psi_1, ..., \psi_m, \phi_1, ..., \phi_m, w, \ddot{\psi}_1, \ddot{\psi}_{r-1}, ..., \ddot{\psi}_{r+1}, ..., \ddot{\psi}_m) = I^r\ddot{\psi}_r \quad (r=0,...,m)$$

$$f_2^s(M_{ij}^k, \psi_1, ..., \psi_m, \phi_1, ..., \phi_m, w, \ddot{\phi}_1, \ddot{\phi}_{s-1}, ..., \ddot{\phi}_{s+1}, ..., \ddot{\phi}_n) = I^s\ddot{\phi}_s \quad (s=0,...,n)$$

$$f_3(M_{ij}^k, \psi_1, ..., \psi_m, \phi_1, ..., \phi_m, w) = I^0\ddot{w} \qquad (5.3\text{-}18)$$

so that the following expression hold for the displacement in time interval Δt [41,51]

$$f_1^r(\tilde{M}_{ij}^k, \tilde{\psi}_1, ..., \tilde{\psi}_m, \tilde{\phi}_1, ..., \tilde{\phi}_m, \tilde{w}, \ddot{\tilde{\psi}}_1, ..., \ddot{\tilde{\psi}}_{r-1}, \ddot{\tilde{\psi}}_{r+1}, ..., \ddot{\tilde{\psi}}_m) = I^r\Delta\ddot{\psi}_r \quad (r=0,..,m)$$

$$f_2^r(\tilde{M}_{ij}^k, \tilde{\psi}_1, ..., \tilde{\psi}_m, \tilde{\phi}_1, ..., \tilde{\phi}_m, \tilde{w}, \ddot{\tilde{\phi}}_1, ..., \ddot{\tilde{\phi}}_{r-1}, \ddot{\tilde{\phi}}_{r+1}, .., \ddot{\tilde{\phi}}_m) = I^r\Delta\ddot{\phi}_r \quad (r=0,...,m)$$

$$f_3(\tilde{M}_{ij}^k, \tilde{\psi}_1, ..., \tilde{\psi}_m, \tilde{\phi}_1, ..., \tilde{\phi}_m, \tilde{w}) = I^0\Delta\ddot{w} \qquad (5.3\text{-}19)$$

in which the incremented terms are demonstrated by a " $\tilde{}$ " symbol. Therefore $(m+n+1)$ coupled and nonlinear equations in terms of $\Delta\psi_i$, $\Delta\phi_i$ and Δw are obtained. The displacements have to satisfy the initial conditions, for $(x=0, L$ and $\theta=0, 2\pi)$

$$\Delta\psi_i = 0 \quad \Delta\phi_i = 0 \quad \Delta w = 0$$
$$\Delta\dot{\psi}_i = \dot{\psi}_i \quad \Delta\dot{\phi}_i = \dot{\phi}_i \quad \Delta\dot{w} = \dot{w} \qquad (5.3\text{-}20)$$

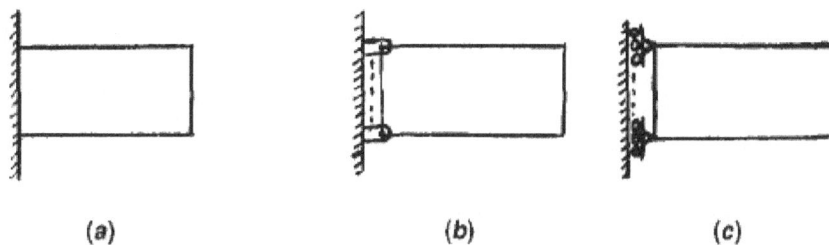

Figure 5.3-2: Some edge conditions of a cylindrical shell.

To demonstrate how the boundary conditions can be incorporated in the proposed formulation, three different edge conditions are discussed below

A - Clamped edge

In this case, the boundary conditions are independent of the z coordinate and can be written as

$$\begin{aligned} \psi_i &= 0 & i &= 0, ..., m \\ \phi_i &= 0 & i &= 0, ..., n \\ w, w_{,x} &= 0 \end{aligned} \qquad (5.3\text{-}21)$$

so that u, v, and w are zero along this edge, Fig. (5.3-2a).

For a simply supported edge, Fig. (5.3-2b), the practical boundary conditions can be expressed as

$$\begin{aligned} w &= 0 & M_x &= 0 \\ \phi_i &= 0 & i &= 0, ..., n \\ \psi_i &= 0 \quad i = 0, ..., m \quad or & i &= 0, 2, ..., m \end{aligned} \qquad (5.3\text{-}22)$$

the last condition corresponds to $u = 0$ and $u = \psi_1 z$, respectively.

C - Longitudinally constrained edge

The boundary conditions for this case, Fig. (5.3-2c) become

$$w, M_x, M_{x\theta}^i, M_{xz}^i = 0 \qquad (5.3\text{-}23)$$

3.5 Numerical Solution

The numerical solution procedure is accomplished through the following steps:

1- The numerical solution begins with discretizing the middle surface of the shell into m×n grid points in the axial and circumferential directions, respectively.
2- The initial deviations of the shell at the grid points are defined and the initial values of the displacement terms (ψ_i, ϕ_i, w) are set to zero.
3- Time is divided into a number of increments.
4- Corresponding increments of mechanical and thermal loads are found.
5- According to the temperature distribution described in the previous step, the mechanical properties in principal direction of the material are determined for each layer and related through Eqs. (5.2-20), (5.2-21) to those in geometrical coordinates of the shell. In this way, the dependency of E_1, E_2, ν_{12} and α_{ij} coefficients on temperature can be expressed as a prescribed function.
6- Derivative terms of ψ_i, ϕ_i and w with respect to the spatial coordinates that appear in Eqs. (5.3-4) and (5.3-12) are approximated by a second-order finite difference method (the central difference method).
7- Based on the displacement terms values (ψ_i, ϕ_i, w, w_0) and their derivative values obtained in the grid points, the strain values are computed.
8- The constitutive equations (5.2-19) to (5.2-21) are considered and the stress components are calculated. Then, using Eqs. (5.3-14) to (5.3-17), the stress resultant terms M_{ij}^k are computed.
9- Derivatives of M_{ij}^k terms in each grid point are approximated by a second-order finite difference method and derivative terms of Eq. (5.3-12) involving multiplication of M_{ij}^k terms (or their derivatives) in the displacement terms (or their derivatives) are substituted using a fourth-order finite difference approximation.
10- In the process of solution, time-invariance for terms appeared in the first sides of Eq. (5.3-19) is assumed during each time step. Indeed, the same idea has been adopted for reducing Eq. (5.3-18) to Eq. (5.3-20). The resulted coupled and nonlinear system of equations is solved using a successive approximation method based on a Gauss-Siedel type of substitution for the unknown variables. Thus, this manner of solution requires several iterations in each time interval. So that, values of $\ddot{\psi}_r$ and $\ddot{\phi}_r$ terms are substituted from the results of the higher equations of the system of equations, i.e. from values corresponding to ($i = 0, ..., r-1$) in Eq. (5.3-19) that are obtained in the present iteration. The remaining terms that correspond to ($i = r, ..., m$), are replaced by the values ob-

tained in the previous iteration of the current time interval. Thus, a set of second order differential equations are derived that can be solved by employing the fourth-order Runge-Kutta method subjected to the initial and boundary conditions given in Eqs. (5.3-20) to (5.3-23).

11- When all equations in each iterative step of the current time interval are solved, the maximum value of the lateral displacements (w_{max}) is determined.

12- In each grid point, the displacement term increments ($\Delta\psi_i, \Delta\phi_i$ and Δw) are added to displacement terms obtained at the end of the previous time interval. To improve the results, solution is continued by using more iterations starting from step 6, until difference of the successive values of w_{max} of the same time interval becomes negligible.

13- The corrected values of $\psi_i, \phi_i, w, \dot\psi, \dot\phi$ and $\dot w$ obtained in this manner are considered as initial values for the next time interval.

14- Beginning from step 4, results corresponding to the next time increments are obtained.

15- Possibility of dynamic buckling occurrence is checked. For this purpose, variations of w_{max} versus time or versus applied load (external pressure, axial load, temperature gradient, etc.) are plotted.

16- In the case of no buckling point, amplitude of the applied loads are increased and calculations are continued starting from step 2.

Buckling load can be determined using one of the two stability criteria explained in Section (5.3-2).

3.6 Results and discussion

In this section, examples of dynamic buckling of composite multi-layer cylindrical shells under mechanical loads are considered. Examples of dynamic buckling of cylindrical shells under thermal loads will be presented in Section (4). As a first example, buckling of a 4-ply laminated imperfect cylindrical shell under uniform axial compression used in references [3] and [15] is reconsidered. The cylinder is assumed as simply supported with radius $R = 19.0\ cm$ and $L/R = 2$. Thickness of each layer is .0135 cm and the laminate staking sequence is $[0°/30°/60°/90°]$. Thus, the laminate is an asymmetric one. The material of each lamina is assumed as Boron/Epoxy, AVCO 550 with the following properties $E_x = 207[GPa]$, $E_\theta = 18.6[GPa]$, $G_{x\theta} = 4.48[GPa]$ and $\nu_{x\theta} = \nu_{xz} = \nu_{z\theta} = .21$. The imperfection is assumed as

$$w_0(x,\theta) = W_0 h.sin(\frac{\pi x}{L}).cos(8\theta) \qquad (5.3\text{-}24)$$

where W_0 is the amplitude coefficient of imperfection and h is the total thickness of the shell.

Results of static and dynamic buckling analysis performed in references [3] and [15] as well as the present results are depicted in Fig. (5.3-3) for comparison purposes. In the earlier results, Love-Kirchhoff assumptions are adopted and von Karman type expressions of strains are chosen. Results of reference [3] are based on Hoff-Simitses criterion (the total potential energy approach) that is explained in reference [44] so that the static and dynamic loads for $W_0 = 0$ are identical. In reference [15], the nonlinear equations of motion are linearized and solved using an incremental method previously used in reference [14]. As seen from this figure, reduction in buckling loads is decreased as the amplitude of imperfection (W_0) is increased. In other words, imperfection sensitivity is more noticeable in small imperfection amplitude values. In comparison with the other results, the present results are lower in the beginning of the buckling curve. This is due to incorporation of the exact influence of the initial imperfections in the strain-displacement relations in the present analysis. On the other hand, due to the difference in the numerical procedures used in solving the equations of motion, the present results are negligibly higher in the end of the curve. Effect of cylinder length and thickness to radius ratios are illustrated in Fig. (5.3-4) for $W_0 = 0.05$.

As a second example, buckling of two imperfect three-layered cross ply ($0°/90°/0°$) cylindrical shells, as those used in reference [33], is investigated for different initial deviations. The imperfection is assumed to have the form

$$w_0 = \sum_{m=1}^{M} \sum_{n=0}^{N} W_0^{mn} sin(\frac{m\pi x}{L})cos(n\theta) \qquad (5.3\text{-}25)$$

where m and n being the numbers of axial half waves and circumferential waves, respectively.

The elastic properties of the shell are

$$
\begin{aligned}
& E_l = 209.5[GPa] && E_t = 7[GPa] \\
& G_{lt} = 3.5[GPa] && G_{tt} = 1.4[GPa] \\
& \nu_{lt} = \nu_{tt} = 0.3 && h_i = h/3 \qquad (5.3\text{-}26)
\end{aligned}
$$

where subscripts l and t stand for directions along and transverse to fibers, respectively. Two three layered thin ($h = .254\ cm$) and moderately thick ($h = 2.54\ cm$) shells with $R = 91.4\ cm$ and $l = 254\ cm$ are considered.

Figure (5.3-5) shows axial load versus axial deflection of the thin shell, for two different amplitudes of initial imperfections. Results of reference [33] and the results of the present analysis are plotted simultaneously for comparison purpose. As mentioned before, dynamic buckling has been occurred in points where abrupt change in slope is noticed in a displacement mode (for example, axial deflection) versus the applied load curve. In Fig. (5.3-5), the axial load is normalized with respect to the static buckling load of the shell. The deformed shape and buckling modes of the cylindrical shell are shown in Fig. (5.3-6). As it may be noticed from this figure, the dominant mode of the shell buckling is appeared with six axial and fourteen circumferential waves. Prior to the buckling point, results of reference [33] are negligibly lower. However, beyond this point, the present results are lower. This is due to the fact that boundary conditions can not be completely satisfied in the displacement-based layer-wise theory proposed by reference [33]. At the end of the present curve, slope increases gradually which is in agreement with the results reported in references [41,51]. The dynamic buckling point correspond to the present analysis is slightly higher.

The postbuckling response for the moderately thick shell is presented in Fig. (5.3-7) and the deformed shape of the shell is illustrated in Fig. (5.3-8). As it can be seen from this figure, in contrast to the thin shell, a buckling mode with two axial and four circumferential waves is the dominant mode for the thick shell. In general, thicker shells tend to buckle at lower axial and circumferential modes.

To compare results of the present theory with those of classical (CL), Mindlin-Riessner (first order shear-deformation) (FOSD), first order shear-deformation with a shear correction factor of 5/6 (FOSDF) and a displacement based higher order shear deformation (HOSD) [5] theories, for thicker shells,some perfect cylinders have the following geometric and material properties are considered

$$R = 19.05[cm] \qquad\qquad R/h = 15$$
$$E_{11} = 206.844 \times 10^9 [Pa] \qquad\qquad E_{22} = 18.6159 \times 10^9 [Pa]$$
$$E_{33} = 18.6159 \times 10^9 [Pa] \qquad\qquad G_{12} = 4.48162 \times 10^9 [Pa]$$
$$G_{13} = 4.48162 \times 10^9 [Pa] \qquad\qquad G_{23} = 2.55107 \times 10^9 [Pa]$$
$$\nu_{11} = 0.21 \qquad\qquad \nu_{13} = 0.21 \qquad\qquad \nu_{23} = 0.45$$

and their stacking sequences are presented in Table (5.3-1).

All laminates are symmetric. Two load cases are considered a uniform axial compression with $(L/R = 1, 5)$ and a lateral pressure on a very long cylinder. Critical loads corresponding to the mentioned theories are presented for axial compression, Table (5.3-2), and external pressure, Table (5.3-3). As it may be noticed from Table (5.3-2), the classical theory results overestimate the buckling loads especially for the shorter length and axial compression. Results of Table (5.3-3) reveals that for long

code no.	stacking sequence
1	$[0^\circ_3]_s$
2	$[0^\circ/90^\circ/0^\circ]_s$
3	$[90^\circ/0^\circ/90^\circ]_s$
4	$[90^\circ_3]_s$
5	$[45^\circ_2/-45^\circ]_s$
6	$[45^\circ/-45^\circ_2]_s$
7	$[-45^\circ/45^\circ/-45^\circ]_s$
8	$[-45^\circ_2/45^\circ]_s$
9	$[30^\circ_2/-60^\circ]_s$

Table 5.3-1: Stacking sequences used in the last example.

Code no. (L/R=1)	CL	FOSD	FOSDF	HOSD	Present
1	82.66	53.76	33.62	34.50	33.56
2	66.64	35.46	29.86	32.40	32.47
3	32.83	23.41	21.81	21.4	21.2
4	14.71	13.82	13.34	13.22	13.14
5	28.41	18.92	17.65	17.33	17.08
6	33.76	21.61	19.99	19.45	19.13
7	36.56	23.42	21.63	21.23	20.98
8	28.41	18.92	17.65	17.33	17.11
9	57.63	28.44	25.52	24.30	23.28

Code no. (L/R=5)	CL	FOSD	FOSDF	HOSD	Present
1	12.52	12.71	12.56	12.45	12.41
2	15.77	14.98	14.77	14.76	14.64
3	15.11	14.01	13.78	13.72	13.59
4	11.38	10.67	10.54	10.54	10.51
5	14.50	12.76	12.49	12.41	12.38
6	17.45	14.93	14.53	14.38	14.34
7	20.66	17.06	16.48	16.12	15.97
8	14.50	12.76	12.49	12.41	12.39
9	12.03	10.66	10.38	10.12	9.92

Table 5.3-2: Critical axial compression in (N/m)$\times 10^{-6}$ for complete fixation of the ends.

Code no. (L/R=1)	CL	FOSD	HOSD	Present
1	1847	1771	1765	1762
2	6687	6274	6191	6148
3	15671	14286	14196	14173
4	20512	18567	18333	18287
5	6226	5805	5743	5711
6	6226	5839	5791	5752
7	6226	5853	5826	5814
8	6226	5805	5743	5707
9	3144	2999	2985	2979

Table 5.3-3: Critical pressure in Pa $\times 10^{-3}$

cylinders, results of the mentioned theories become closer and the difference becomes higher for thicker cylinders. Comparison of results of Table (5.3-2) reveals that incorporating the influence of transverse shear deformations in composite materials is necessary even for symmetric staking sequences of plies.

4 Higher-Order Deformation Theory

In this section, based on the energy form of the three-dimensional theory of elasticity a hybrid higher-order theory for thermal as well as mechanical dynamic buckling analysis of cylindrical shells is developed that includes the effect of all stress and strain components. Therefore, the transverse normal stress and strain that were ignored in the previous theory, are automatically considered in this theory. It is proved that transverse normal stress and strain have significant influence on the buckling loads, especially for antisymmetric staking sequences of the layers [19]. Besides, variation of the radial component of the displacements through the thickness of the shell is considered. Love-Timoshenko assumption used for evaluating the functionals in the Sect.(3) is not used here. Investigations by Eslami and Shariyat [36] show that a theory of this type, while posses almost the accuracy of the 3-D elasticity analysis in describing the global laminate response, have more tendency to converge even though it contains full nonlinear terms related to the complete nonlinear strain-displacement relations. This in turn leads to a remarkable saving in the computation time.

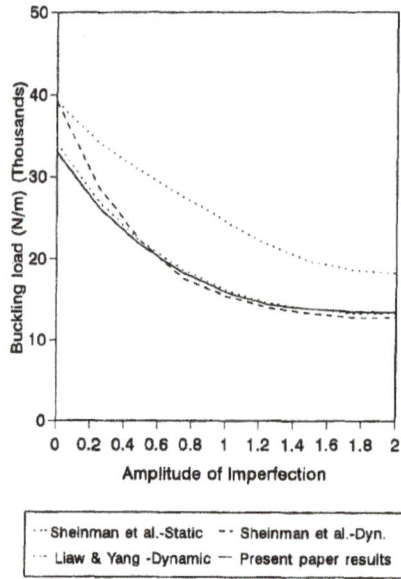

Figure 5.3-3: Effect of imperfection amplitude on the dynamic buckling load.

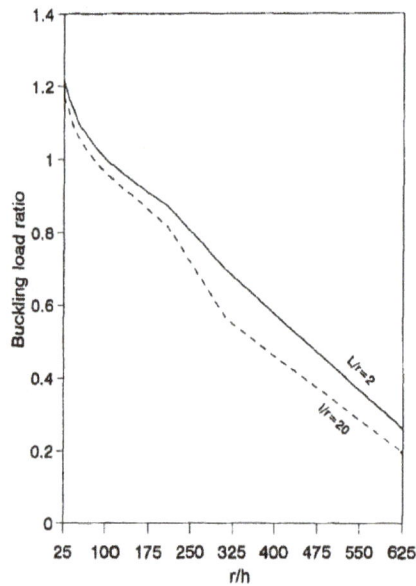

Figure 5.3-4: Effect of thickness to radius and length to radius ratios on the dynamic buckling load of the shell.

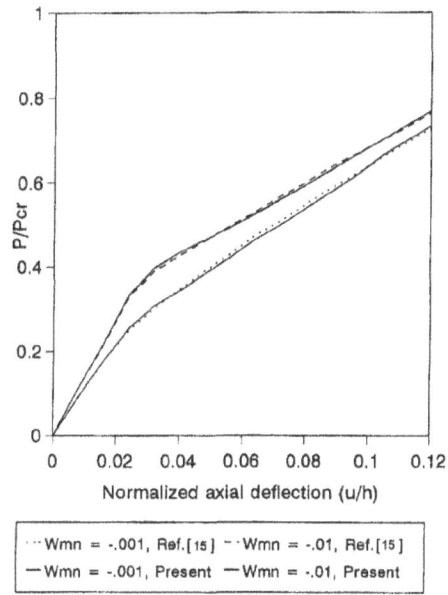

Figure 5.3-5: Comparison of the postbuckling analysis results of a thin shell. The illustrated results correspond to $(x = l/2, \ z = -h/2)$.

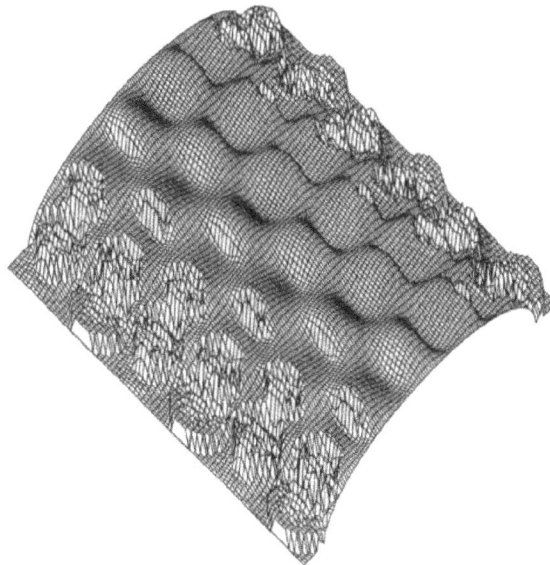

Figure 5.3-6: Buckling mode of the thin cylindrical shell.

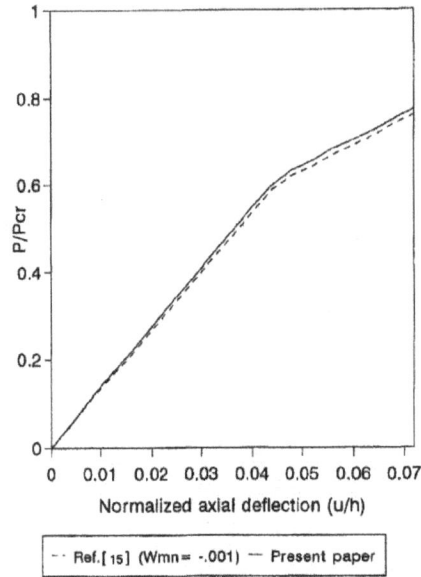

Figure 5.3-7: Comparison of the postbuckling analysis results of a moderately thick shell. The illustrated results correspond to $(x = l/2,\ z = -h/2)$.

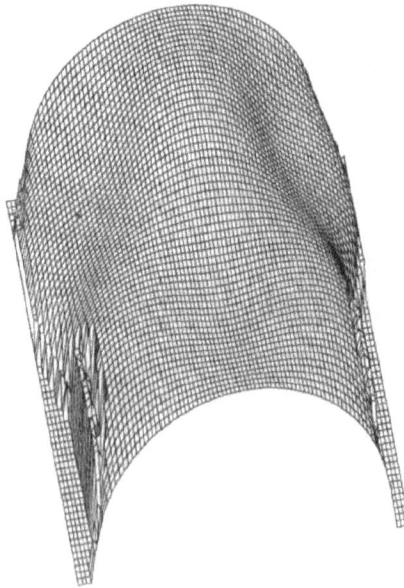

Figure 5.3-8: Buckling mode of the thick cylindrical shell.

4.1 Development of Hamilton's functional

The equations of motion can be derived through using Hamilton's principle. From Eq. (5.3-6) it can be proved written as

$$\int_{t_1}^{t_2}\{\int_S\int_{r_{in}}^{r_{out}}(\sigma_{ij}\delta\epsilon_{ij}+\rho\ddot{u}_i\delta u_i)dr.dS-\delta\Omega\}dt=0 \qquad (5.4\text{-}1)$$

where ρ is the density per unit volume and Ω is the work done by the external forces. According to Brush and Almroth [52], for a circular cylindrical shell under external fluid pressure, this work is equal to

$$\Omega=p\int_{\bar{S}}[\bar{w}+\frac{1}{2\bar{r}}(\bar{v}^2-\bar{v}\bar{w}_{,\theta}+\bar{v}_{,\theta}\bar{w}+\bar{w}^2)]d\bar{S} \qquad (5.4\text{-}2)$$

where $\bar{S}$ is the external surface of the shell and $(\bar{u},\bar{v},\bar{w})$ are the displacement components of the external surface of the shell. If the external pressure is a dead loading type, only the first term of the above integral will appear. The increments of the strain components can be readily found from Eq. (5.2-12). These expressions after some rearrangements are

$$\delta\epsilon_{rr}=u_{,r}\delta u_{,r}+v_{,r}\delta v_{,r}+[1+(w+w_0)_{,r}]\delta w_{,r}$$

$$\delta\epsilon_{\theta\theta}=\frac{1}{r^2}\{[v-(w+w_0)_{,\theta}]\delta v+(r+v_{,\theta}+w+w_0)\delta w+u_{,\theta}\delta u_{,\theta}$$

$$+(r+v_{,\theta}+w+w_0)\delta v_{,\theta}+[(w+w_0)_{,\theta}-v]\delta w_{,\theta}\}$$

$$\delta\epsilon_{xx}=(1+u_{,x})\delta u_{,x}+v_{,x}\delta v_{,x}+(w+w_0)_{,x}\delta w_{,x}$$

$$\delta\epsilon_{r\theta}=\frac{1}{2r}\{-[1+(w+w_0)_{,r}]\delta v+v_{,r}\delta w+u_{,\theta}\delta u_{,r}+(r+v_{,\theta}+w+w_0)\delta v_{,r}$$

$$+[(w+w_0)_{,\theta}-v]\delta w_{,r}+u_{,r}\delta u_{,\theta}+v_{,r}\delta v_{,\theta}+[1+(w+w_0)_{,r}]\delta w_{,\theta}\}$$

$$\delta\epsilon_{rx}=\frac{1}{2}\{(1+u_{,x})\delta u_{,r}+v_{,x}\delta v_{,r}+(w+w_0)_x\delta w_{,r}+u_{,r}\delta u_{,x}+v_{,r}\delta v_{,x}$$

$$+[1+(w+w_0)_{,r}]\delta w_{,x}\}$$

$$\delta\epsilon_{\theta,x}=\frac{1}{2r}\{-(w+w_0)_{,x}\delta v+v_{,x}\delta w+(1+u_{,x})\delta u_{,\theta}+v_{,x}\delta v_{,\theta}+(w+w_0)_{,x}\delta w_{,\theta}$$

$$+u_{,\theta}\delta u_{,x}+(r+v_{,\theta}+w+w_0)\delta v_{,x}+[(w+w_0)_{,\theta}-v]\delta w_{,x}\} \qquad (5.4\text{-}3)$$

Substituting Eqs. (5.4-3) and (5.4-2) into Eq. (5.4-1) leads to

$$\int_{t_1}^{t_2}\{\int_S[\int_{r_{in}}^{r_{out}}(\rho\ddot{u}\delta u$$

$$-\frac{1}{r}\{\frac{1}{r}\sigma_\theta[(w+w_0)_{,\theta}-v]+\sigma_{r\theta}[1+(w+w_0)_{,r}]+\sigma_{\theta x}(w+w_0)_{,x}-\rho r\ddot{v}\}\delta v$$

$$+\frac{1}{r}\{\frac{1}{r}\sigma_\theta(r+v_{,\theta}+w+w_0)+\sigma_{r\theta}v_{,r}+\sigma_{\theta x}v_{,x}+\rho r\ddot{w}\}\delta w$$

252CHAPTER 5. DYNAMIC BUCKLING OF CYLINDRICAL SHELLS

$$+\{\sigma_r u_{,r} + \frac{1}{r}\sigma_{r\theta} u_{,\theta} + \sigma_{rx}(1 + u_{,x})\}\delta u_{,r}$$

$$+\{\sigma_r v_{,r} + \frac{1}{r}\sigma_{r\theta}(r + v_{,\theta} + w + w_0) + \sigma_{rx} v_{,x}\}\delta v_{,r}$$

$$+\{\sigma_r[1 + (w + w_0)_{,r}] + \frac{1}{r}\sigma_{r\theta}[(w + w_0)_{,\theta} - v] + \sigma_{rx}(w + w_0)_{,x}\}\delta w_{,r}$$

$$+\frac{1}{r}\{\frac{1}{r}\sigma_\theta u_{,\theta} + \sigma_{r\theta} u_{,r} + \sigma_{\theta x}(1 + u_{,x})\}\delta u_{,\theta}$$

$$+\frac{1}{r}\{\frac{1}{r}\sigma_\theta(r + v_{,\theta} + w + w_0) + \sigma_{r\theta} v_{,r} + \sigma_{\theta x} v_{,x}\}\delta v_{,\theta}$$

$$+\frac{1}{r}\{\frac{1}{r}\sigma_\theta[(w + w_0)_{,\theta} - v] + \sigma_{r\theta}[1 + (w + w_0)_{,r}] + \sigma_{\theta x}(w + w_0)_{,x}\}\delta w_{,\theta}$$

$$+\{\sigma_x(1 + u_{,x}) + \sigma_{rx} u_{,r} + \frac{1}{r}\sigma_{\theta x} u_{,\theta}\}\delta u_{,x}$$

$$+\{\sigma_x v_{,x} + \sigma_{rx} v_{,r} + \frac{1}{r}\sigma_{\theta x}(r + v_{,\theta} + w + w_0)\}\delta v_{,x}$$

$$+\{\sigma_x(w + w_0)_{,x} + \sigma_{rx}[1 + (w + w_0)_{,r}] + \frac{1}{r}\sigma_{\theta x}[(w + w_0)_{,\theta} - v]\}\delta w_{,x}].dr].dS$$

$$-\int_{\bar{S}} \frac{p}{2\bar{r}}\{(2\bar{v} - \bar{w}_{,\theta})\delta\bar{v} + [2\bar{r} + (\bar{v}_{,\theta} + 2\bar{w})]\delta\bar{w} + \bar{w}\delta\bar{v}_{,\theta} - \bar{v}\delta\bar{w}_{,\theta}\}.d\bar{S}\}.dt = 0$$

$$(5.4\text{-}4)$$

In derivation of the above expressions, as it is common, only the external work due to the lateral forces are appeared in Hamilton's functional and the work of the external loads applied on the two ends of the shell is not considered explicitly. The effect of these loads may be accounted for when imposing the boundary conditions. If the compression and shear stresses applied at the ends of the shell are denoted by q and τ, respectively, the following terms should be added to the first side of Eq. (5.4-4), in order to explicitly incorporate this work

$$-\int_{t_1}^{t_2} \int_0^{2\pi} \int_{r_{in}}^{r_{out}} r(q\delta u_l + \tau_t \delta v_l).dr.d\theta.dt \qquad (5.4\text{-}5)$$

where u_l and v_l are the longitudinal and circumferential displacements corresponding to the end edge of the shell ($x = l$). In this case, only the displacement boundary conditions must be satisfied. Almost in all references integration across the thickness is carried over using the variable z which is the point distance from the reference surface. In these approaches, Love-Timoshenko [31] (i.e. $r = \bar{r} + z \cong \bar{r}$) is used. To avoid this assumption, r is chosen as the integration variable of the current work.

Integrating by parts in the x and θ directions leads to the following expression for Eq. (5.4-4)

$$\int_{t_1}^{t_2}\{\int_S[\int_{r_{in}}^{r_{out}}(\{\rho\ddot{u}-[[\frac{1}{r}\sigma_\theta u_{,\theta}+\sigma_{r\theta}u_{,r}+\sigma_{\theta x}(1+u_{,x})]_{,\theta}$$

$$-[r\sigma_x(1+u_{,x})+r\sigma_{rx}u_{,r}+\sigma_{\theta x}u_{,\theta}]_{,x}\}\delta u$$

$$+\{r\sigma_r u_{,r}+\sigma_{r\theta}u_{,\theta}+r\sigma_{rx}(1+u_{,x})\}\delta u_{,r}$$

$$-\{\frac{1}{r}\sigma_\theta[(w+w_0)_{,\theta}-v]+\sigma_{r\theta}[1+(w+w_0)_{,r}]+\sigma_{\theta x}(w+w_0)_{,x}-\rho\ddot{v}\}$$

$$+[\frac{1}{r}\sigma_\theta(r+v_{,\theta}+w+w_0)+\sigma_{r\theta}v_{,r}+\sigma_{\theta x}v_{,x}]_{,\theta}$$

$$+[r\sigma_x v_{,x}+r\sigma_{rx}v_{,r}+\sigma_{\theta x}(r+v_{,\theta}+w+w_0)]_{,x}\}\delta v$$

$$+[r\sigma_r v_{,r}+\sigma_{r\theta}(r+v_{,\theta}+w+w_0)+r\sigma_{rx}v_{,x}\}\delta v_{,r}$$

$$+\{\frac{1}{r}\sigma_\theta(r+v_{,\theta}+w+w_0)+\sigma_{r\theta}v_{,r}+\sigma_{\theta x}v_{,x}+\rho r\ddot{w}$$

$$-[\frac{1}{r}\sigma_\theta[(w+w_0)_{,\theta}-v]+\sigma_{r\theta}[1+(w+w_0)_{,r}]+\sigma_{\theta x}(w+w_0)_{,x}]_{,\theta}$$

$$-[r\sigma_x(w+w_0)_{,x}+r\sigma_{rx}[1+(w+w_0)_{,r}]+\sigma_{\theta x}[(w+w_0)_{,\theta}-v]]_{,x}\}\delta w$$

$$+[r\sigma_r[1+(w+w_0)_{,r}]+\sigma_{r\theta}[(w+w_0)_{,\theta}-v]+r\sigma_{rx}(w+w_0)_{,x}]\delta w_{,r}$$

$$-p(\bar{v}-\bar{w}_{,\theta})\delta\bar{v}+[\bar{r}+(\bar{v}_{,\theta}+\bar{w})]\delta\bar{w}]d\theta.dx\}.dt=0$$

$$(5.4\text{-}6)$$

4.2 The governing equations

A high-order Mindlin-Riessner type shear deformation theory is used to establish the governing equations of the imperfect cylindrical composite shell. The proposed solution has the following form

$$u_i(r,\theta,x,t)=\sum_{k=0}^{n}[U_i(\theta,x,t)]_k.r^k \qquad (5.4\text{-}7)$$

the upper limit can be chosen to be identical for all of the displacement components. Thus, one may write

$$u=\sum_{k=0}^{n}U_k.r^k \qquad v=\sum_{k=0}^{n}V_k.r^k \qquad w=\sum_{k=0}^{n}W_k.r^k \quad (5.4\text{-}8)$$

In this equation, U_k, V_k, and W_k are coefficients of the proposed power series solution and are determined after incorporating the initial and boundary conditions in the governing equations. It is proved that incorporating the effect of transverse shear and normal strains require at least a third order expressions for the axial and circumferential displacement components and at least a second order expression for the radial component [19]. However, choosing interpolating polynomials of order higher than 3 causes local extrema between the precision points. The incremental form of Eq. (5.4-8) is

$$\delta u = \sum_{k=0}^{n} \delta U_k.r^k \qquad \delta v = \sum_{k=0}^{n} \delta V_k.r^k \qquad \delta w = \sum_{k=0}^{n} \delta W_k.r^k$$

(5.4-9)

Substitution of Eqs. (5.4-8) and (5.4-9) into Eq. (5.4-6) and noting that U_k, V_k, and W_k are arbitrary and nonzero coefficients, leads to the equations

$$\sum_{j=0}^{n} \int_{r_{in}}^{r_{out}} ([\sigma_\theta \sum_{k=0}^{n} U_{k,\theta} r^{j+k-1} + \sigma_{r\theta} \sum_{k=0}^{n} U_k.kr^{j+k-1} + \sigma_{\theta x}(r^j + \sum_{k=0}^{n} U_{k,x} r^{j+k})]_{,\theta}$$

$$+[\sigma_x(r^{j+1} + \sum_{k=0}^{n} U_{k,x} r^{j+k+1}) + \sigma_{rx} \sum_{k=0}^{n} kU_k r^{j+k} + \sigma_{\theta x} \sum_{k=0}^{n} U_{k,\theta} r^{j+k}]_{,x}$$

$$-\rho \sum_{k=0}^{n} \ddot{U}_k r^{j+k+1} - j[\sigma_r \sum_{k=0}^{n} k.U_k r^{j+k-1} + \sigma_{r\theta} \sum_{k=0}^{n} U_{k,\theta} r^{j+k-1}$$

$$+\sigma_{rx}(r^j + \sum_{k=0}^{n} U_{k,x} r^{j+k})]).\delta U_j.dr = 0$$

$$\sum_{j=0}^{n} \{ \int_{r_{in}}^{r_{out}} (\sigma_\theta \sum_{k=0}^{n} [(W_k + W_{0,k}),_\theta - V_k].r^{j+k-1} + \sigma_{r\theta}[\sum_{k=0}^{n}(W_k + W_{0,k})k.r^{j+k-1}$$

$$+r^j] + \sigma_{\theta x} \sum_{k=0}^{n}(W_k + W_{0k}),_x r^{j+k} - \rho \sum_{k=0}^{n} \ddot{V}_k r^{j+k+1}$$

$$+[\sigma_\theta[r^j + \sum_{k=0}^{n}(V_{k,\theta} + W_k + W_{0k})r^{j+k-1}] + \sigma_{r\theta} \sum_{k=0}^{n} k.V_k r^{j+k-1} + \sigma_{\theta x} \sum_{k=0}^{n} V_{k,x} r^{j+k}]_{,\theta}$$

$$+[\sigma_x \sum_{k=0}^{n} V_{k,x} r^{j+k+1} + \sigma_{rx} \sum_{k=0}^{n} k.V_k r^{j+k} + \sigma_{\theta x}[r^{j+1} + \sum_{k=0}^{n}(V_{k,\theta} + W_k + W_{0k})r^{j+k}]]_{,x}$$

$$-j[\sigma_r \sum_{k=0}^{n} k.V_k r^{j+k-1} + \sigma_{r\theta}[r^j + \sum_{k=0}^{n}(V_{k,\theta} + W_k + W_{0k})r^{j+k-1}]$$

$$+\sigma_{rx} \sum_{k=0}^{n} V_{k,x} r^{j+k}])\delta V_j.dr + p \sum_{k=0}^{n}(V_k - W_{k,\theta})r_{out}^{j+k}.\delta V_j\} = 0$$

$$\sum_{j=0}^{n} \{ \int_{r_{in}}^{r_{out}} (\sigma_\theta[r^j + \sum_{k=0}^{n}(V_{k,\theta} + W_k + W_{0k})r^{j+k-1}] + \sigma_{r\theta} \sum_{k=0}^{n} kV_k r^{j+k-1} + \sigma_{\theta x} \sum_{k=0}^{n} V_{k,x} r^{j+k}$$

$$+\rho \sum_{k=0}^{n} \ddot{W}_k r^{j+k+1} - [\sigma_\theta \sum_{k=0}^{n}[(W_k + W_{0k}),_\theta - V_k]r^{j+k-1} + \sigma_{r\theta}[\sum_{k=0}^{n}(W_k$$

$$+W_{0k}).kr^{j+j-1} + r^j] + \sigma_{\theta x} \sum_{k=0}^{n}(W_k + W_{0k}),_x r^{j+k}]_{,\theta}$$

$$-[\sigma_x \sum_{k=0}^{n}(W_k + W_{0k}),_x r^{j+k+1} + \sigma_{rx}[r^{j+1} + \sum_{k=0}^{n}(W_k + W_{0k}).k.r^{j+k}]$$

$$+\sigma_{\theta x} \sum_{k=0}^{n}[(W_k + W_{0k}),_\theta - V_k]r^{j+k}]_{,x} + j[\sigma_r[r^j + \sum_{k=0}^{n} k(W_k$$

4. *HIGHER-ORDER DEFORMATION THEORY* 255

$$+W_{0k})r^{j+k-1}] + \sigma_{r\theta}\sum_{k=0}^{n}[(W_k + W_{0k}),_\theta - V_k]r^{j+k-1} + \sigma_{rx}\sum_{k=0}^{n}(W_k$$

$$+W_{0k}),_x r^{j+k}]).\delta W_j.dr - p[r_{r_{out}}^{j+1} - \sum_{k=0}^{n}(V_{k,\theta} - W_k)r_{out}^{j+k}].\delta W_j\} = 0 \tag{5.4-10}$$

In the above equation, the initial imperfection function w_0 can be expressed in power series form

$$w_0 = \sum_{k=0}^{n} W_{0k}.r^k \tag{5.4-11}$$

Defining the following symbols

$$M^l = \int_{r_{in}}^{r_{out}} \sigma.r^l.dr \qquad I^l = \int_{r_{in}}^{r_{out}} \rho.r^l.dr \tag{5.4-12}$$

and noting that $\delta U_j, \delta V_j$, and δW_j ($j = 0, ..., n$) are independent and arbitrary values, Eq. (5.4-10) results in the following governing equations

$$\sum_{k=0}^{n}[(M_\theta^{j+k-1}U_{k,\theta} + k.M_{r\theta}^{j+k-1}U_k + M_{\theta x}^{j+k}U_{k,x}),_\theta + (M_x^{j+k+1}U_{k,x}$$

$$+k.M_{rx}^{j+k}U_k + M_{\theta x}^{j+k}U_{k,\theta}),_x - I^{j+k+1}\ddot{U}_k - j(k.M_r^{j+k-1}U_k$$

$$+M_{r\theta}^{j+k-1}U_{k,\theta} + M_{rx}^{j+k}U_{k,x})] + M_{\theta x,\theta}^j + M_{x,x}^{j+1} - jM_{rx}^j = 0$$

$$\sum_{k=0}^{n}\{M_\theta^{j+k-1}[(W_k + W_{0,k}),_\theta - V_k] + kM_{r\theta}^{j+k-1}(W_k + W_{0,k}) + M_{\theta x}^{j+k}(W_k + W_{0k}),_x$$

$$-I^{j+k+1}\ddot{V}_k + [M_\theta^{j+k-1}(V_{k,\theta} + W_k + W_{0k}) + kM_{r\theta}^{j+k-1}.V_k + M_{\theta x}^{j+k}V_{k,x}],_\theta$$

$$+[M_x^{j+k+1}V_{k,x} + kM_{rx}^{j+k}.V_k + M_{\theta x}^{j+k}(V_{k,\theta} + W_k + W_{0k})],_x - j[kM_r^{j+k-1}V_k$$

$$+M_{r\theta}^{j+k}(V_{k,\theta} + W_k + W_{0k}) + M_{rx}^{j+k}V_{k,x}] + p(V_k - W_{k,\theta})r_{out}^{j+k}\}$$

$$+M_{r\theta}^j + M_{\theta,\theta}^j + M_{\theta x,x}^{j+1} - jM_{r\theta}^j = 0$$

$$\sum_{k=0}^{n}\{M_\theta^{j+k-1}(V_{k,\theta} + W_k + W_{0k}) + kM_{r\theta}^{j+k-1}V_k + M_{\theta x}^{j+k}V_{k,x} + I^{j+k+1}\ddot{W}_k$$

$$-[M_\theta^{j+k-1}[(W_k + W_{0k}),_\theta - V_k] + kM_{r\theta}^{j+k-1}(W_k + W_{0k}) + +M_{\theta x}^{j+k}(W_k + W_{0k}),_x],_\theta$$

$$-[M_x^{j+k+1}(W_k + W_{0k}),_x + kM_{rx}^{j+k}(W_k + W_{0k}) + M_{\theta x}^{j+k}[(W_k + W_{0k}),_\theta - V_k]],_x$$

$$+j[kM_r^{j+k-1}(W_k + W_{0k}) + M_{r\theta}^{j+k-1}[(W_k + W_{0k}),_\theta - V_k] + M_{rx}^{j+k}(W_k + W_{0k}),_x]$$

$$+pr_{out}^{j+k}(V_{k,\theta} - W_k)\} + M_\theta^j - M_{r\theta,\theta}^j - M_{rx,x}^{j+1} + jM_r^j - pr_{out}^{j+1} = 0 \tag{5.4-13}$$

Thus, there are three governing equations associated with each j value ($j = 0, ..., n$). Therefore, a set composed of $3(n + 1)$ governing equations in terms of U_j, V_j, and W_j is obtained.

Evaluation of the M expressions

Substituting Eq. (5.2-19) into Eq. (5.4-12) gives the following equation for the stress resultants expressions

$$\{M^l\} = \sum_{s=0}^{N-1} \{M_s^{*l} - (M_T^l)_s\} \tag{5.4-14}$$

where s is the layer counter, N is the number of layers and

$$\{M_s^{*l}\} = [\bar{Q}]^{(s)} \int_{r_s}^{r_{s+1}} \{\epsilon\} r^l.dr \qquad \{(M_T^l)_s\} = [\bar{Q}]^{(s)} \{\alpha^{(s)}\} \int_{r_s}^{r_{s+1}} \Delta T.r^l.dr \tag{5.4-15}$$

Thus, the stress resultant expressions can be found after evaluating the following integrals

$$\int_{r_s}^{r_{s+1}} \epsilon_{ij}.r^l.dr = e_{ij}^l|_{r_s}^{r_{s+1}} \tag{5.4-16}$$

and substituting their values in Eq. (5.4-12). Substituting Eqs. (5.4-8) and (5.4-11) into Eq. (5.4-12), the final form of the e_{ij}^l terms of Eq. (5.4-16) are

$$e_{rr}^l = \sum_{k=0}^n \frac{k}{k+l} W_k r^{k+l} + \frac{1}{2}\{\sum_{j=0}^n \sum_{k=0}^n \frac{kj}{k+j+l-1}(W_k W_j + 2W_k W_{0j} + V_k V_j$$

$$+U_k U_j)r^{k+j+l-1}\}$$

$$e_{\theta\theta}^l = \sum_{k=0}^n \frac{1}{k+l}(V_{k,\theta} + W_k)r^{k+l} + \frac{1}{2}\{\sum_{j=0}^n \sum_{k=0}^n [(V_{k,\theta} + W_k + W_{0k})(V_{j,\theta} + W_j + W_{0j})$$

$$+(W_{k,\theta} + W_{0k,\theta} - V_k)(W_{j,\theta} + W_{0j,\theta} - V_j) + U_{k,\theta}U_{j,\theta} - W_{0k}W_{0j} - W_{0k,\theta}W_{0j,\theta}]$$

$$\cdot\frac{1}{k+j+l-1}.r^{k+j+l-1}\}$$

$$e_{xx}^l = \sum_{k=0}^n \frac{k}{k+l+1}U_{k,x}r^{k+l+1} + \frac{1}{2}[\sum_{j=0}^n \sum_{k=0}^n \frac{1}{k+j+l+1}(W_{k,x}W_{j,x} + 2W_{k,x}W_{0j,x}$$

$$+V_{k,x}V_{j,x} + U_{k,x}U_{j,x})r^{k+j+l+1}]$$

$$e_{r\theta}^l = \frac{1}{2}\sum_{k=0}^n \frac{1}{k+l}(kV_k + W_{k,\theta} - V_k)r^{k+l} + \sum_{j=0}^n \sum_{k=0}^n \frac{j}{2(k+j+l-1)}[(W_{k,\theta} - V_k)W_j$$

$$-V_k W_{0j} + W_j W_{0k,\theta} + W_{k,\theta}W_{0j} + (V_{k,\theta} + W_k + W_{0k})V_j + U_j U_{k,\theta}]r^{k+j+l-1}\}$$

$$e_{rx}^l = \frac{1}{2}\sum_{k=0}^n \frac{1}{k+l+1}W_{k,x}r^{k+l+1} + \sum_{j=0}^n \sum_{k=0}^n \frac{j}{2(k+j+l)}[(W_{k,x}W_j + W_{0k,x}W_j$$

$$+W_{0j}W_{k,x} + V_j V_{k,x} + U_j U_{k,x})r^{k+j+l}] + \frac{1}{2}\sum_{k=0}^n \frac{k}{k+l}U_k r^{k+l}$$

$$e_{\theta x}^l = \frac{1}{2}\{\sum_{k=0}^n \frac{1}{k+l+1}V_{k,x}r^{k+l+1} + \sum_{k=0}^n \frac{1}{k+l}U_{k,\theta}r^{k+l} + \sum_{j=0}^n \sum_{k=0}^n \frac{1}{k+j+l}$$

$$\cdot[V_{k,x}(V_{j,\theta} + W_j + W_{0j}) + U_{k,\theta}U_{j,x} + (W_k + W_{0k})_{,x}(W_{j,\theta} - V_j) + W_{k,x}W_{0j,\theta}]$$

$$\cdot r^{k+j+l}\} \tag{5.4-17}$$

4.3   Numerical solution scheme

The nonlinear equations of motion (5.4-13) may be rewritten in the following form

$$f_m(M_{ij}, U_k, V_k, W_k, \ddot{U}_{k\&k\neq m}) = I^{2m+1}\ddot{U}_m$$
$$g_m(M_{ij}, U_k, V_k, W_k, \ddot{V}_{k\&k\neq m}) = I^{2m+1}\ddot{V}_m \qquad m = 0, ..., n$$
$$h_m(M_{ij}, U_k, V_k, W_k, \ddot{W}_{k\&k\neq m}) = I^{2m+1}\ddot{W}_m \qquad (5.4\text{-}18)$$

Nonlinear systems are usually treated by means of an incremental solution procedure. Thus, the applied loads are increased incrementally and the increments of displacements, strains, and stresses are computed. Following this procedure, the perturbed form of Eq. (5.4-18) is expressed as

$$f_m(\tilde{M}_{ij}, \tilde{U}_k, \tilde{V}_k, \tilde{W}_k, \Delta\ddot{U}_{k\&k\neq m}) = I^{2m+1}\Delta\ddot{U}_m$$
$$g_m(\tilde{M}_{ij}, \tilde{U}_k, \tilde{V}_k, \tilde{W}_k, \Delta\ddot{V}_{k\&k\neq m}) = I^{2m+1}\Delta\ddot{V}_m \qquad m = 0, ..., n$$
$$h_m(\tilde{M}_{ij}, \tilde{U}_k, \tilde{V}_k, \tilde{W}_k, \Delta\ddot{W}_{k\&k\neq m}) = I^{2m+1}\Delta\ddot{W}_m \qquad (5.4\text{-}19)$$

where the incremented quantities are indicated by a " ~ " symbol. These equations should be solved simultaneously in each time interval Δt. This is accomplished by substituting the partial derivatives with respect to x and θ by the corresponding finite difference expressions and by employing the fourth order Runge-Kutta method for time marching during an iterative solution scheme. To this end, the middle surface of the shell is discretized into a two dimensional mesh composed of circumferential and axial nodes, and the thermal and mechanical loads are applied incrementally. Since the temperature dependency of the mechanical properties in principal directions of the material ($E_i, \nu_{ij}, \alpha_{ij}$) may be expressed by prescribed functions, these values in the geometrical coordinates of the shell are determined through Eqs. (5.2-20) and (5.2-21). The strain values at the grid points are calculated from Eq. (5.2-5) through substituting the partial derivatives of the displacement components defined in Eqs. (5.4-8) and (5.4-9) by a finite difference approximation. In the beginning of the solution, U_k, V_k, and W_k values are set to zero and are modified through the proceeding stages of the same time interval iterations. The stress components and, consequently, the expressions for M are determined from Eqs. (5.2-19) and (5.4-14) to (5.4-17). Therefore, the modified distribution of the displacement components is achieved by simultaneously solving Eqs. (5.4-19), employing the fourth order Runge-Kutta method subjected to the initial and boundary conditions. In the process of solution, time-invariance of displacement coefficients is assumed during each time interval. A Gauss-Seidel type iterative method of solution that includes all of the foregoing derivations to establish the required equations is implicitly adopted. In the case of stress, force, and moment boundary conditions, corresponding values should be substituted directly in the formulations. At the end of each time interval, the maximum value of a certain displacement mode (say, w_{max}) is determined and the corrected values of U_k, V_k, W_k, $\dot{U}_k$, $\dot{V}_k$, and $\dot{W}_k$ are considered as initial values for the next time increment. The steps mentioned above are continued until the final time is reached. Buckling occurrence is investigated using the generalized concept of dynamic buckling proposed by Budiansky or increasing amplitudes criterion described in Section 2.3 of this chapter.

4.4   Results and discussions

In all examples studied in this section, loads are assumed to be step function with infinite duration. Static buckling loads are determined using the generalized criteria proposed by Budiansky [42] and employing the dynamic relaxation method for the resulted governing equations.

To compare the accuracy of the results of the present theory with that of the existing classical theories and the higher-order shear-deformation and three dimensional approaches previously introduced by the author [32,36], an example of dynamic buckling of multilayered composite circular cylindrical shells under mechanical load is considered. For this purpose, dynamic buckling of a 4-ply laminated imperfect cylindrical shell under uniform axial compression is investigated. The cylinder is assumed as simply supported with $R = 19[cm]$ and $L/R = 2$, where R and L are the shell radius and length, respectively. Thickness of each layer is 0.0135 [cm] and the staking sequence is considered to be $[0°/30°/60°/90°]$. Each lamina is constructed from Boron/Epoxy, AVCO 550 material whose properties are as follows

$$E_x = 207[GPa] \qquad\qquad E_\theta = E_x = 18.6[GPa]$$
$$G_{x\theta} = 4.48[GPa] \qquad\qquad \nu_{x\theta} = \nu_{xz} = \nu_{z\theta} = .21$$

The initial geometric imperfection of the shell is assumed to have the following form

$$w_0(x,\theta) = W_0 h.sin(\frac{\pi_x}{L}).cos(8\theta)$$

where W_0 and h are the amplitude of imperfection and the total thickness of the shell, respectively.

Results of the references [15] and [3] as well as results of the present approach are depicted in Fig. (5.4-1). Results of references [15] and [3]

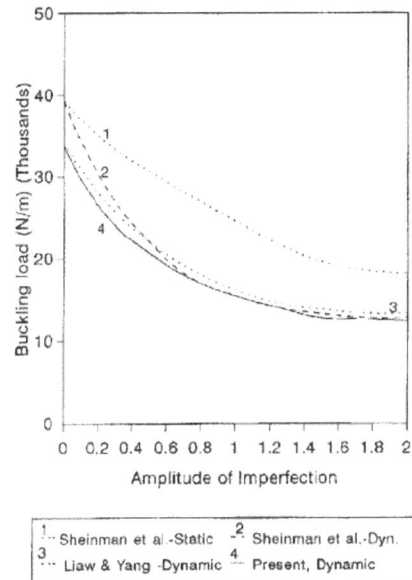

Figure 5.4-1: Effect of imperfection amplitude on the dynamic buckling load.

are based on equivalent single-layer energy approaches. Reference [15] consumes the Love-Kirchhoff assumptions and the nonlinear equations of motion are linearized before solution. Results of reference [3] are based on Hoff-Simitses total potential approach criterion for buckling detection which is described in details in reference [44]. Though the results of the exact three dimensional approach introduced by the author are more conservative, it is observed that the required run-time for solving the present governing equations is considerably less than that of the 3-D approach and the results are much more stable. Indeed, discretizing the shell in three perpendicular directions, causes a high sensitivity to mesh or element size or element numbers and to time increment and even to the values of the mechanical properties. Therefore, the convergency is reached within a specific value of Δt for fixed values of the other parameters. For this reason, the 3-D elasticity method is usually not self-started.

From Fig. (5.4-1), it is easily seen that the predicted buckling loads of composite cylindrical shells with small initial geometrical imperfection amplitude have significant differences for different theories. In Fig. (5.4-2), results of the present theory are compared with those obtained using the higher-order shear-deformation (HOSD), the general layer-wise (GLW) and the three dimensional elasticity theories previously proposed by the author [32,35,36]. As mentioned before, the present results are

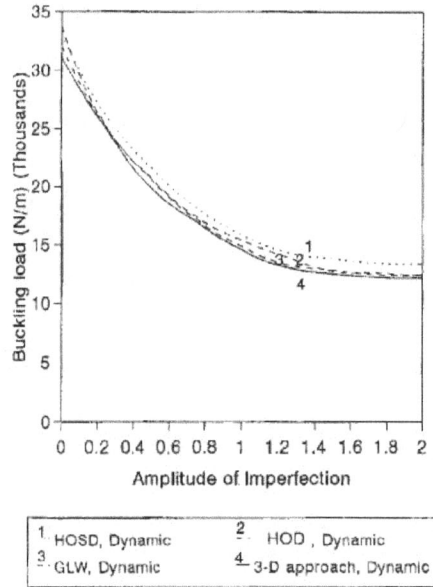

Figure 5.4-2: Comparison of the result of the higher-order shear deformation equivalent-layer theory, general layer-wise theory, 3-D elasticity approach, and the present theory.

very close to the results of GLW and 3-D approaches with remarkable saving in run-time, and more stability in solution.

Thermal buckling is investigated for cylindrical shells constructed from materials with the following mechanical and thermal properties

$$E_{11} = 1.5 \times 10^5 [N/cm^2] \qquad E_{22} = E_{33} = 0.1 E_{11}$$
$$G_{12} = G_{13} = G_{23} = 0.5 E_{22} \qquad \nu_{12} = \nu_{13} = \nu_{23} = 0.25$$
$$\alpha_1 = \alpha_2 = \alpha_3 = 10^{-6} [C^{\circ -1}] \qquad L^2/Rh = 400$$

It is assumed that the shells are subjected to uniform temperature rise throughout. Static and dynamic thermal loads are determined and the static critical loads are compared with those of reference [21]. In reference [21], static thermal buckling analysis of the composite cylindrical shells, is accomplished through extending the linear Semiloof shell element formulated by Irons. This method is based on the classical assumptions and infinitesimal strains. The predicted static critical temperature (T_{cr}) proposed by reference [21] for angle-ply cylindrical shells with axisymmetric buckling modes, along with the results of the static and dynamic buckling analysis developed in the current section, are illustrated in Fig. (5.4-3). For this purpose, two staking sequence ($[\phi/-\phi/\phi]$ and $[\phi/-\phi/\phi/-\phi/\phi]$) are used. As it may be noted, the critical temperature is considerably dependent on the inclination angle of the fibers (ϕ), which is measured

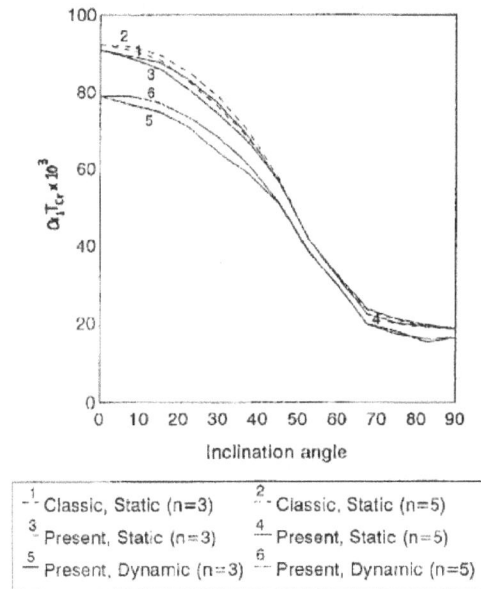

Figure 5.4-3: Comparison of the classical and the present critical temperatures of the symmetric, angle-ply, cylindrical shells.

from the cylinder cross section. Its maximum is attained in ($\phi = 90°$), where the least axial strength of the shell is obtained. Indeed, in the cylindrical shells, the predominant stress is the axial one and circumferential stress is developed mainly in the axial edges. It is shown (for example, in references [14]) that in the case of mechanical buckling, the critical loads vary symmetrically about ($\phi = 45°$). As is shown in Fig. (5.4-4), due to the effect of the nonlinearity in the strain-displacement expressions the differences between curves corresponding to these two approaches is larger in this case. In Fig. (5.4-4), staking sequences are taken to be ($[\phi/-\phi], [\phi/-\phi/\phi/-\phi]$ and $[\phi/-\phi/\phi/-\phi/\phi/-\phi]$). Figures (5.4-3) and (5.4-4) show that increasing number of the layers, while fixing the overall thickness of the shell, leads to a higher strength for the shell. This variation is more pronounced for the two-layer shell. This conclusion is in agreement with the report of reference [48] for mechanical buckling. Linear analysis done by Eslami [49] et al. reveals that the critical buckling temperatures for isotropic materials are independent of modulus of elasticity and are inversely proportional to the coefficient of thermal expansion. Whereas, in the nonlinear analysis introduced in this section, all of the mechanical properties are coupled together. Thus, the critical load is a function of all of mechanical and thermal properties of the materials. On the other hand, as the orthotropy degree G_{ij}/E_1 increases, the discrepancy between results of the classical and the more

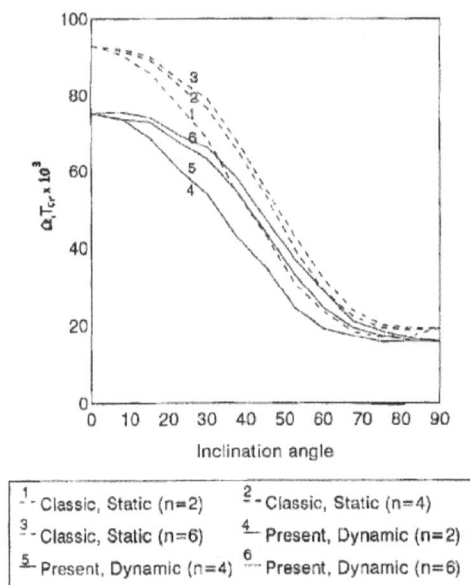

Figure 5.4-4: Comparison of the classical and the present critical temperature of the antisymmetric, angle-ply, cylindrical shells.

accurate theories (the present theory, layer-wise theory and the 3-D approach) increases.

The influence of the thermal expansion coefficients ratio (α_2/α_1) is illustrated in Fig. (5.4-5). It is seen that for lower ratios of the thermal expansion coefficients, fiber orientation play a significant role in the thermal buckling load. Whereas, for higher ratios, the fiber orientation effect diminishes.

5   The Generalized Layer-wise Theory

In this section a layer-wise theory that posses full three dimensional modeling capability and accurately satisfy both displacement and stress boundary conditions, is developed. For this purpose, all components of strain and stress are considered and the exact expressions of strain-displacement components are employed. No simplification other than the Kantrowich type of solution is made. One of the major advantages of the proposed theory compared to the 3-D elasticity analysis, presented in Section 2, is that the solution of the resulted full nonlinear governing equations of the present analysis shows more convergency. For this reason, some authors [8-12] used results of the other theories as starting values in their solutions.

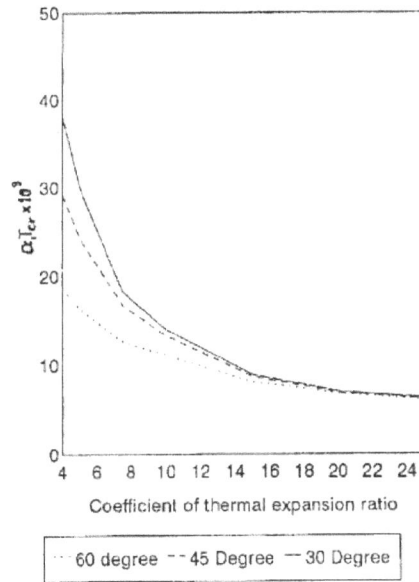

Figure 5.4-5: Variation of the dynamic buckling temperature with coefficient of thermal expansion ratio for the antisymmetric shells ($n = 4$).

5.1 Layer-wise formulations

We used a Kantrovich type of approximation in deriving the governing equations of the imperfect, multilayered composite shell. So that, variations of the displacement components through the shell thickness is considered to be independent of their variations in the x and θ directions. Furthermore, displacement variation trends within the k-th layer is assumed to be different from that of the neighboring layers. Therefore, we can write

$$u_i^{(k)}(r, \theta, x, t) = < \phi(r) >^{(k)} \{U_i(\theta, x, t)\} \qquad (k = 1, ..., N)$$
$$(5.5\text{-}1)$$

where $< \phi(r) >^{(k)}$ is the shape function row matrix of the k-th layer and $\{U_i(\theta, x, t)\}$ is the displacement values vector of grid points chosen through the thickness of the layer to enable approximation of the displacement components variations in this direction. Then, if $(n + 1)$ grid points are adopted, an alternative form of Eq. (5.5-1) will be

$$u_i^{(k)} = \sum_{j=0}^{n^{(k)}} (U_i^{(k)})_j . \phi_j^{(k)}(r) \qquad (k = 1, ..., N) \qquad (5.5\text{-}2)$$

Because $(n + 1)$ points are selected in each layer, if the approximation functions $\phi_j(r)$ are assumed to be polynomials, these polynomials will be of order n. A more suitable interpolation polynomial in the form of Eq. (5.5-2) is the Lagrange interpolating Polynomial, in this situation

$$\phi_j(r) = \prod_{\substack{i=0 \\ i \neq j}}^{n} \frac{r - r_i}{r_j - r_i} \tag{5.5-3}$$

hence

$$u_{i,r} = \sum_{j=0}^{n} \phi_{j,r}.(U_i)_j = \sum_{j=0}^{n} \sum_{s=0}^{n} \frac{1}{r_j - r_s} \prod_{\substack{i=0 \\ i \neq j \\ i \neq s}}^{n} \frac{r - r_i}{r_j - r_i} (U_i)_j \tag{5.5-4}$$

and

$$\delta u_i^{(k)} = \sum_{j=0}^{n^{(k)}} \delta(U_i^{(k)})_j . \phi_j^{(k)}(r) \qquad (k = 1, ..., N) \tag{5.5-5}$$

where $\delta(U_i^{(k)})_j$ are values of δu_i in the grid points. As it may be noticed, each point of the shell has three degrees of freedom $(U_j^{(k)}, V_j^{(k)},$ and $W_j^{(k)})$. It is evident that these terms are repeated in the interpolating polynomials of the neighboring layers for grid points located at the common interface. Therefore, if Eq. (5.5-5) is substituted in Eq. (5.4-6) and rearranging the terms with respect to $\delta U_j^{(k)}, \delta V_j^{(k)},$ and $\delta W_j^{(k)}$ and noting that these values are arbitrary and nonzero, the resulted governing equations will be: for internal points of each layer

$$\int_{r_k}^{r_{k+1}} (\{[\frac{1}{r}\sigma_\theta u_{,\theta} + \sigma_{r\theta} u_{,r} + \sigma_{\theta x}(1 + u_{,x})]_{,\theta} + [r\{\sigma_x(1 + u_{,x}) + r\sigma_{rx} u_{,r}$$

$$+\sigma_{\theta x} u_{,\theta}\}]_{,x}] - \rho \ddot{u}\}\phi_j^{(k)} + \{r\sigma_r u_{,r} + \sigma_{r\theta} u_{,\theta} + r\sigma_{rx}(1 + u_{,x})\}\phi_{j,r}^{(k)}).dr = 0$$

$$\int_{r_k}^{r_{k+1}} (\{\frac{1}{r}\sigma_\theta[(w + w_0)_{,\theta} - v] + \sigma_{r\theta}[1 + (w + w_0)_{,r}] + \sigma_{\theta x}(w + w_0)_{,x} - \rho r \ddot{v}\}$$

$$+[\frac{1}{r}\sigma_\theta(r + v_{,\theta} + w + w_0) + \sigma_{r\theta} v_{,r} + \sigma_{\theta x} v_{,x}]_{,\theta} + [r\sigma_x v_{,x} + r\sigma_{rx} v_{,r} + \sigma_{\theta x}(r +$$

$$v_{,\theta} + w + w_0)]_{,x}\}\phi_j^{(k)} - [r\sigma_r v_{,r} + \sigma_{r\theta}(r + v_{,\theta} + w + w_0) + r\sigma_{rx} v_{,x}\}\phi_{j,r}^{(k)}).dr = 0$$

$$\int_{r_k}^{r_{k+1}} (\{\frac{1}{r}\sigma_\theta(r + v_{,\theta} + w + w_0) + \sigma_{r\theta} v_{,r} + \sigma_{\theta x} v_{,x} + \rho r \ddot{w}$$

$$-[\frac{1}{r}\sigma_\theta[(w + w_0)_{,\theta} - v] + \sigma_{r\theta}[1 + (w + w_0)_{,r}] + \sigma_{\theta x}(w + w_0)_{,x}]_{,\theta}$$

$$-[r\sigma_x(w+w_0)_{,x} + r\sigma_{rx}[1+(w+w_0)_{,r}] + \sigma_{\theta x}[(w+w_0)_{,\theta} - v]]_{,x}\}\phi_j^{(k)}$$

$$+[r\sigma_r[1+(w+w_0)_{,r}] + \sigma_{r\theta}[(w+w_0)_{,\theta} - v] + r\sigma_{rx}(w+w_0)_{,x}]\phi_{j,r}^{(k)}).dr = 0$$

$$(5.5\text{-}6)$$

and for grid points located in the interfaces

$$\int_{r_{k-1}}^{r_k} (\{[\frac{1}{r}\sigma_\theta u_{,\theta} + \sigma_{r\theta} u_{,r} + \sigma_{\theta x}(1+u_{,x})]_{,\theta} + [r\{\sigma_x(1+u_{,x}) + r\sigma_{rx}u_{,r}$$

$$+\sigma_{\theta x}u_{,\theta}\}]_{,x}] - \rho\ddot{u}\}\phi_{n(k-1)}^{(k-1)} + \{r\sigma_r u_{,r} + \sigma_{r\theta}u_{,\theta} + r\sigma_{rx}(1+u_{,x})\}\phi_{n(k-1),r}^{(k-1)}).dr$$

$$+\int_{r_k}^{r_{k+1}} (\{\rho\ddot{u} - [[\frac{1}{r}\sigma_\theta u_{,\theta} + \sigma_{r\theta}u_{,r} + \sigma_{\theta x}(1+u_{,x})]_{,\theta} + [r\{\sigma_x(1+u_{,x}) +$$

$$r\sigma_{rx}u_{,r} + \sigma_{\theta x}u_{,\theta}\}]_{,x}] - \rho\ddot{u}\}\phi_0^{(k)} + \{r\sigma_r u_{,r} + \sigma_{r\theta}u_{,\theta} + r\sigma_{rx}(1+u_{,x})\}\phi_{0,r}^{(k)}).dr = 0$$

$$\int_{r_{k-1}}^{r_k} (\{\frac{1}{r}\sigma_\theta[(w+w_0)_{,\theta} - v] + \sigma_{r\theta}[1+(w+w_0)_{,r}] + \sigma_{\theta x}(w+w_0)_{,x} - \rho r\ddot{v}\}$$

$$+[\frac{1}{r}\sigma_\theta(r+v_{,\theta}+w+w_0) + \sigma_{r\theta}v_{,r} + \sigma_{\theta x}v_{,x}]_{,\theta} + [r\sigma_x v_{,x} + r\sigma_{rx}v_{,r} + \sigma_{\theta x}(r+$$

$$v_{,\theta}+w+w_0)]_{,x}\}\phi_{n(k-1)}^{(k-1)} - [r\sigma_r v_{,r} + \sigma_{r\theta}(r+v_{,\theta}+w+w_0) + r\sigma_{rx}v_{,x}\}\phi_{n(k-1),r}^{(k-1)}).dr$$

$$\int_{r_k}^{r_{k+1}} (\{\frac{1}{r}\sigma_\theta[(w+w_0)_{,\theta} - v] + \sigma_{r\theta}[1+(w+w_0)_{,r}] + \sigma_{\theta x}(w+w_0)_{,x} - \rho r\ddot{v}\}$$

$$+[\frac{1}{r}\sigma_\theta(r+v_{,\theta}+w+w_0) + \sigma_{r\theta}v_{,r} + \sigma_{\theta x}v_{,x}]_{,\theta} + [r\sigma_x v_{,x} + r\sigma_{rx}v_{,r} + \sigma_{\theta x}(r+$$

$$v_{,\theta}+w+w_0)]_{,x}\}\phi_0^{(k)} - [r\sigma_r v_{,r} + \sigma_{r\theta}(r+v_{,\theta}+w+w_0) + r\sigma_{rx}v_{,x}\}\phi_{0,r}^{(k)}).dr = 0$$

$$\int_{r_{k-1}}^{r_k} (\{\frac{1}{r}\sigma_\theta(r+v_{,\theta}+w+w_0) + \sigma_{r\theta}v_{,r} + \sigma_{\theta x}v_{,x} + \rho r\ddot{w}$$

$$-[\frac{1}{r}\sigma_\theta[(w+w_0)_{,\theta} - v] + \sigma_{r\theta}[1+(w+w_0)_{,r}] + \sigma_{\theta x}(w+w_0)_{,x}]_{,\theta}$$

$$-[r\sigma_x(w+w_0)_{,x} + r\sigma_{rx}[1+(w+w_0)_{,r}] + \sigma_{\theta x}[(w+w_0)_{,\theta} - v]]_{,x}\}\phi_{n(k-1)}^{(k-1)}$$

$$+[r\sigma_r[1+(w+w_0)_{,r}] + \sigma_{r\theta}[(w+w_0)_{,\theta} - v] + r\sigma_{rx}(w+w_0)_{,x}]\phi_{n(k-1),r}^{(k-1)}).dr$$

$$+\int_{r_k}^{r_{k+1}} (\{\frac{1}{r}\sigma_\theta(r+v_{,\theta}+w+w_0) + \sigma_{r\theta}v_{,r} + \sigma_{\theta x}v_{,x} + \rho r\ddot{w}$$

$$-[\frac{1}{r}\sigma_\theta[(w+w_0)_{,\theta} - v] + \sigma_{r\theta}[1+(w+w_0)_{,r}] + \sigma_{\theta x}(w+w_0)_{,x}]_{,\theta}$$

$$-[r\sigma_x(w+w_0)_{,x} + r\sigma_{rx}[1+(w+w_0)_{,r}] + \sigma_{\theta x}[(w+w_0)_{,\theta} - v]]_{,x}\}\phi_0^{(k)}$$

$$+[r\sigma_r[1+(w+w_0)_{,r}] + \sigma_{r\theta}[(w+w_0)_{,\theta} - v] + r\sigma_{rx}(w+w_0)_{,x}]\phi_{0,r}^{(k)}).dr = 0$$

$$(5.5\text{-}7)$$

and for grid points lay on the external surface of the shell, the following expressions

$$+p(v^N - w_{,\theta}^N)\phi_j^N(r_{out})$$

$$-p[1 + (v_{,\theta}^N + w^N)\phi_j^N(r_{out}) \qquad (5.5\text{-}8)$$

to the second and third equations of Eq. (5.5-6), respectively. Values have the superscript "N" correspond to points located on the outer surface of the shell.

5.2 Numerical solution and buckling criteria

The governing equations (5.5-6) and (5.5-7) include multiplications of the stress terms in the displacement expressions. Thus, remembering that the stresses are nonlinear functions of the displacements, the resulted equations are coupled and are highly nonlinear functions of the displacements. Nonlinear problems of mechanics are often solved by adopting an incremental formulation. By this incremental solution procedure, the real time-variant system is approximated in a step by step way assuming time-invariance within each time step.

Rearrangement of Eqs. (5.5-6) and (5.5-7) can led to equations of the following type ($m = 0, ..., n^{(k)}$)

$$f^m(\sigma_{ij}, U_i, w_0, U_{i,x}, w_{0,x}, U_{i,\theta}, w_{0,\theta}, \ddot{U}_i|_{i \neq s}) = I_m^k \ddot{(U_s)}_m \quad s = 1,3$$
$$g^m(\sigma_{ij}, U_i, w_0, U_{i,x}, w_{0,x}, U_{i,\theta}, w_{0,\theta}, \ddot{U}_i|_{i \neq s}) = (I^k + I^{k-1})\ddot{(U_s)}_m \quad s = 1,3$$
$$h^m(\sigma_{ij}, U_i, w_0, U_{i,x}, w_{0,x}, U_{i,\theta}, w_{0,\theta}, \ddot{U}_i|_{i \neq s}) = I^k \ddot{(U_s^N)}_m \quad s = 1,3 \quad (5.5\text{-}9)$$

for each layer or interface, where

$$I_m^k = \int_k^{k+1} r.\rho.\phi_m^{(k)}(r).dr \qquad (5.5\text{-}10)$$

and U_1, U_2, and U_3 are equivalent to U, V, and W respectively. Therefore, the following expression hold for the displacements in time interval Δt [41,51]

$$f^m(\tilde{\sigma}_{ij}, \tilde{U}_i, w_0, \tilde{U}_{i,x}, w_{0,x}, \tilde{U}_{i,\theta}, w_{0,\theta}, \ddot{U}_i|_{i \neq s}) = I^k(\Delta \ddot{U}_s)_m \quad s = 1,3$$
$$g^m(\tilde{\sigma}_{ij}, \tilde{U}_i, w_0, \tilde{U}_{i,x}, w_{0,x}, \tilde{U}_{i,\theta}, w_{0,\theta}, \ddot{U}_i|_{i \neq s}) = (I^k + I^{k-1})(\Delta \ddot{U}_s)_m \quad s = 1,3$$
$$h^m(\tilde{\sigma}_{ij}, \tilde{U}_i, w_0, \tilde{U}_{i,x}, w_{0,x}, \tilde{U}_{i,\theta}, w_{0,\theta}, \ddot{U}_i|_{i \neq s}) = I^k(\Delta \ddot{U}_s^N)_m \quad s = 1,3 (5.5\text{-}11)$$

in which values correspond to the end of the time interval are demonstrated by a " $\tilde{\ }$ " symbol. Therefore $3[\sum_{k=1}^N (n^{(k)} + 1)]$ coupled and nonlinear equations in terms of Δu, Δv, and Δw are obtained. The displacements have to satisfy the initial conditions, for ($x = 0, L$ and $\theta = 0, 2\pi$)

$$\Delta U_m = 0 \quad \Delta V_m = 0 \quad \Delta W_m = 0$$
$$\Delta \dot{U}_m = \dot{U}_m \quad \Delta \dot{V}_m = \dot{V}_m \quad \Delta \dot{W}_m = \dot{W}_m \qquad (5.5\text{-}12)$$

The numerical solution procedure is accomplished through the following steps

1- The numerical solution begins with discretizing each the internal and external surfaces as well as the interfaces into m×n grid points in the axial and circumferential directions, respectively. If more precision is required, each layer can be divided into many virtual layers with the same mechanical properties.

2- The initial deviations of the shell at the grid points are defined and the initial values of the displacement terms U_i are set to zero.

3- Time is incremented and the corresponding temperature increase is calculated and corresponding increments of mechanical and thermal loads are found.

4- Starting form point lay near the fixed edges, derivative terms of U_i with respect to the spatial coordinates that appear in Eqs. (5.2-12), (5.5-6) and (5.5-7) are approximated by a second-order finite difference method (the central difference method).

5- Based on the displacement terms values (U_i, w_0) and their derivative values obtained in the grid points of each layer, the strain values are computed from Eq. (5.2-12). In this regard, the displacement components and their derivatives are determined from Eq. (5.5-6) and its equivalent form for derivative values, respectively.

6- The constitutive equations are considered and using Eqs. (5.2-19) to (5.2-21), the stress components are calculated. It is evident that this step is not necessary for the boundary points if the stress boundary conditions are prescribed.

7- Derivative terms of Eq. (5.5-10) involving multiplication of the stress components in the displacement terms (or their derivatives) are substituted using a fourth-order finite difference approximation.

8- In the process of solution, time-invariance for terms appeared in the first sides of Eq. (5.5-10) is assumed during each time step. Thus, values of $\ddot{U}_i$ terms of the first side of the equations are that are calculated substituted from the results of the upper equations of the present step of solution, and the remaining terms are replaced by the values obtained in the previous iterative step of the current time interval. Thus, a set of second order differential equations are derived that can be solved by employing the fourth-order Runge-Kutta method subjected to initial conditions appeared in Eq. (5.5-11).

9- When all equations in each iterative step of the current time interval are solved, the maximum value of a displacement component (for example, the lateral displacement w_{max}) is determined.

10- In each grid point, the displacement term increments (ΔU_i) are added to displacement terms obtained at the end of the previous time interval. To improve the results, solution is continued by using more iterations starting from step 4, until difference of the successive values of w_{max} of the same time interval becomes negligible.

11- The corrected values of $U_i, \dot{U}_i$ obtained in this manner are considered as initial values for the next time interval.

12- Beginning from step 3, results corresponding to the next time step are obtained.

13- Possibility of dynamic buckling occurrence is checked. For this purpose, variations of w_{max} versus time or versus applied load (external pressure, axial load, temperature gradient, etc.) are plotted.

14- In the case of no buckling point, amplitude of the applied loads are increased and calculations are continued starting from step 2.

5.3   Results and discussion

Almost all examples of the well-known references in the field of dynamic buckling of composite multilayered cylindrical shells are restricted to studying buckling under mechanical loads. Thus, since the aims of the following examples are comparison among results of the various theories proposed so far and verification of the results obtained using the developed formulations, the adopted examples are consistent with buckling under mechanical loads. Besides, to avoid occurrence of local extrema of the displacement field within each layer, second-order Lagrange interpolating polynomial $\phi_j^{(k)}$ are adopted.

To compare results of the present theory with those of the higher shear deformation, equivalent layer theories, some perfect cylinders of reference [5] have the following geometric and material properties are considered

$$R = 19.05[cm] \qquad\qquad R/h = 15$$
$$E_{11} = 206.844 \times 10^9[Pa] \qquad E_{22} = 18.6159 \times 10^9[Pa]$$
$$E_{33} = 18.6159 \times 10^9[Pa] \qquad G_{12} = 4.48162 \times 10^9[Pa]$$
$$G_{13} = 4.48162 \times 10^9[Pa] \qquad G_{23} = 2.55107 \times 10^9[Pa]$$
$$\nu_{11} = 0.21 \qquad \nu_{13} = 0.21 \qquad \nu_{23} = 0.45$$

and their stacking sequences are presented in Table (5.5-1). In reference

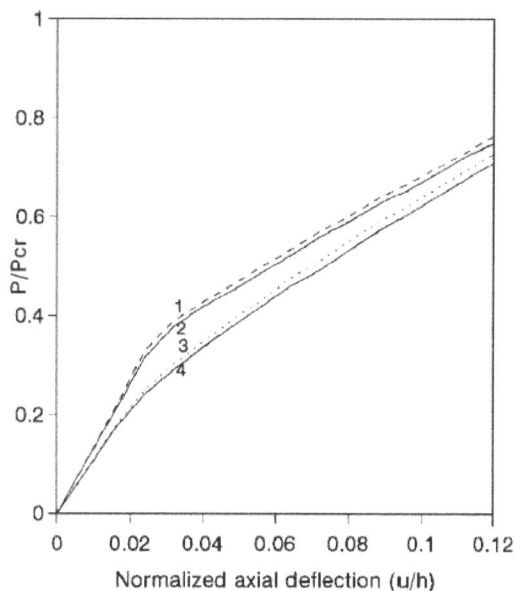

Figure 5.5-1: Comparison of the post buckling analysis result of a thin shell. The illustrated results correspond to the internal surface at t $x = 1$: $1 = W^{mn} = -0.001$, Ref. 33: $2 = W^{mn} = -0.001$, present study: $3 = W^{mn} = -0.01$, Ref. 33; $4 = W^{mn} = -0.01$, present study.

Code no.	stacking sequence
1	$[0_3^\circ]_s$
2	$[0^\circ/90^\circ/0^\circ]_s$
3	$[90^\circ/0^\circ/90^\circ]_s$
4	$[90_3^\circ]_s$
5	$[45_2^\circ/-45^\circ]_s$
6	$[45^\circ/-45_2^\circ]_s$
7	$[-45^\circ/45^\circ/-45^\circ]_s$
8	$[-45_2^\circ/45^\circ]_s$
9	$[30_3^\circ/-60^\circ]_s$

Table 5.5-1: Stacking sequences used

[5], static buckling of perfect circular cylindrical shells using an equivalent layer approach is investigated by means of a displacement-based formulation. Two load cases are considered: a uniform axial compression with $(L/R = 1, 5)$ and a lateral pressure on a very long cylinder. Critical loads corresponding to classical theory (CL), first order shear deformation theory (FOSD), higher order shear deformation, equivalent layer theory (HOSDEL) as well as the present generalized layer-wise theory (GLW) are presented for axial compression, Table (5.5-2), and external pressure, Table (5.5-3). As it may be noticed from Table (5.5-2), classical theory results overestimate the buckling loads especially for the shorter length. Results of Table (5.5-3) reveals that for long cylinders, results of the mentioned theories become closer and the difference becomes higher for thicker cylinders.

Another comparison is made with a conventional displacement-based layer-wise propose by Reddy [33]. For this purpose, postbuckling of two imperfect multilayered cylindrical shells, as those presented in reference [33], is investigated for different initial deviations. The imperfection is assumed to have the form

$$w_0 = \sum_{m=1}^{M} \sum_{n=0}^{N} W_0^{mn} sin(\frac{m\pi x}{l}) cos(n\theta) \qquad (5.5\text{-}13)$$

where m and n being the numbers of axial half waves and circumferential waves, respectively.

The mechanical properties of the shell are

$$E_l = 209.5[GPa] \qquad E_t = 7[GPa]$$
$$G_{lt} = 3.5[GPa] \qquad G_{tt} = 1.4[GPa]$$
$$\nu_{lt} = \nu_{tt} = 0.3 \qquad h_i = h/3$$

where subscripts l and t stand for directions along and transverse to fibers, respectively. Two three layered thin ($h = .254$ [cm]) and moderately thick ($h = 2.54$ [cm]) shells with $R = 91$ [cm] and $l = 254$ [cm] are considered.

Figure (5.5-1) shows axial load versus axial deflection curve of the thin shell, for two different amplitudes of initial geometrical imperfections. Results of reference [33] and results of the present analysis are plotted simultaneously for comparison purpose. As mentioned before, dynamic buckling occurs in points where abrupt change in displacement modes (for example, axial deflection) due to small increase in the applied loads is noticed. In Fig. (5.5-1), the axial load is normalized with respect to the static buckling load of the shell. The deformed shape and buckling

Code no. (L/R=1)	CL	FOSD	HOSDEL	GLW
1	82.66	53.76	34.50	32.38
2	66.64	35.46	32.40	31.26
3	32.83	23.41	21.40	21.13
4	14.71	13.82	13.22	13.04
5	28.41	18.92	17.33	16.92
6	33.76	21.61	19.45	18.86
7	36.56	23.42	21.23	20.74
8	28.41	18.92	17.33	16.92
9	57.63	28.44	24.30	22.31

Code no. (L/R=5)	CL	FOSD	HOSDEL	GLW
1	12.52	12.71	12.45	12.37
2	15.77	14.98	14.76	14.52
3	15.11	14.01	13.72	13.48
4	11.38	10.67	10.54	10.49
5	14.50	12.76	12.41	12.35
6	17.45	14.93	14.38	14.32
7	20.66	17.06	16.12	15.86
8	14.50	12.76	12.41	12.35
9	12.03	10.66	10.12	9.89

Table 5.5-2: Critical axial compression in $(N/m) \times 10^{-6}$ for complete fixation of the ends.

Code no. (L/R=1)	CL	FOSD	HOSDEL	GLW
1	1847	1771	1765	1758
2	6687	6274	6191	6126
3	15671	14286	14196	14147
4	20512	18567	18333	18218
5	6226	5805	5743	5689
6	6226	5839	5791	5714
7	6226	5853	5826	5804
8	6226	5805	5743	5689
9	3144	2999	2985	2974

Table 5.5-3: Critical pressure in Pa $\times 10^{-3}$

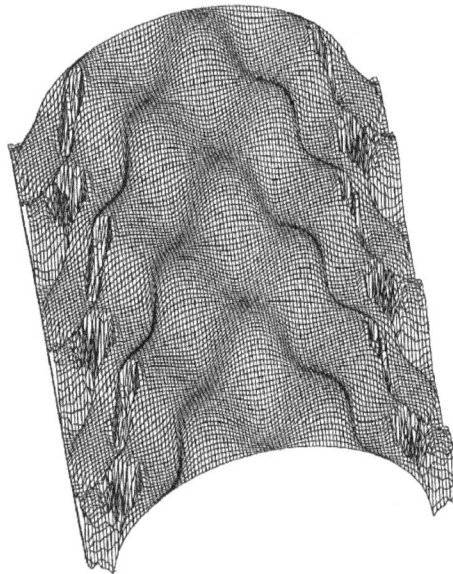

Figure 5.5-2: Buckling mode of the cylindrical shell.

modes of the cylindrical shell are shown in Fig. (5.5-2). The major disadvantage of the analysis presented in reference [33] is that generally, the boundary conditions are not completely satisfied. On the other hand, due to incorporation of exact strain-displacement expressions in the present analysis, the present results are lower. At the end of the present curve, slope increases gradually which is in agreement with the results reported in references [41,51].

The postbuckling response for the moderately thick shell is presented in Fig. (5.5-3) and the deformed shape of the shell is illustrated in Fig. (5.5-4). Comparison of Figs. (5.5-2) and (5.5-4), reveals that in contrast to the thin shell, lower buckling modes are more magnified so that in the limit (for very thick vessels), buckling accomplishes in an axisymmetric manner [18]. The difference between the predicted critical loads of these theories, is more for the moderately thick shell.

Finally, a comparison with the latest 3-D analysis proposed by reference [12] is done. The proposed formulations of the above reference are suitable for static buckling of thick perfect single layer circular cylindrical shells under mechanical loads and are based on infinitesimal strains assumption (linear strain-displacement expressions). In deriving the final results, one of the triple Fourier series (which means separation of variables in all coordinate directions, whereas in the present analysis, independency is taken to be in radial direction only) is chosen, boundary conditions are satisfied in average and the results are obtained using a finite difference method.

Table (5.5-4) gives a comparison of the predicted loads for a very long shell, under external pressure. The critical pressures are normalized with respect to the classical theory (Donnell's shell theory) results. Two cases of material properties are considered: Circumferential reinforcement ($E_2 \gg E_1, E_3$) and axial reinforcement ($E_3 \gg E_1, E_2$). Results of the preliminary 3-D analysis is denoted by (P3D).

Now the buckling of a 4-ply laminated imperfect cylindrical shell under uniform axial compression of references [3] and [15] is reexamined. The cylinder is considered to be simply supported with radius $R = 19.0$ [cm] and $L/R = 2$. Thickness of each layer is .0135 [cm] and the laminate construction is $[0°/30°/60°/90°]$. The material of each lamina is assumed as Boron/Epoxy, AVCO 550 with the following properties

$$E_x = 207[GPa], \qquad E_\theta = 18.6[GPa], \qquad G_{x\theta} = 4.48[GPa]$$
$$\nu_{x\theta} = \nu_{xz} = \nu_{z\theta} = .21$$

The initial geometric imperfections of the shell are defined as

$$w_0(x,\theta) = W_0 h.sin(\frac{\pi x}{L}).cos(8\theta) \qquad (5.5-14)$$

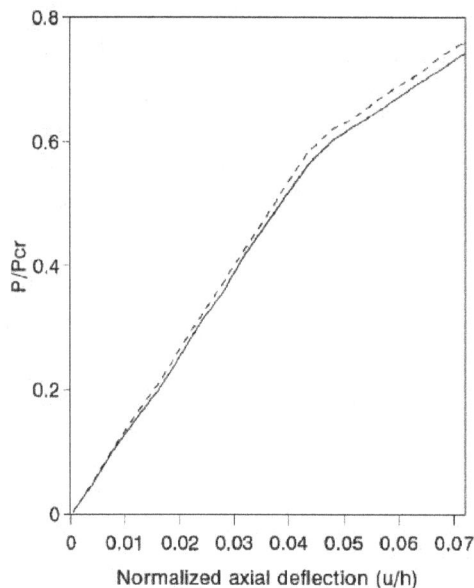

Figure 5.5-3: Comparison of the post buckling analysis result of a moderately thick shell. The illustrated results correspond to the internal surface at $x = 1$: - - =Ref. 33 ($W^{mn} = -0.001$), and ——=present paper.

where W_0 is the amplitude of imperfection and h is the total thickness of the shell.

Results of static and dynamic buckling analysis performed by references [15] and [3] as well as the present results are depicted in Fig. (5.5-5). In the earlier results, an equivalent layer approach is adopted and von Karman type expressions of strains are chosen. Results of reference [3] are based on Hoff-Simitses criterion (the total potential energy approach) that is explained in detail in reference [44], so that the static and dynamic loads for $W_0 = 0$ are identical. In reference [15], Love-Kirchhoff assumptions are employed and the nonlinear equations of motion are linearized and solved using an incremental method previously described in reference [14]. As seen from this figure, the reduction in buckling loads is diminished as imperfection amplitude increases. In other words, imperfection sensitivity is more noticeable in small imperfection amplitudes. In comparison with the other results, the present formulation gives smaller critical loads. This is due to incorporation of the exact strain-displacement relations of the imperfect shell in the relatively exact nonlinear analysis presented here which is compatible with the exact 3-d elasticity analysis.

Figure (5.5-6) shows the effect of various impulsive loads (rectangular, triangular, and parabolic impulsive loads) and their time duration on the

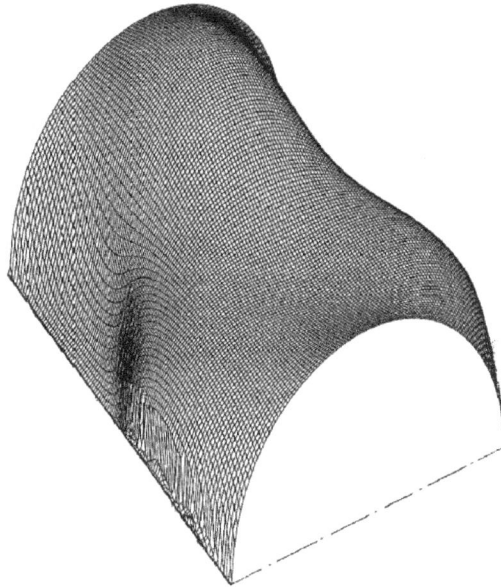

Figure 5.5-4: Buckling mode of the thick cylindrical shell.

predicted critical loads. For this purpose, the following nondimensional time is used

$$t^* = \hat{\tau}/\hat{\tau}_0 \qquad (5.5\text{-}15)$$

where $\bar{\tau}$ is the pulse duration and $\hat{\tau}_0$ is the free vibration period of the shell. In Fig. (5.5-6), P_{cr} is the static buckling load. These results indicate that shell stiffness increases for a short duration of loading, specially if the pulse duration is comparable to the hoop breathing mode. For short time duration, the dynamic buckling loads are larger that the static buckling load. It is believed that this phenomenon is due to the stress wave revibration between the impacted and fixed ends of the shell [37-39????]. For a given amplitude, the step load has a maximum curve area. Thus, as may be expected, this type of loading has the worst influence on the strength of the shell. The critical loads corresponding to the triangular impulsive loads are the largest. The second example of the static analysis is reconsidered, and the dynamic buckling due to the dynamic load is investigated. The results of the analysis are shown in Fig. (5.5-7). The reduction in the dynamic buckling load for the thick shell is more that that of the thin shell. The reason is that the effects of transverse shear and normal stresses are more significant in thick shells.

Comparison of the above numerical examples, verify the accuracy and conservation of the present theory results. As it may be expected, the difference among the results of the above mentioned theories is higher

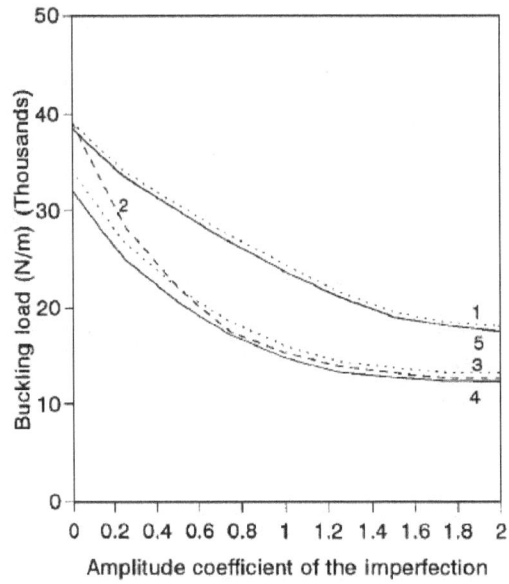

Figure 5.5-5: Effect of imperfection amplitude on the dynamic buckling load: 1=Sheinman Liaw (static); 2=Sheinman et al. (dynamic); 3=Liaw and yang (dynamic); 4=present study (dynamic); and 5=present study (static)

for buckling under dynamic loads.

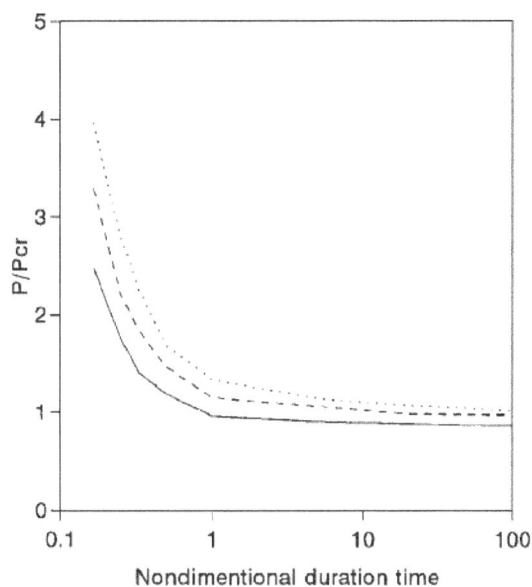

Figure 5.5-6: Effect of different types of impulsive loads and various loading durations on the buckling load of the cylindrical shell: —rectangular impulse; - - - ,parabolic impulse; and ...,triangular impulse.

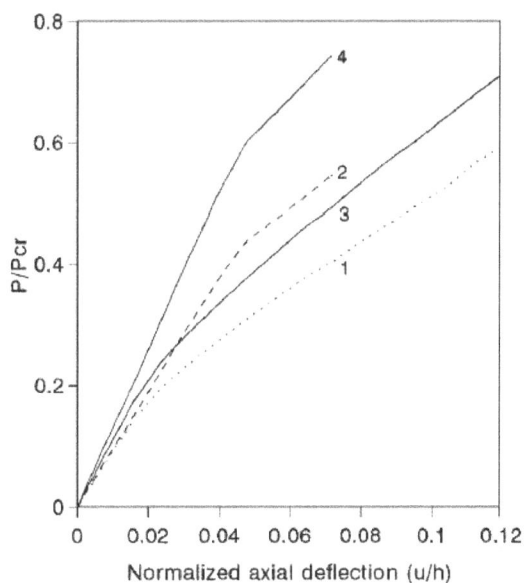

Figure 5.5-7: Comparison between the dynamic and the static buckling results of the thin and thick shells.the illustrated results correspond to the internal surface at $x = 1$ $W^{mn} = -0.01$, 1=thin shell, dynamic; 2=thick sell, dynamic; 3=thin shell, static; and 4=thick shell, static.

Geometry	FOSD	HOSDEL [37]	P3D	GLW
Circumferential reinforcement				
h=6.35[mm], h/R=0.03,L/R=100	0.9668	0.9637	0.9694	0.9608
h=12.7[mm], h/R=0.07,L/R=100	0.9050	0.8933	0.9148	0.8892
Axial reinforcement				
h=6.35[mm], h/R=0.03,L/R=100	0.9822	0.9822	0.9817	0.9816
h=12.7[mm], h/R=0.07,L/R=100	0.9588	0.9556	0.9605	0.9528

Table 5.5-4: Comparison of various theories results for cylindrical shells under external pressure

6   References

1 - Booton, M., and Tennyson, R.C., " Buckling of Imperfect Anisotropic Circular Cylinder under Combined Loading ", AIAA J., Vol. 17, pp. 278-287, 1979.

2 - Jones, R.M., and Hennemann, J.C.F., " Effect of Prebuckling Deformation on Buckling of Laminated Composite Circular Shells ", AIAA J., Vol. 18, pp. 110-115, 1980.

3 - Shienman, I., Shaw, I.D., and Simitses, G.J., " Nonlinear Analysis of Axially Loaded Laminated Cylindrical Shells ", Comput. Struct., Vol. 16, pp. 131-137, 1983.

4 - Simitses, G.J., Shaw, I.D., and Sheinman, I., " Stability of Imperfect Laminated Cylinders: A Comparison Between Theory and Experiment ", AIAA J., Vol. 23, pp. 1086-1092, 1985.

5 - Simitses, G.J., and Anastasiadis, J.S., " Shear Deformable Theories for Cylindrical Laminates - Equilibrium and Buckling with Applications ", AIAA J., Vol. 30, No. 3, pp. 826-834, 1992.

6 - Shen, H.S., and Williams, F.W., " Postbuckling Analysis of Stiffened Laminated Panels Loaded in Compression ", Int. J. Solids Struct., Vol. 3o, No. 12, pp. 1589-1630, 1993.

7 - Song, Y., Sheinman, I., and Simitses, G.J., " Thermoelastoviscoplastic Buckling Behavior of Cylindrical Shells ", J. Eng. Mech., Vol. 121, No. 1, pp. 62-70, 1995.

8 - Kardomateas, G.A., " Buckling of Thick Orthotropic Cylindrical Shells under External Pressure ", J. Appl. Mech., Vol. 60, pp. 195-202, 1993.

9 - Kardomateas, G.A., " Stability Loss in Thick Transversely Isotropic Cylindrical Shells under Axial Compression ", J. Appl. Mech., Vol. 60, pp. 506-513, 1993.

10 - Kardomateas, G.A., and Chung, C.B., " Buckling of Thick Orthotropic Cylindrical Shells under External Pressure Based on Non-Planar Equilibrium Modes ", Int. J. Solids Struct., Vol. 31, No. 16, pp. 2195-2210, 1994.

11 - Kardomateas, G.A., " Bifurcation of Equilibrium in Thick Orthotropic Cylindrical Shells under Axial Compression ", J. Appl. Mech., Vol. 62, pp. 43-52, 1995.

12 - Kardomateas, G.A., and Philobos, M.S., " Buckling of Thick Orthotropic Cylindrical Shells under Combined External Pressure and Axial Compression ", AIAA J., Vol. 33, No. 10, pp. 1946-1953, 1995.

13 - Eslami, M.R., and Shariyat, M., 1996, " Elastic-Plastic and Creep Buckling of Imperfect Cylinders under Mechanical and Thermal Loading ", ASME Trans., J. Press. Vess. Tech., Vol. 119, No. 1, pp. 27-36, 1997.

14 - Saigal, S., Yang, T.Y., and Kapania, R.K., " Dynamic Buckling of Imperfection-Sensitive Shell Structures ", AIAA J., Vol. 24, No. 10, pp. 718-724, 1987.

15 - Liaw, D.G., and Yang, T.Y., " Symmetric and Asymmetric Dynamic Buckling of Laminated Thin Shells with the Effect of Imperfection and Damping ", J. Compos. Mater., Vol. 24, pp. 188-207, 1990.

16 - Argento, A., and Scott, R.A., " Dynamic Instability of Layered Anisotropic Circular Cylindrical Shells, Part I; Theoretical Development ", J. Sound and Vib., Vol. 162, No. 2, pp. 311-322, 1993.

17 - Kasagi, A., and Sridharan, S., " Imperfection Sensitivity Layered

Composite cylinders ", J. Eng. Mech., Vol. 121, No. 7, pp. 810-818, 1995.

18 - Donnell, L.H., *Beams, Plates, and Shells*, Mc Graw-Hill, New York, 1976.

19 - Chang, J.S., and Leu, S.Y., " Thermal Buckling Analysis of Anti-symmetric Angle-Ply Laminates Based on a Higher-Order Displacement Field ", Compos. Sci. and Tech., Vol. 41, pp. 109-128, 1991.

20 - Radhamohan, S.K., and Venkataramana, J., " Thermal Buckling of Orthotropic Cylindrical Shells ", AIAA J., Vol. 13, No. 3, pp. 397-399, 1975.

21 - Thangaratnam, R.K., Palaninathan, R., and Ramachandran, J., " Thermal Buckling of Laminated Composite Shells ", AIAA J., Vol. 28, No. 5, pp. 859-860, 1990.

22 - Ma, S.F., and Wilcox, M.W., " Thermal Buckling of Axisymmetric Angle-Ply Laminated Cylindrical Shells ", Composite Eng., Vol. 1, No. 3, pp. 183-192, 1991.

23 - Huang, N.N, and Tauchert, T.R., " Large Deflection of Laminated Cylindrical and Doubly Curved Panels under Thermal Loading ", Comput. Struct., Vol. 41, No. 2, pp. 303-312, 1991.

24 - Kossira, H., and Haupt, M., " Buckling of Laminated Plates and Cylindrical Shells Subjected to Combined Thermal and Mechanical Loads ", (ed: Jullien, J.F.) in: *Buckling of Shell Structures, on Land, in the Sea and in the Air*, pp. 201-212, 1991.

25 - Weller, T., and Patlashenko, I., " Postbuckling of Infinite Length Cylindrical Panels under Combined Thermal and Pressure Loading ", Int. J. Solids Struct., Vol. 30, No. 12, pp. 1649-1662, 1993.

26 - Thornton, E.A., " Thermal Buckling of Plates and Shells ", Appl. Mech. Rev., Vol. 46, No. 10, pp. 485-506, 1993.

27 - Zukas, J.A., " Effect of Transverse and Normal Shear Strains in Orthotropic Shells ", AIAA J., Vol. 12, No. 12, pp. 1753-1755, 1974.

28 - Reddy, J.N., and Liu, C.F., " A Higher Order Shear Deformation Theory of Laminated Elastic Shells ", Int. J. Eng. Science, Vol. 23, No.

3, pp. 319-330, 1985.

29 - Stein, M., " Nonlinear Theory for Plates and Shells Including the Effect of Transverse Bending ", AIAA J., Vol. 24, No. 9, pp. 1537-1544, 1986.

30 - Dennis, S.T., and Palazotto, A.N., " Transverse Shear Deformation in Orthotropic Cylindrical Pressure Vessels using a Higher-Order Shear Theory ", AIAA J., Vol. 27, No. 10, pp. 1441-1447, 1989.

31 - Barbero, E.J., Reddy, J.N., and Teply, J.L., " General Two-Dimensional Theory of Laminated Cylindrical Shells ", AIAA J., Vol. 28, No. 3, pp. 544-553, 1990.

32 - Eslami, M.R., Shakeri, M., and Shariyat, M., " A High-Order Theory for Dynamic Buckling and Post buckling Analysis of Laminated Cylindrical Shells ", Trans, ASME, J. Pressure Vessel Tech., Vol. 121, pp. 94-102, 1998.

33 - Reddy, J.N., and Savoia, M., " Layer-Wise Shell Theory for Post-buckling of Laminated Circular Cylindrical Shells ", AIAA J., Vol. 30, No. 8, pp. 2148-2154, 1992.

34 - Robbins, J.R., and Reddy, J.N., " Modelling of Thick Composites using a Layer-wise Laminate Theory ", Int. J. Num. Meth. Eng., Vol. 36, pp. 655-677, 1993.

35 - Eslami, M.R., Shariyat, M., and Shakeri, M., Dynamic Buckling of Analysis of Composite Cylindrical Shells under Mechanical Loads Based on Layerwise Theory, AIAA Journal, Vol. 36, No. 10, pp. 1874-1882, 1998.

36 - Eslami, M.R., and Shariyat, M., " Dynamic Buckling and Post-buckling of Imperfect Orthotropic Cylindrical Shells under Mechanical and Thermal Loads Based on the Three-Dimensional Theory of Elasticity, Trans ASME, J. Appl. Mech, Vol. 66, pp. 476-484, 1999.

37 - Washizu, K., *Variational Methods in Elasticity and Plasticity*, Pergamon Press, London, 1982.

38 - Flugge, W., *Tensor Analysis and Continuum Mechanics*, Springer-Verlag, Berlin, 1972.

39 - Vinson, J.R., and Sierakowski, R.L., *The Behavior of Structures Composed of Composite Materials*, Chapter 5, Martinus Nijhoff Publishers, 1987.

40 - Eslami, M.R., and Shariyat, M., " On Thermal Dynamic Buckling Analysis of Imperfect Laminated Cylindrical Shells ", ZAMM, Vol. 80, No. 3, pp. 171-182, 2000.

41 - Shariyat, M., Eslami, M.R., and Shakeri, M., " Elastic-Plastic Dynamic Buckling Analysis of Imperfect Cylindrical Shells ", Amirkabir Journal of Science and Tech. Vol. 11, No. 42, pp. 35-46, 1999.

42 - Budiansky, B., " Theory of Buckling and Postbuckling Behavior of Elastic Structures ", Advances in Appl. Mech., pp. 1-65, 1974.

43 - Simitses, G.J., Tabiei, A., and Anastasiadis, J.S., " Buckling of Moderately Thick, Laminated Cylindrical Shells under Lateral Pressure ", Compos. Eng., Vol. 3, No. 5, pp. 409-417, 1993.

44 - Simitses, G.J., *Dynamic Stability of Suddenly Loaded Structures*, Springer-Verlag, Berlin, 1990.

45 - Lindberg, H.E., " Dynamic Buckling of Cylindrical Shells form Oscillating Waves Following Axial Impact ", Int. J. Solids Struct., Part 1, Vol. 23, No. 6, pp. 669-692, 1987.

46 - Lindberg, H.E., " Dynamic Pulse Buckling of Imperfection Sensetive Shells ", Trans. ASME, J. Appl. Mech., Vol. 58, pp. 743-748, 1991.

47 - Zimcik, D.G., and Tennyson, R.C., " Stability of Circular Cylindrical Shells under Transient Axial Impulsive Loading ", AIAA J., Vol. 18, No. 6, pp. 691-699, 1980.

48 - Ganapathi, M., and Varadan, T.K., " Dynamic Buckling of Laminated Anisotropic Spherical Caps ", J. Appl. Mech., Vol. 62, pp. 13-19, 1995.

49 - Eslami, M.R., Ziaii, A.R., and Ghorbanpour, A., " Thermoelastic Buckling of Thin Cylindrical Shells Based on Improved Stability Equations ", J. Thermal Stresses, Vol. 19, No. 4, PP. 299-316, 1996.

50- Hinton, H., *Static and Dynamic Analysis of Plates and Shells; Theory, Software and Applications*, Springer-Verlag, Berlin, 1992.

51 - Gillat, R., Feldman, E., and Aboudi, J., " Axisymmetric Response of Nonlinearly Elastic Cylindrical Shells to Dynamic Axial Load ", Int. J. Impact Eng., Vol. 13, No. 4, pp. 545-554, 1993.

52 - Brush, D.O., and Almroth, B.O., *Buckling of Bars, Plates and Shells*, Mc Graw-Hill, New York, 1975.

Index

سرشناسه	: اسلامی، محمدرضا، ۱۳۲۳
	Eslami, Mohammad Reza
عنوان و نام پدیدآور	Thermo-Mechanical Buckling of Composite Plates and Shells /M.R Eslami
مشخصات نشر	: تهران: دانشگاه صنعتی امیرکبیر (پلی‌تکنیک تهران)، ۱۳۸۹=۲۰۱۰م
مشخصات ظاهری	: [۲۸۴] ص: نمودار
شابک	: 978-964-463-384-3
وضعیت فهرست‌نویسی	: فیپا
یادداشت	: انگلیسی
یادداشت	: کتابنامه: ص. ۲۷۷--۲۸۲
یادداشت	: نمایه
آوانویسی	: ترمو -- مکنیکال
موضوع	: مهندسی مکانیک
شناسه افزوده	: دانشگاه صنعتی امیرکبیر (پلی‌تکنیک تهران)
شناسه افزوده	: Amirkabir University of Technology
رده‌بندی کنگره	: ۱۳۸۹ ۴ت۵الف/TJ۱۴۵
رده‌بندی دیویی	: ۶۲۱
شماره کتابشناسی ملی	: ۱۸۲۱۶۰۶

انتشارات دانشگاه صنعتی امیرکبیر
(پلی‌تکنیک تهران)

عنوان کتاب	:	Thermo-Mechanical Buckling of Composite Plates and Shells
تألیف	: پروفسور محمدرضا اسلامی	
ناشر	: انتشارات دانشگاه صنعتی امیرکبیر (پلی‌تکنیک تهران)	
لیتوگرافی، چاپ و صحافی	: انتشارات دانشگاه صنعتی امیرکبیر (پلی‌تکنیک تهران)	
چاپ اول	: پاییز ۱۳۸۸	
شابک	: ۹۷۸-۹۶۴-۴۶۳-۳۸۴-۳	ISBN : 978-964-463-384-3

آدرس مرکز پخش: خیابان ولیعصر، روبروی خیابان بزرگمهر، فروشگاه کتاب مرکز نشر
دانشگاه صنعتی امیرکبیر (پلی‌تکنیک تهران) – تلفن: ۶۶۴۹۸۸۶۸
وبسایت: http://publication.aut.ac.ir

Thermo-Mechanical Buckling

of

Composite Plates and Shells

تأليف:

پروفسور محمدرضا اسلامی

استاد دانشکده مهندسی مکانیک

دانشگاه صنعتی امیرکبیر (پلی‌تکنیک تهران)

پاییز ۱۳۸۸